Asymmetric Gearing

Asymmetric Gearing

Alexander L. Kapelevich

CRC Press
Taylor & Francis Group
Boca Raton London New York

CRC Press is an imprint of the
Taylor & Francis Group, an **informa** business

CRC Press
Taylor & Francis Group
6000 Broken Sound Parkway NW, Suite 300
Boca Raton, FL 33487-2742

Printed on acid-free paper

International Standard Book Number-13: 978-1-138-55444-3 (Hardback)

Visit the Taylor & Francis Web site at
http://www.taylorandfrancis.com

and the CRC Press Web site at
http://www.crcpress.com

Contents

Preface

Why Asymmetric?

It is well known that the opposite gear tooth flanks are functionally different for the majority of gear drives, where the load and duration of its application are much greater for one tooth flank than for the other one. Such gear drives are often called unidirectional. First of all, it should be understood that this does not mean that these unidirectional drives cannot operate in the opposite direction; they usually can. Second, there is a difference between the rotation direction and the load transmission direction—in some cases, these two directions do not coincide. For example, a gravity or spring loaded gear drive output shaft may rotate in both directions, while the direction of loading and loaded gear tooth flanks remain the same. Another example: when a vehicle is moving through flat terrain or uphill, its transmission conveys power from the engine to the wheels. When it is moving downhill driven by gravity, its transmission is typically downshifted to slow the vehicle down. In this case, the vehicle transmission conveys power from the wheels to the engine. In both cases the direction of the wheels' rotation is the same regardless of terrain, but directions of the power transmission and loaded tooth flanks are different. However, this kind of transmission can typically be classified as a unidirectional one, because for most vehicles braking by downshifting accounts for a small fraction of the vehicle's operating life, and the gear tooth load is usually lower than it is in driving mode.

Despite a functional asymmetry of the unidirectional gear drive opposite tooth flanks, they have symmetric profiles for practically all existing gear transmissions. This makes the gears uniquely different from the other rotational power transmitting machine components. For instance, like a gear tooth flank a gas turbine blade also transmits rotational power when a gas flow load is applied to its high pressure surface. However, the high pressure surface of the turbine blade has a concave airfoil profile, which is very different from an opposite convex profile of the suction surface. This turbine blade airfoil profile asymmetry is necessary for achieving maximum performance and efficiency. The modern asymmetric turbine blade profile is the result of evolution from the flat symmetric profile blades of wind or water mill turbines that were used centuries ago.

Charles H. Logue, in his "American Machinist Gear Book" dated 1922, describing the benefits of gears with asymmetric buttress teeth wrote: "... it is a great wonder that this tooth is not more extensively used." Now, almost a hundred years later, it is still a mystery why the modern unidirectional gear drive teeth are not designed asymmetric to take advantage of functional asymmetry of their flank operation for performance maximization.

This can be partially explained by a fact that modern (also known as traditional) gear design, which describes a gear tooth macrogeometry as a result of a generating (tooling) rack's conjugate motion around the gear center, has been the dominant gear design method for over a hundred years. Applications of alternative gear design methods are hardly known and used today.

Another likely reason for overlooking asymmetric gearing is the overstandardization of modern gear design. There are very few applications for gears with nonstandard (typically non-involute) tooth profiles, like in cycloidal gears, Wildhaber and Novikov gears, convoloid gears, etc. But even those nonstandard gears generally have symmetric teeth.

All existing gear design standards define and utilize a symmetric involute tooth profile. The vast majority of gears are designed with the standard 20° pressure angle and tooth proportions. In some demanding applications, like, for example, in aerospace and automotive industries, the pressure angle and tooth proportions are altered to provide higher transmission load carrying capacity and improve other performance characteristics. Nevertheless, these gears are designed by the same rack generating design methodology that has not evolved for many years.

Traditional gear design (combined with the standardization of tooling rack parameters) was a major factor in both enhancing gear drive characteristics and in increasing the productivity of gear fabrication. However, over the last several decades, this design method reached its limits in gear performance improvement. Since then, the main contributors to gear drive development have been achievements in gear materials, manufacturing technology and tooling, thermal treatment, tooth surface engineering and coatings, tribology and lubricants, testing technology and diagnostics, simulation software, and gear tooth microgeometry—which defines deviation from the nominal involute surface to achieve optimal tooth contact localization for higher load capacity and lower transmission error.

The design objective of asymmetric gears is to improve the performance of the primary drive tooth flanks at the expense of the performance of the opposite coast flanks, which are unloaded or lightly loaded during a relatively short work period. Asymmetric teeth make it possible to simultaneously increase the contact ratio and operating pressure angle beyond the limits achievable with conventional symmetric gears. A key advantage of asymmetric gears is drive flank contact stress reduction, which results in higher power transmission density. Bending stress in these gears is also minimized, because of an optimized root fillet. Another important advantage is the ability to design the coast tooth flanks independently from the drive tooth flanks, i.e., managing tooth stiffness while keeping a desirable drive flank pressure angle and contact ratio. This allows for an increase in tooth tip deflection, thus softening a tooth engagement impact and reducing gear noise and vibration.

The book describes involute asymmetric tooth gears. It further elaborates on the alternative Direct Gear Design® approach that was developed to maximize the performance of custom gear drives and described in detail in my previous book, "Direct Gear Design."

This book comprehensively defines all aspects of asymmetric gearing from its development history to the latest implementations, including the tooth geometry analysis, parameter limits and area of existence definition, tooth and mesh parameter optimization, stress analysis and rating, some design specifics, gear fabrication and inspection, and analytical and experimental comparison to the best specimens of symmetric tooth gears. Some segments of this book were recently published in technical magazines and presented at international gear conferences.

This book is addressed, first of all, to my dear colleagues—gear engineers. We should not stand aside letting gear tooth geometry be defined using the old-fashioned traditional gear design with all its self-limitations delegating gear drive improvements to material experts, technologists, tribologists, and other field professionals (who, by the way, are doing a brilliant job in gear drive characteristic enhancement). The goal of this book is to show readers the great potentials of asymmetric gearing and inspire engineers to utilize these advantages in their designs to maximize gear drive performance.

Acknowledgments

This book would not have been possible were it not for the collaboration, inspiration, and support from my family, friends, colleagues. I express deep gratitude to:

- Dr. Yuriy V. Shekhtman for his constant assistance in software development and many years of friendship
- Robert L. Errichello (Geartech), Yuriy Gmirya (Lockheed Martin), Dr. Anders Flodin (Höganäs AB), Prof. Robert L. Mott (University of Dayton, OH), Prof. Karsten Stahl (FZG), and Prof. Aizoh Kubo (Gear Technologies Co.) for support and encouragement in the writing of this book
- Dr. Franz J. Joachim (Joachim Gear Consulting), Dr. Michael Andersson (Höganäs AB), Dr. Daniel Kadach (FZG), Sergej Lieder (ZF), Dr. Andreas Langheinrich (Scholz), Dr. Mehr and Dr. Winkel (Liebherr), and Christian Weihmann (Profiroll Technologies) for their help in finding the reference literature about asymmetric gears for this book
- Prof. Victor E. Starzhinsky (MPRI), Prof. Veniamin I. Goldfarb (Izhevsk State Technical University), and Prof. Stephen P. Radzevich (Ricardo) for an opportunity to write the chapters about asymmetric gearing for their books
- my colleagues from Russia and Ukraine: Dr. Viacheslav M. Ananiev (CIAM), Dr. Igor L. Glikson and Alexander A. Stupakov (Motor Sich) for sharing their expertise and friendship
- Dmitry Karfagenskiy for editing assistance and Alexander A. Shumskiy for his help in graphics work
- American Gear Manufacturer's Association for the privilege to be its member and to work in its Aerospace gearing committee
- Jonathan Plant, Ryan Farrar, and Edward Curtis from CRC Press and Alexandra Andrejevich from codeMantra for the book publication preparation

Alexander L. Kapelevich

Author

Dr. Alexander L. Kapelevich is a gear design consultant at AKGears, LLC. He holds a Master's Degree in Mechanical Engineering at Moscow Aviation Institute and a Ph.D. Degree in Mechanical Engineering at Moscow State Technical University. Dr. Kapelevich has 40 years of gear transmission research and design experience. He began his career working in the Russian Aviation Industry, where he was involved in R&D, software development, testing, and failure investigation for aerospace gear transmissions. Living in the USA since 1994, he developed the Direct Gear Design® methodology for custom gear transmissions which have been implemented in various fields such as aerospace, automotive, agriculture, defense, robotics, racing, and many others. His specialty is the gear drive architecture, planetary systems, gear tooth geometry optimization, gears with asymmetric teeth, and gear transmission performance maximization. Dr. Kapelevich is the author of the book titled "Direct Gear Design" and many technical articles. He is a member of American Gear Manufacturers Association.

1

History of Asymmetric Gears and Modern State of Art

While gears with asymmetric teeth (or asymmetric gears) have been known for many years, their history is not sufficiently recorded in modern gear literature. For this reason, some modern gear researchers have concluded that asymmetric tooth gears were discovered just several decades ago. This chapter, although it may not describe the complete history of asymmetric gearing, this chapter should clarify this matter and shed light upon the origin of asymmetric gearing. It also describes a current state of art based on recent publications about asymmetric gears.

Apparently the first asymmetric tooth gears were invented by Mother Nature and applied in the leg joints of the tiny planthopper insect Issus Coleoptratus (Figure 1.1a). According to British biologists Malcolm Burrows and Gregory Sutton [1], who discovered that juveniles of the species have an intricate gearing system that locks their back legs together, this insect has its leg joints with "10 to 12 gear teeth, asymmetric in their form, looking like shark fins on just a portion of a circle. Each tooth is about 50 microns in width and 15 to 30 microns in length and about 30 microns apart from its neighbors." The gears (Figure 1.1b) operate "when the animal ratchets its legs to cock its body for the spring. This meshing prevents the animal from spinning to one side as the animal would do if the legs did not fire at precisely the same time."

People have used gears for many centuries. The first, so-called lantern gears, simple rectangular or cylindrical tooth profiles (Figure 1.2) were replaced with more sophisticated cycloid profiles. In the seventeenth century, the Dutch mathematician Christiaan Huygens studied the circle involute, which is the path traced out by a point on a straight line that rolls around a circle, in order to apply it to his first pendulum clock. In the mid-eighteenth century, the Swiss scientist Leonard Euler introduced the involute of a circle for the gear tooth flank profiles. An important feature of the involute gear profiles is producing the theoretically constant rotational velocities' ratio. Both the cycloid and involute gears, as they are known today, usually have symmetric tooth profiles.

It is not clear when the first asymmetric tooth gears were invented. Although the asymmetric tooth wheels were used as a part of the ratchet mechanisms (Figure 1.3) centuries ago, for example, in the catapults invented in Ancient Greece in the fourth century BC. Teeth of the ratchet wheel have an asymmetric buttress shape with moderate slope on one flank and a much

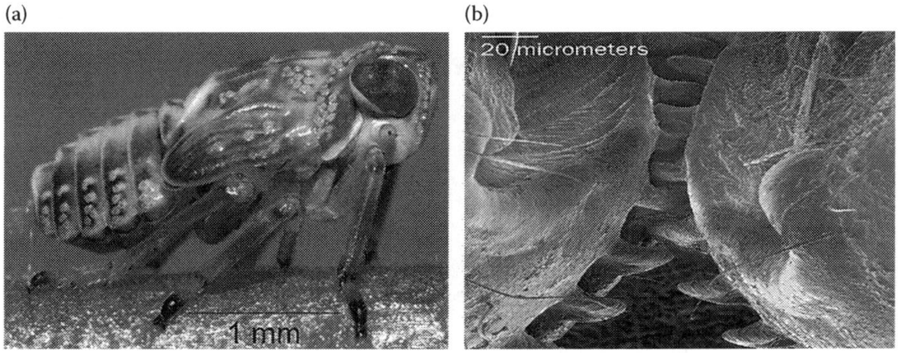

FIGURE 1.1
Issus Coleoptratus (a) with the asymmetric teeth gears (b) in its leg joints. (From Burrows, M and Sutton, G. 2013. Interacting gears synchronize propulsive leg movements in a jumping insect. Science, September, 1254–1256. With permission.)

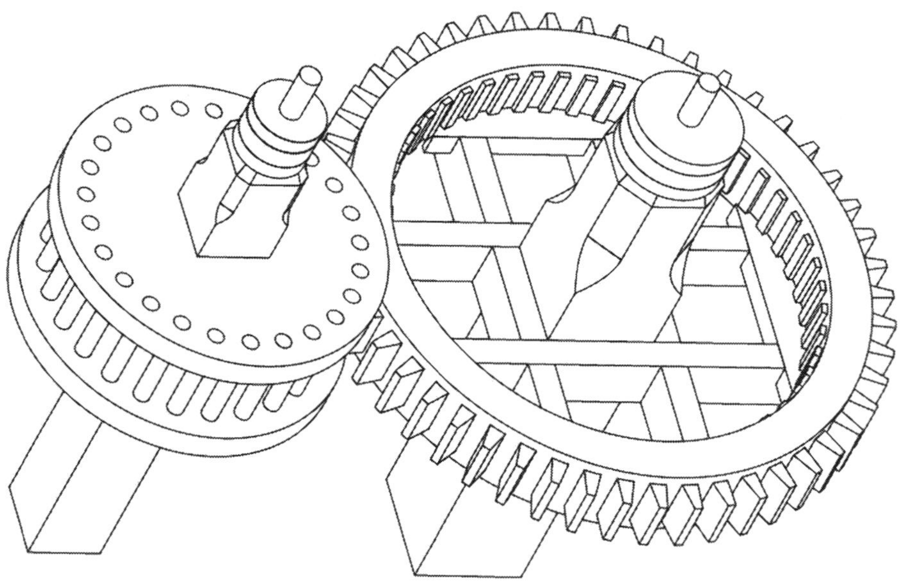

FIGURE 1.2
Ancient lantern gear drive. (Redrawn from Willis, R., 1841, *Principles of Mechanism*, London: John W. Parker, p. 43.)

steeper slope on the other flank. Since the ratchet asymmetric tooth wheels were known for many centuries, perhaps it is safe to presume that the asymmetric tooth gears might have been used much earlier than it was recorded in gear literature.

Similar to the ratchet wheels, the first asymmetric gears had a buttress tooth shape with low pressure angle at the drive tooth flanks, and with supporting (buttressing) coast flanks with high pressure angle.

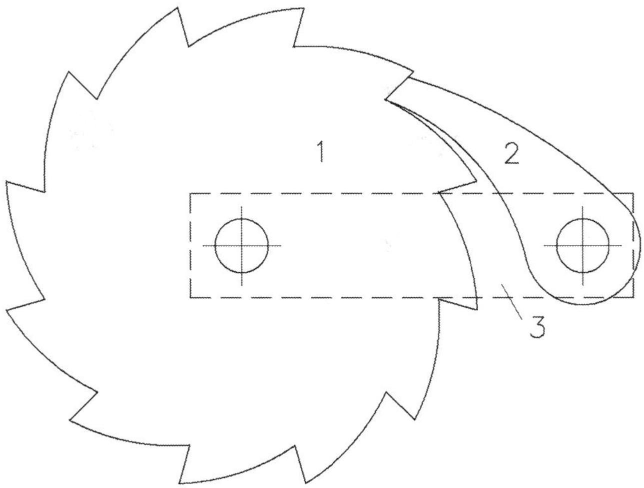

FIGURE 1.3
Ratchet mechanism; 1: asymmetric tooth wheel, 2: pawl, 3: base.

Apparently, a first documented description of asymmetric buttress gears was offered by the Cambridge Professor Robert Willis in his book "Principles of Mechanism" [2]: "In all the figures of teeth hitherto given the teeth are symmetrical, so that they will act whether the wheels be turned one way or the other. If a machine be of such a nature that the wheels are only required to turn in one direction, the strength of the teeth may be doubled by an alteration of form exhibited in Figure 1.4 (Figure 64 in [2], A.K.). This represents of circumference of a pair of wheels of which the lowest is the driver, and

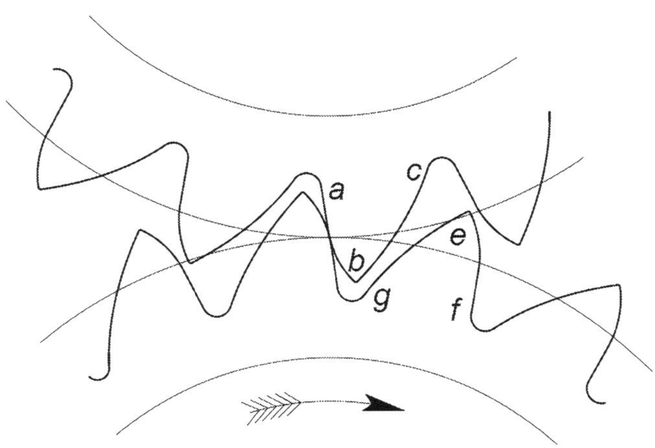

FIGURE 1.4
Asymmetric gear mesh. (Redrawn from Willis, R. 1841, *Principles of Mechanism*, London: John W. Parker, West Strand, Cambridge: J. & J.J. Deighton., p. 138.)

always moves in the direction of the arrow, consequently the right side of its teeth and the left side of the follower's teeth are the only portions that are ever called into action; and these sides are formed as usual (the epicycloid profile, A.K.). But the back of each tooth, both in the driver and follower, is proposed to be bounded by an arc of an involute, as *eg* or *cb*. The bases of these involutes being proportional to the pitch circles, they will during the motion be sure to clear each other, because, geometrically speaking, they would, if the wheels moved the reverse way, work together correctly; but the inclination of their common normal to the line of centers is too great for the transmission of pressure. The effect of this shape is to produce a very strong root, by taking away matter from the extremity of the tooth where the original form has more than is required for strength, and adding it to the root." In the second addition of his book "Principles of Mechanism" [3] Prof. Willis mentioned that this tooth form was "suggested by me in 1838."

In 1894, the German mechanical engineer-scientist Franz Reuleaux elaborated upon asymmetric tooth shape proportions in his book "The Constructor" [4]. He wrote: "By combining *evolute* and epicycloids—using the two curves for opposite sides of the same tooth—a profile of great strength is obtained. This form is of especial service for heavy-duty driving when motion is constantly in the same direction. From the peculiar form these have been called thumb-shaped teeth." As Robert Willis earlier, Reuleaux also used the epicycloid drive tooth flanks and the involute coast flanks, but he established the 53° coast pressure angle (Figure 1.5).

In 1908, Henry J. Spooner, Professor of the Polytechnic School of Engineering in London, describing, as he called them, the "Gee's buttress teeth" wrote in his book [5]: "In cases where a pair of wheels run always in the same direction the teeth may be strengthened by making their backs in the form of a buttress the driving faces of the teeth being of the usual form. In this way, it is claimed that they can be made 35 per cent, stronger than ordinary teeth. The back faces may be described by smaller rolling circles or by involutes of considerable obliquity, but whatever curve is used for them, there is an obvious increase of obliquity which makes this form of gearing quite unsuitable for use in cases where backlash is likely to occur, as severe stresses upon the teeth, rims, and journals are caused by the wedging action of the back of the teeth." He also added in the footnote: "These teeth were first suggested by Willis, in 1838" and "If all curves be involutes, a large base circle for the working sides *mn* (Figure 1.6) is required, and a small base circle for the opposite sides." His reference to Prof. Willis is apparently related to the buttress teeth in general, because in that book the buttress teeth have the epicycloid drive flank profile. So, Prof. Spooner's book [5] might present a first reference to the asymmetric teeth with both flanks formed by the involutes.

In 1910 Charles H. Logue, in his "American Machinist Gear Book" [6] describes buttress involute asymmetric tooth gears (Figure 1.7): "It is apparent that the object is to obtain a strong tooth or a pair of gears operating continuously in one direction. This is accomplished by increasing the angle of

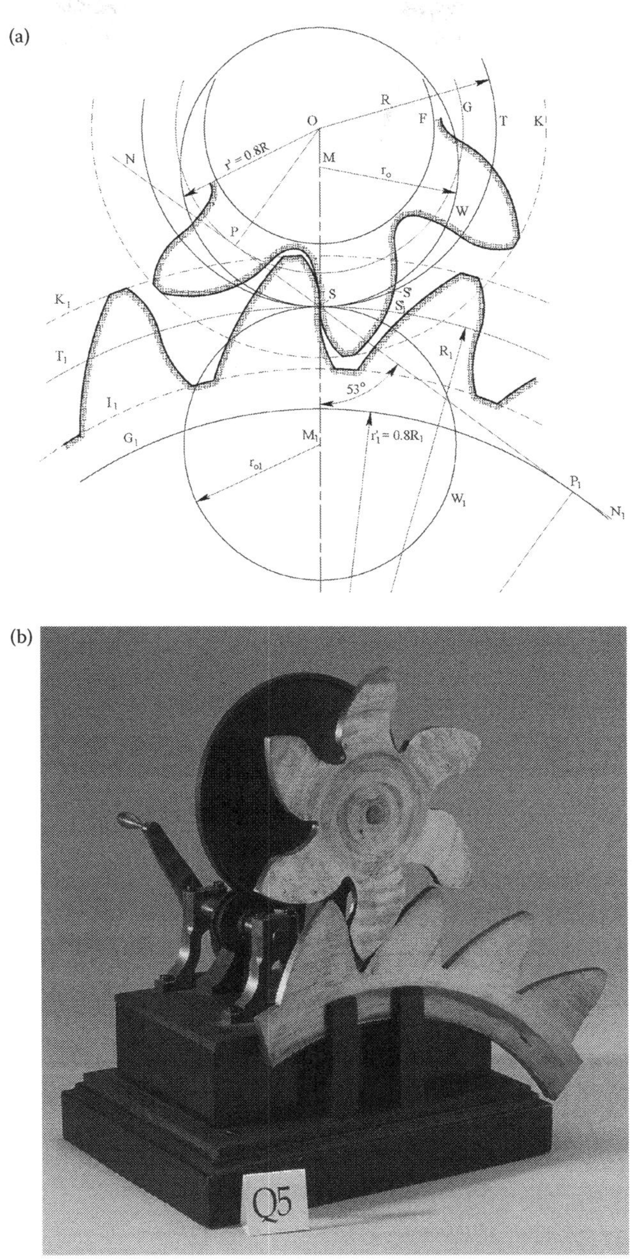

FIGURE 1.5
The Reuleaux's thumb-shaped tooth gears. Gear (a) mesh schematics. (Redrawn from Reuleaux, F., *The Constructor. A Hand-Book of Machine Design*, Philadelphia: H.H. Suplee, 1894, p. 134.) Gear (b) model from Reuleaux Collection of Kinematic Mechanisms of Cornell University. (From Jon Reis Photography, Q5 thumb shaped gear teeth profiles, Cornell University Library, http://kmoddl.library.cornell.edu/model_metadata.php?m=111. Copyright © Jon Reis. With permission.)

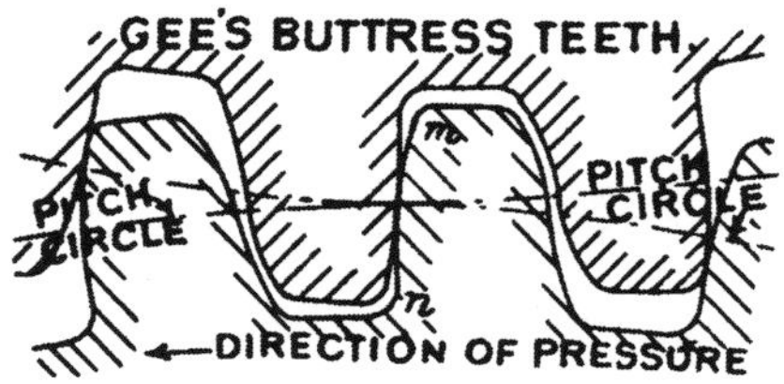

FIGURE 1.6
The Gee's buttress tooth gears. (Redrawn from Spooner, H.S., 1908, *Machine Drawing and Design for Beginners*. Longmans, Green, and Co., p. 170.)

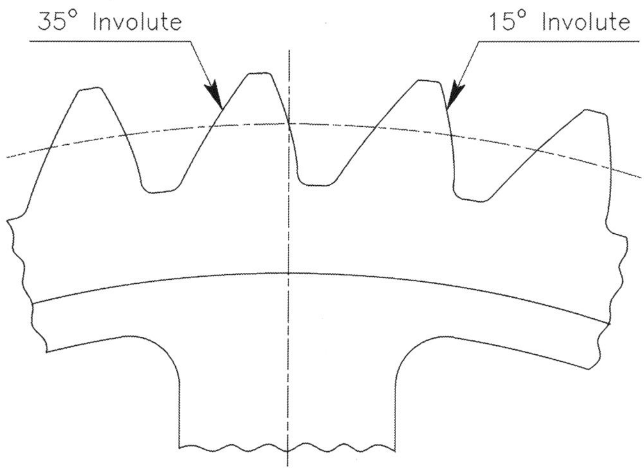

FIGURE 1.7
Buttressed tooth (a) and buttressed teeth in contact (b). (Redrawn from Logue, C.H., 1910, *American Machinist Gear Book*, McGraw-Hill Book Company, p. 30.)

obliquity of the back of the tooth, the face of the tooth being any angle desired. If the back of the tooth is correctly formed the gears will operate satisfactorily in either direction although with an increased pressure on their bearings when using the back face of the teeth owing to the increased obliquity of action. For many purposes, there is no objection to this, and it is a great wonder that this tooth is not more extensively used. Of course, there must be a limit to the angle of the back of the tooth. For practical purposes the curve at the top of tooth at the back should not extend farther than the center line of the tooth; for an addendum of I/p or 0.6866p, this will occur at an angle of about 32 degrees. A greater angle than this will subject the tooth to breakage at the point."

In 1912 the Austrian inventor Kais Konigl patented a process for manufacture of gear wheels with the asymmetric buttress teeth [7]. The loaded tooth flank was cycloidal and unloaded tooth flank had an involute profile.

In 1917 Prof. O.A. Leutwiler applied involute profiles for both drive and coast flanks of the buttress or, as he called them, hook-tooth gears (Figure 1.8).

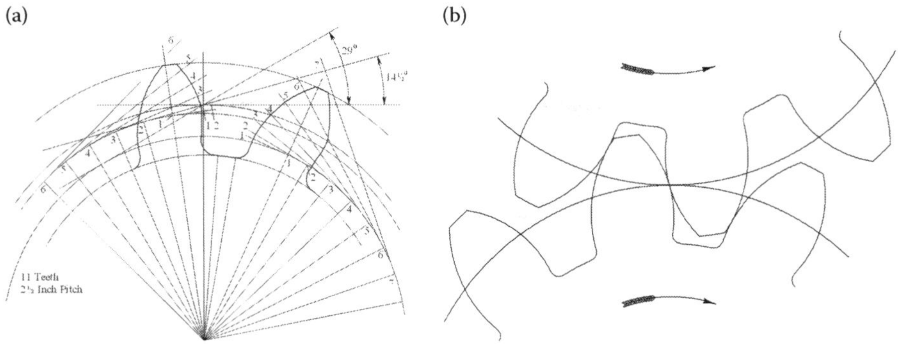

FIGURE 1.8
Buttress or hook-tooth gear. (Redrawn from Leutwiler, O.A., 1917, *Element of Machine Design*, London: McGraw-Hill Book Company, p. 134.)

He suggested the 15° pressure angle for drive flanks and the 35° pressure angle for coast flanks. He wrote [8]: "The buttress or hook-tooth gear can be used in cases where the power is always transmitted in the same direction. The load side of the tooth has the usual standard profile, while the back side of the tooth has a greater angle of obliquity. To compare its strength with that of the standard tooth, use the following method: Make a drawing of the two teeth and measure their thicknesses at the tops of the fillets; then the strength of the hook tooth is to the standard as the square of the tooth thickness is to the square of the thickness of the standard tooth."

Prof. Gustav Niemann and Dr. Heinz Glaubitz have compared the standard 20° pressure angle gears, nonstandard 30° pressure angle gears, internal tooth gears, and asymmetric tooth gears using the photoelastic tooth models and also pulsator tooth testing [9]. They studied two types of asymmetric gears, with the 40° drive pressure angle and the 15° drive coast pressure angle, and with the buttress teeth with the 15° drive pressure angle and the 40° drive coast pressure angle (Figure 1.9).

Authors of earlier and some modern publications about asymmetric gear geometry chose the buttress tooth form that has drive flanks with low pressure angle and the coast supporting flanks with high pressure angle, because it provides noticeable root bending stress reduction in comparison with the standard symmetric tooth form. However, in comparison to modern nonstandard optimized symmetric tooth gears, the buttress asymmetric tooth gears do not show significant difference in root bending stress. Although, it is well known that gear transmission power density depends mainly upon the tooth flank surface durability, which is defined by the contact stress level and scuffing resistance. From this point, the application of a high pressure angle for drive tooth flanks and a low pressure angle for coast tooth flanks is more promising. In addition, this tooth form provides lower stiffness and better gear mesh impact dampening.

In most publications, gear researchers defined asymmetric gear tooth geometry traditionally generated by the preselected asymmetric gear rack

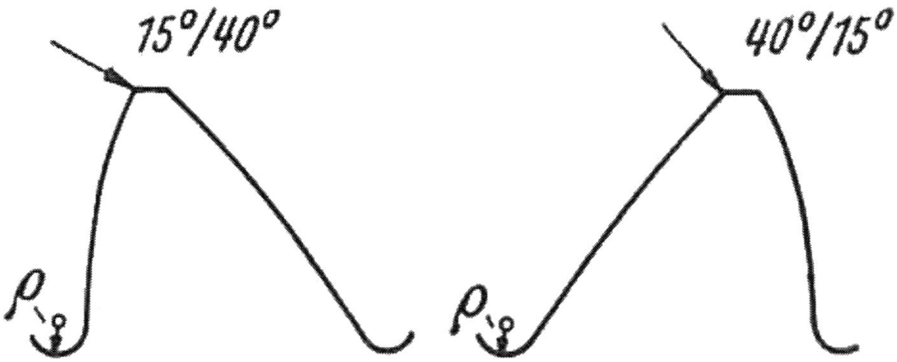

FIGURE 1.9
(a) Buttressed tooth with the low pressure angle drive flank; (b) Asymmetric tooth with the high pressure angle drive flank. (Redrawn from Niemann, G., Glaubitz, H., 1950. Zahnfußfestigkeit geradverzahnter Stirnräder aus Stahl, VDI-Zeitschrift, Band 92: p. 923.)

(Figure 1.10), which is typically modified from the standard symmetric rack by increasing the pressure angle of one flank. The opposite flank and other rack tooth proportions remain unchanged. This approach, although simple and accustomed for standard symmetric tooth gears, and also convenient for gear machining, is not the best choice for asymmetric tooth gear design. First of all, there are no standards or standard software for gears with asymmetric teeth. Second, asymmetric tooth gears are typically considered for custom demanding applications, where required performance cannot be achieved by using the best nonstandard optimized symmetric tooth gears. This makes essential optimizing asymmetric tooth geometry (see Chapter 5), which cannot be delivered by using an asymmetric tooling rack modified from a standard symmetric one or designed from scratch.

Direct Gear Design® is an alternative (to the traditional rack generation) gear geometry definition method. It is based on the theory of so-called generalized parameters authored by Prof. Edgar B. Vulgakov [10] that separates

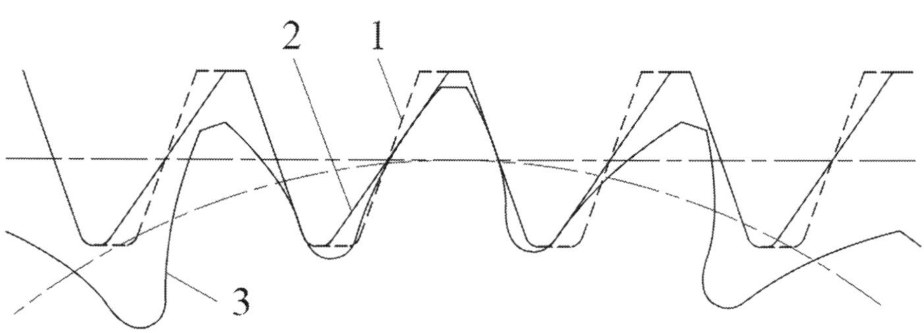

FIGURE 1.10
Asymmetric gear rack generation: 1: standard symmetric generating rack; 2: modified asymmetric generating rack profile; 3: gear profile.

gear geometry from fabrication technology and tooling parameters. This theory demonstrates how an involute gear tooth, a gear, and a gear mesh could be defined, analyzed, and optimized without using a generating rack. In the 1970s Prof. Vulgakov had initiated analytical and experimental research of asymmetric tooth gears for aviation gear drives, including helicopter and aero-engine gearboxes. Some results of this research were presented in his books and articles [10–13]. He applied his theory of generalized parameters to asymmetric tooth gears, defining their geometry without using rack generation parameters. According to his approach, an asymmetric tooth is constructed by two halves of symmetric teeth with different base circles (Figure 1.11). Later such asymmetric tooth presentation was utilized by several other gear researchers. In order to achieve the maximum operating pressure angle and contact ratio, the drive tooth flank uses one half of the symmetric tooth, which has the pointed tip and smaller base circle. A necessary tooth tip land is provided by the coast flank that is one half of the other symmetric tooth, which has a greater base circle and a large tooth tip land. Although this approach might look logical, the maximum drive flank transverse contact ratio and pressure angle of such asymmetric gears cannot be greater than for symmetric gears with the pointed tooth tips. This limitation does not allow for realization of all asymmetric tooth performance improvement potentials.

An alternative approach for the asymmetric tooth construction was utilized in the dissertation [14]. It presents an asymmetric tooth formed with two involutes of two different base circles. Figure 1.12 shows that such an asymmetric tooth has a much longer active involute flank than a symmetric

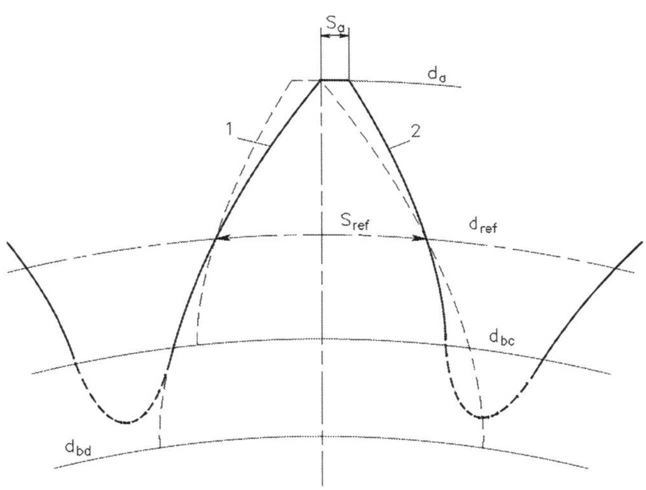

FIGURE 1.11
Asymmetric tooth constructed with two halves of the symmetric teeth; 1: drive flank from base diameter d_{bd}, 2: coast flank from base diameter d_{bc}, S_{ref}: tooth thickness at reference diameter d_{ref}, S_a: thickness at tooth tip diameter d_a.

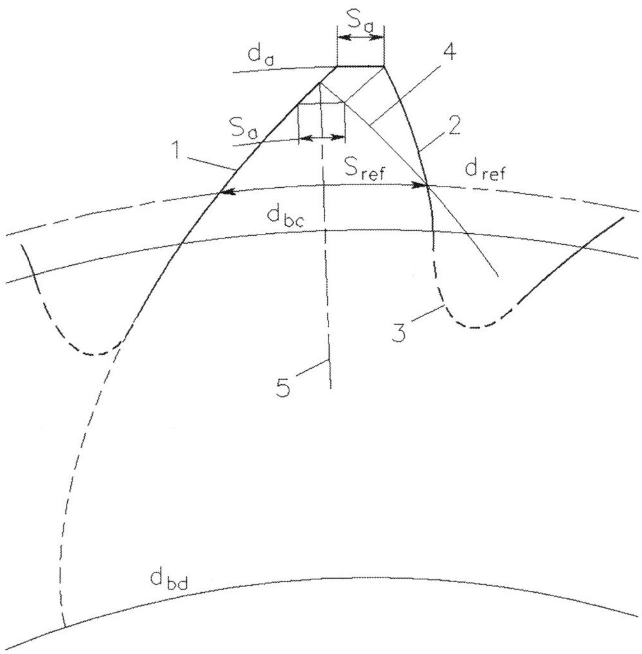

FIGURE 1.12
Asymmetric tooth constructed with two involutes; 1: drive flank of the base diameter d_{bd}, 2: asymmetric tooth coast flank of the base diameter d_{bc}, 3: root fillet, 4: coast flank of the symmetric tooth, 5: axis of the symmetric tooth formed by flanks 1 and 4, S_{ref}: tooth thicknesses at the reference diameter d_{ref}, S_a: tooth thicknesses at the outer diameter d_a.

tooth with identical drive pressure angle and tooth thicknesses at the reference and tip diameters. This simultaneously enables a high drive pressure angle and required contact ratio in the asymmetric gear mesh a way beyond the limitations of the asymmetric teeth constructed by two halves of different symmetric teeth (Figure 1.11). Such an approach addresses all possible asymmetric tooth profiles—from a virtually symmetric tooth shape to a tooth shape with extremely high asymmetry.

This approach to asymmetric gear design was first experimentally implemented in the gearbox of the electric generator driven by the 9I56 gas turbine engine produced by the Kaluga Dvigatel Plant (Kaluga, Russia), described in Section 11.1. Later this asymmetric gear tooth profile definition method was utilized in a number of gear test specimens (Figure 1.13), which were comprehensively tested to explore potentials of asymmetric gearing for aviation gear drives. Testing of these asymmetric tooth gears along with the similar benchmark symmetric gears demonstrated substantial benefits of asymmetric gears in load carrying capacity and also proved the possibility of noise and vibration level reduction. This experimental confirmation of the analytically predicted advantages of asymmetric gears

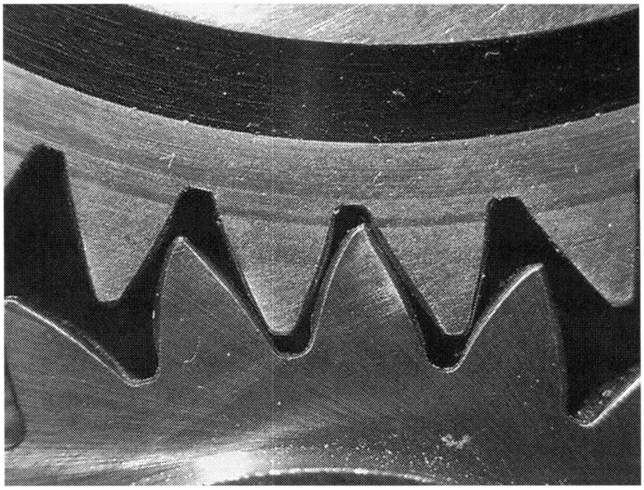

FIGURE 1.13
Asymmetric tooth gear test specimens. (Kapelevich A.L. Direct Gear Design Drives Performance. Gear Solutions, January 2004, 28–29. With permission.)

allowed them to be recommended for implementation in aviation gear transmissions. Apparently the first such implementation of asymmetric tooth gears was in the TV7–117S turboprop engine gearbox, described in detail in Section 11.2.

The inspection parameter definition of the asymmetric tooth gears was described in the article [15]. The possibility of application of asymmetric teeth in air compressor rotors was explored in the study [16]. In 1984, Prof. Bolotovsky with coauthors published the article [17], in which increasing of the coast pressure angle is described as a way to increase the asymmetric tooth bending strength. Authors analyzed the blocking contour of an asymmetric gear pair, defined a measurement over pins for an asymmetric tooth gear, and demonstrated the automotive gear pump with the asymmetric teeth with the increased flow rate.

The end of the twentieth century and beginning of the twenty-first could be definitely characterized as a period of an explosion of interest in asymmetric gearing. It attracts attention of more and more gear scientists and engineers, and becomes a subject of intensive research and development in over 20 countries, including China, Germany, India, Italy, Japan, Russia, Spain, Sweden, Turkey, Ukraine, USA, and others. Figure 1.14 shows an exponential growth of publications related to asymmetric gears. Unfortunately, a format of this book does not allow to mention of all modern explorers of the asymmetric tooth gears and their contributions in development of asymmetric gearing.

Table 1.1 presents references related to each research topic of the asymmetric tooth gears publications.

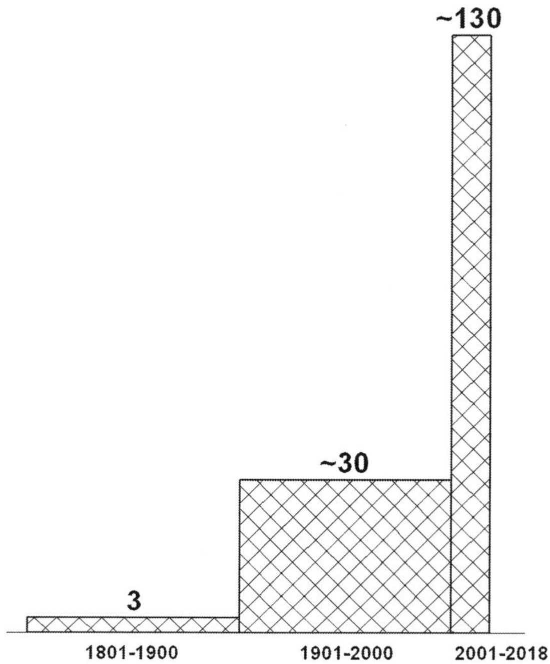

FIGURE 1.14
Grown of number of publications about symmetric gearing.

TABLE 1.1
Asymmetric Gear Publication References

Topics			Reference Number
Introduction and historical overview			1–6, 9, 19, 20–31, 40, 100, 107, 120, 124, 129, 134, 138, 149
Tooth type	Buttress		2–8, 17, 18, 26, 30–32, 38, 43, 57
	Cycloidal and involute		2–7
	Involute		8–164
Gear Type	Spur	Conventional	2–15, 17–25, 27–38, 40, 42–44, 46–48, 50–58, 60–75, 77–84, 86–106, 108, 109, 111–117, 119–128, 130–135, 137–158, 160–164
		High contact ratio	14, 18, 26, 64, 77, 107
	Helical		41, 45, 49, 59, 76, 85, 159
	Bevel		110, 119, 129
	Face, face-worm		39, 136
	Self-locking		53, 61, 107
Area of existence and blocking contour			12, 14, 17, 29, 79, 107

(Continued)

TABLE 1.1 (*Continued*)

Asymmetric Gear Publication References

Topics			Reference Number
Design approach	Traditional (rack generation)		17, 25, 27, 28, 32, 38, 42, 43, 47, 54, 60, 66, 73, 75, 76, 81, 85, 86, 91, 92, 96, 97, 99, 101, 106, 111, 112, 114, 120, 128, 137, 140, 143, 149, 153, 154, 158, 162, 163
	Direct Gear Design		10–14, 21, 24, 29, 33–35, 37, 40, 41, 44, 52, 58, 67–69, 71, 74, 79, 89, 100, 104, 105, 107, 126, 133, 144–148, 155, 157, 159, 160, 164
Optimization	Macrogeometry	Tooth flank	46, 60, 82, 87, 104, 107, 133, 144, 156
		Tooth root	44, 46, 66, 68, 73, 90, 93, 107, 117, 118, 129, 141, 144, 151, 157, 160, 161
	Microgeometry		36, 39, 128, 164
Stress analysis, math modeling, simulation	Modified Lewis equation		57, 70, 72, 99, 100, 116, 120, 131
	Photoelasticity		9, 14
	Finite element analysis		14, 30, 31, 38, 43, 45, 51, 56, 58, 60, 65, 82–84, 87, 92, 98, 100, 101, 107, 108, 113–115, 122, 125, 129, 132, 141–143, 145, 154, 159
	Boundary element analysis		91, 106, 112, 130
	Dynamic analysis		64, 65, 103, 113
Inspection			14, 15, 94, 107
Manufacturing technology, and tooling			14, 20, 54, 67, 74, 88, 100, 107, 126, 136–139, 142, 152, 153, 155, 156
Materials	Powder metal alloys		117, 118, 139, 150, 151, 156
	Polymers		40, 41, 50, 55, 70, 72, 80, 121, 135, 142
Testing			9, 14, 18, 26, 32, 38, 43, 46, 61, 89, 95, 100, 105, 107, 109, 113, 121, 127, 130, 142, 150, 151, 156
Application	Gearboxes		14, 34, 35, 62, 82, 95, 102, 107, 110, 119, 123, 125, 150, 151, 156
	Pumps and compressors		16, 17, 23, 41, 69, 102, 125, 127, 158

A growing number of publications, and a wide spectrum of the research topics, clearly indicate that asymmetric gearing definitely became one of the modern trends in advanced gear transmissions.

The following chapters of the book describe practically all aspects of asymmetric gearing from the tooth geometry analysis to applications of asymmetric tooth gears.

2

Macrogeometry of Asymmetric Tooth Gears

As it was mentioned in the Preface, the traditional rack generating gear design with its embedded performance limitations cannot be considered as a suitable design tool for asymmetric gear design. Besides, there are no standards for asymmetric gear design based on the rack generation, for asymmetric generating racks or tooling. A subject of this chapter is macrogeometry of involute asymmetric gears, analysis of the gear tooth and mesh parameters utilizing the Direct Gear Design method [107]. All these parameters, unless otherwise specified, are defined for spur gears or for helical gears in the transverse section. The following equations are defined, assuming that a normal backlash is equal to zero, unless otherwise specified.

2.1 Involute and Involute Function

Involute of a circle (Figure 2.1) is the path traced out by a point on the straight line that rolls around a circle.

Figure 2.1 shows the involute of a circle with a radius $R = 1$. It starts at the point A. A normal line to the involute at the point X is tangent to the circle with the center O at the point B. The angle α_x between the lines OX and OB is called a profile angle. The length of the arc AB is equal to the line BX, which is equal to $L = R \times \tan\alpha_x$. If the radius $R = 1$, $L = \tan\alpha_x$. Then the roll angle between lines OA and OB is equal to $\phi_x = L/R = \tan\alpha_x$ (in radians). This makes the involute angle between the lines OA and OX, also called the involute function, equal to

$$inv\alpha_x = \tan\alpha_x - \alpha_x \tag{2.1}$$

where the profile angle α_x is in radians.

An involute function along with common trigonometric functions is used for an involute gear analysis, including for the gears with asymmetric teeth.

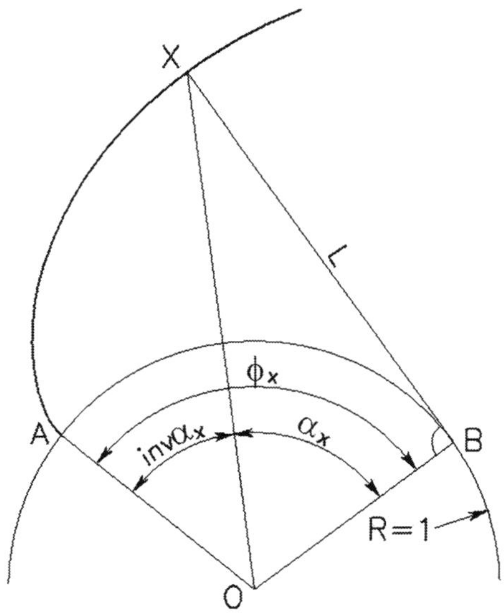

FIGURE 2.1
Involute of a circle.

2.2 Tooth Parameters

Two involute flanks of an asymmetric tooth (see Figures 2.2 and 2.3) are unwounded from two different base diameters d_{bd} and d_{bc}. In the following equations the symbol "*d*" is used for the drive flank and the symbol "*c*" is used for the coast flank of an asymmetric tooth. The involute tooth flanks are intersected at the point with the profile angles ν_d and ν_c.

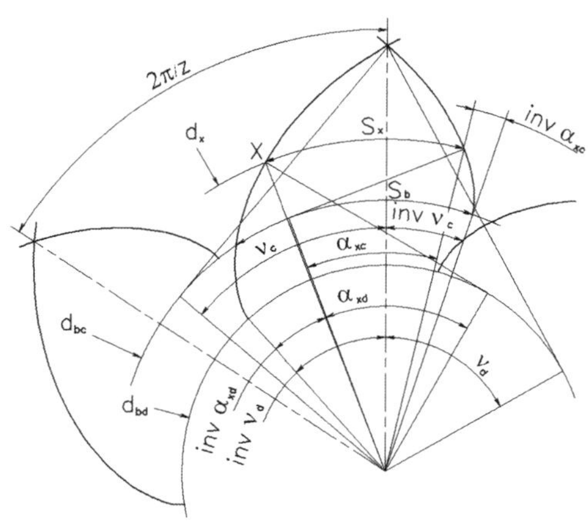

FIGURE 2.2
Involute flanks of the external asymmetric gear teeth.

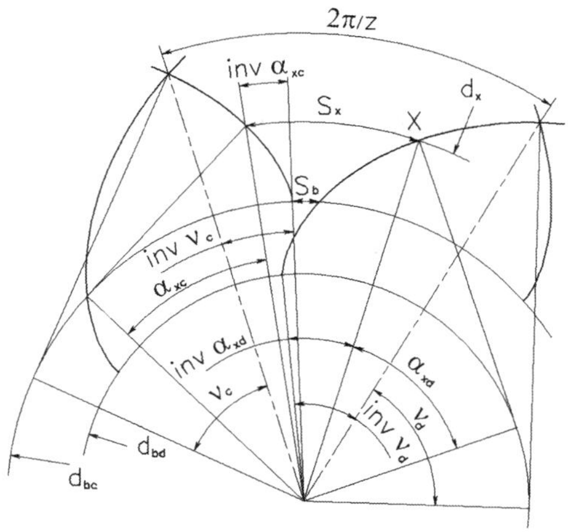

FIGURE 2.3
Involute flanks of the internal asymmetric gear teeth.

Diameter d_x at the point X of the drive flank can be defined as follows:

$$d_x = d_{bd}/\cos\alpha_{xd} = d_{bc}/\cos\alpha_{xc}. \quad (2.2)$$

Then the tooth asymmetry factor K is

$$K = d_{bc}/d_{bd} = \cos\alpha_{xc}/\cos\alpha_{xd} = \cos\nu_c/\cos\nu_d, \quad (2.3)$$

where α_{xd} and α_{xc} are the drive and coast flank involute profile angles at diameter d_x.

For most applications of asymmetric tooth gears, unless otherwise specified, the drive flank profile angle α_{xd} is greater than the coast flank profile angle α_{xc}. This means $d_{bd} < d_{bc}$ and asymmetry factor $K > 1.0$. For symmetric tooth gears $K = 1.0$.

At the coast base circle diameter d_{bc} the coast flank profile angle $\alpha_{bc} = 0$ and the drive flank profile angle α_d from (2.3) is

$$\alpha_d = \arccos(1/K). \quad (2.4)$$

The base tooth thickness of the asymmetric tooth can be defined only at the coast flank base circle d_{bc} (see Figures 2.2 and 2.3):

- for external tooth

$$S_b = \frac{d_{bc}}{2} \times inv(\nu_d) + inv(\nu_c) - inv(\arccos(1/K)), \quad (2.5)$$

- for internal tooth

$$S_b = \frac{d_{bc}}{2} \times 2\pi/z - inv(\nu_d) - inv(\nu_c) + inv\left(\arccos(1/K)\right), \tag{2.6}$$

where z is the number of teeth.

The relations between the drive and coast profile angles, described by Equation (2.3), are shown in Figure 2.4 at the different values of asymmetry factor K.

The base circular pitches are as follows:

- for the drive tooth flanks

$$p_{bd} = \frac{\pi d_{bd}}{z}, \tag{2.7}$$

- for the coast tooth flanks

$$p_{bc} = \frac{\pi d_{bc}}{z}. \tag{2.8}$$

From (2.3), the relation between the drive and coast tooth flank base circular pitches is

$$K = \frac{p_{bc}}{p_{bd}}. \tag{2.9}$$

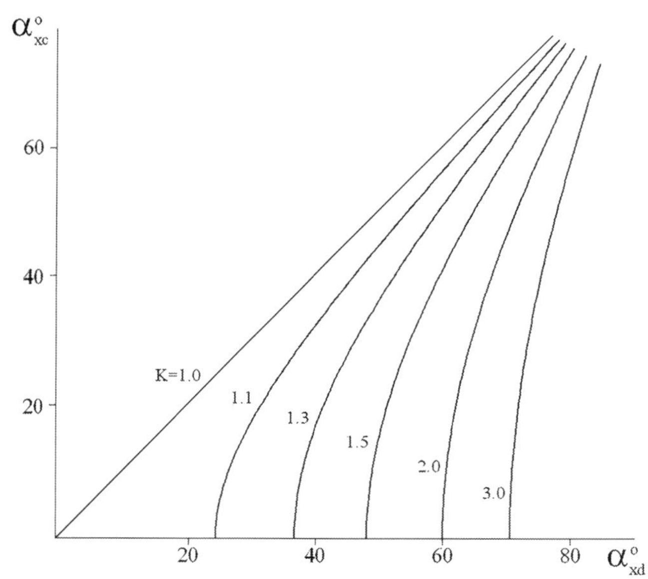

FIGURE 2.4
Relations between the drive and coast profile angles at the different values of an asymmetry factor K.

The tooth thickness at diameter d_x is

- for external tooth

$$S_x = \frac{d_x}{2} \times \left(inv(\nu_d) + inv(\nu_c) - inv(\alpha_{xd}) - inv(\alpha_{xc})\right) \tag{2.10}$$

or

$$S_x = \frac{d_{bd}}{2\cos\alpha_{xd}} \times \left(inv(\nu_d) + inv(\nu_c) - inv(\alpha_{xd}) - inv(\alpha_{xc})\right), \tag{2.11}$$

- for internal tooth

$$S_x = \frac{d_x}{2} \times \left(\frac{2\pi}{z} - inv(\nu_d) - inv(\nu_c) + inv(\alpha_{xd}) + inv(\alpha_{xc})\right) \tag{2.12}$$

or

$$S_x = \frac{d_{bd}}{2\cos\alpha_{xd}} \times \left(\frac{2\pi}{z} - inv(\nu_d) - inv(\nu_c) + inv(\alpha_{xd}) + inv(\alpha_{xc})\right). \tag{2.13}$$

An asymmetric tooth profile must include a tip land and tip radii, and a root fillet between the teeth (shown in the dash lines in Figures 2.5–2.7). The tooth tip diameter d_a is

$$d_a = d_{bd}/\cos\alpha_{ad} = d_{bc}/\cos\alpha_{ac}, \tag{2.14}$$

where α_{ad} and α_{ac} are the drive and coast profile angles at the diameter d_a.

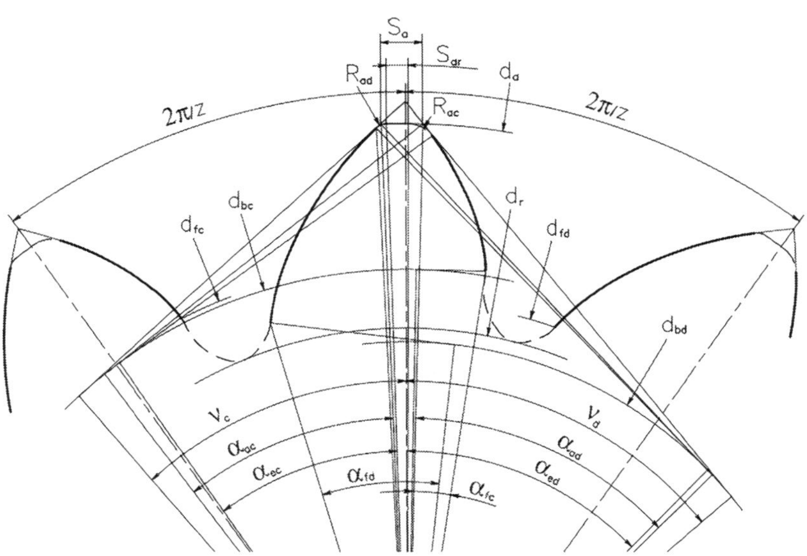

FIGURE 2.5
External asymmetric gear teeth.

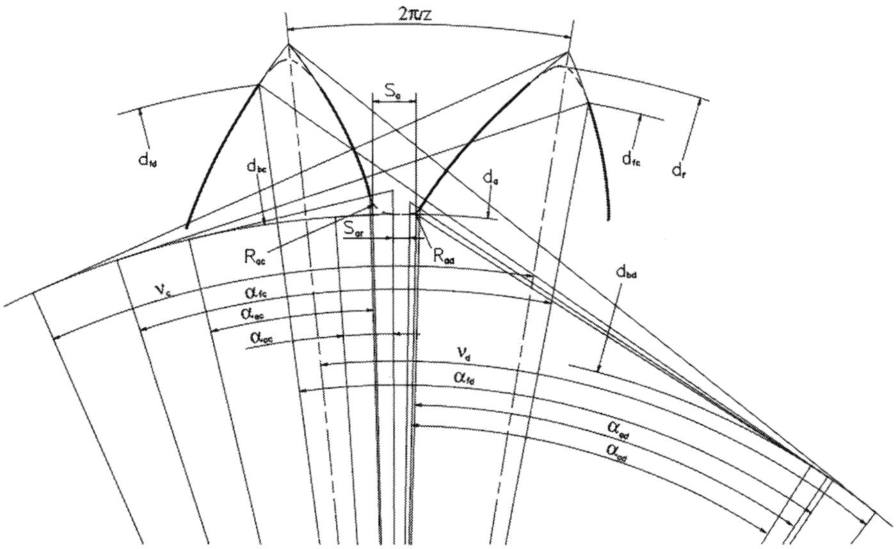

FIGURE 2.6
Internal asymmetric gear teeth.

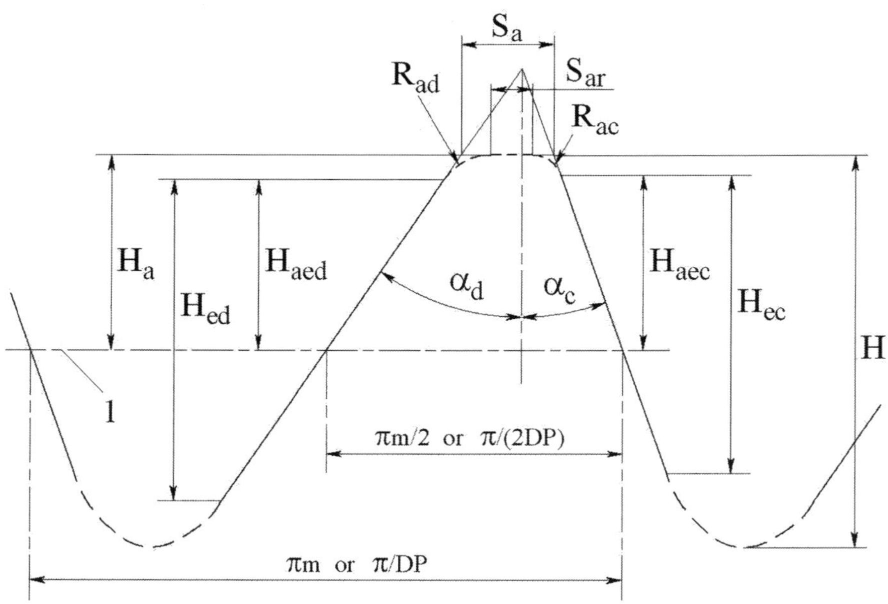

FIGURE 2.7
Asymmetric gear rack teeth; 1: nominal pitch line; α_d and α_c: rack profile angles; H_a: nominal tooth addendum; H_{aed} and H_{aec}: nominal effective tooth addendums; H_{ed} and H_{ec}: nominal effective tooth profile depths; H: whole tooth depth; m: module (in metric system); DP: diametral pitch (in English system).

The virtual tooth tip land S_a (Figure 2.5) of the external gear tooth is defined considering tooth tip radii equal to zero from Equation (2.10) or (2.11):

$$S_a = \frac{d_a}{2} \times \left(inv(\nu_d) + inv(\nu_c) - inv(\alpha_{ad}) - inv(\alpha_{ac}) \right) \tag{2.15}$$

or

$$S_a = \frac{d_{bd}}{2\cos\alpha_{ad}} \times \left(inv(\nu_d) + inv(\nu_c) - inv(\alpha_{ad}) - inv(\alpha_{ac}) \right), \tag{2.16}$$

- for internal tooth (Figure 2.6) from Equation (2.12) or (2.14)

$$S_a = \frac{d_a}{2} \times \left(2\pi/z - inv(\nu_d) - inv(\nu_c) + inv(\alpha_{ad}) + inv(\alpha_{ac}) \right) \tag{2.17}$$

or

$$S_a = \frac{d_{bd}}{2\cos\alpha_{ad}} \times \left(2\pi/z - inv(\nu_d) - inv(\nu_c)) + inv(\alpha_{ad}) + inv(\alpha_{ac}) \right). \tag{2.18}$$

The profile angles α_{ed} and α_{ec} are the effective involute angles at the tooth tip. They are defined considering the tooth tip radii R_{ad} and R_{ac} for external gears as follows:

- for drive flanks

$$\alpha_{ed} = \arctan\left(\tan\left(\arccos\left(\frac{d_{bd}}{d_a \mp 2R_{ad}} \right) \right) \pm \frac{2R_{ad}}{d_{bd}} \right), \tag{2.19}$$

- for coast flanks

$$\alpha_{ec} = \arctan\left(\tan\left(\arccos\left(\frac{d_{bc}}{d_a \mp 2R_{ac}} \right) \right) \pm \frac{2R_{ac}}{d_{bc}} \right). \tag{2.20}$$

In Equations (2.19) and (2.20), the top sign (+ or –) is for external gears and the bottom sign (+ or –) is for internal gears.

The tooth tip land S_{ar} of an external gear tooth is defined considering the tip radii R_{ad} and R_{ac}:

$$S_{ar} = \frac{d_a}{2} \times \left(inv(\nu_d) + inv(\nu_c) - \tan\alpha_{ed} - \tan\alpha_{ec} + \arctan\left(\tan\alpha_{ed} - \frac{2R_{ad}}{d_{bd}} \right) \right.$$
$$\left. + \arctan\left(\tan\alpha_{ec} - \frac{2R_{ac}}{d_{bc}} \right) \right). \tag{2.21}$$

The tooth tip land S_{ar} of the internal gear tooth is

$$S_{ar} = \frac{d_a}{2} \times \left(\frac{2\pi}{z} - inv(v_d) - inv(v_c) + \tan\alpha_{ed} + \tan\alpha_{ec} - \arctan\left(\tan\alpha_{ed} + \frac{2R_{ad}}{d_{bd}} \right) - \arctan\left(\tan\alpha_{ec} + \frac{2R_{ac}}{d_{bc}} \right) \right). \quad (2.22)$$

The root diameter d_r is defined as a result of the root fillet profile optimization (see Chapter 5).

The asymmetric gear rack tooth is shown in Figure 2.7. The tooth tip land S_{ar} of the rack tooth is

- in the metric system

$$S_{ar} = \frac{\pi m}{2} - H_{aed}\tan\alpha_d - R_{ad}\cos\alpha_d - H_{aec}\tan\alpha_c - R_{ac}\cos\alpha_c \quad (2.23)$$

or

- in the English system

$$S_{ar} = \frac{\pi}{2DP} - H_{aed}\tan\alpha_d - R_{ad}\cos\alpha_d - H_{aec}\tan\alpha_c - R_{ac}\cos\alpha_c, \quad (2.24)$$

where the module m (in millimeters, metric system) is the gear rack axial pitch divided by π and the diametral pitch DP (in 1/inch., English system) is the number π divided by the gear rack axial pitch.

The nominal gear rack tooth addendum H_a is

$$H_a = H_{aed} + R_{ad}(1 - \sin\alpha_d) = H_{aec} + R_{ac}(1 - \sin\alpha_c). \quad (2.25)$$

The virtual tooth tip land S_{ar} of the gear rack tooth is

- in the metric system

$$S_a = \frac{\pi m}{2} - H_a(\tan\alpha_d + \tan\alpha_c) \quad (2.26)$$

or

- in the English system

$$S_a = \frac{\pi}{2DP} - H_a(\tan\alpha_d + \tan\alpha_c). \quad (2.27)$$

2.3 Gear Mesh Characteristics

In this chapter, the gear geometry considers the gear tooth tip radii and the normal mesh backlash equal to zero. Effects of the tooth tip radii and backlash are taken into account in the pitch factor analysis (Chapter 2.4) and the tolerance analysis (Chapter 8).

In traditional gear design, the module m in the metric system (or diametral pitch DP in the English system) is a scale factor for defining the gear tooth size. Prof. E.B. Vulgakov [10] proposed to use a gear base circle diameter as a scale factor. In Direct Gear Design the nominal transverse operating module m_w in the metric system (or the nominal transverse operating diametral pitch DP_w in the English system) is used as a gear tooth scale factor to make it easy to understand for engineers, who are familiar with the traditional gear design. This also allows the nominal operating gear mesh center distance to be directly defined as follows:

$$a_w = \frac{m_w(z_2 \pm z_1)}{2} \text{ or } a_w = \frac{z_2 \pm z_1}{2DP_w}, \tag{2.28}$$

where a "+" sign is for external gear mesh and a "–" sign is for internal gear mesh; indexes "1" and "2" are related to parameters of the pinion with number of teeth z_1 and of the gear with number of teeth z_2 consequently. A pinion typically (but not always) has fewer teeth than a gear and is the driving element of the gear pair.

The operating pitch diameters of mating gears are as follows:

$$d_{w1,2} = m_w z_{1,2} \text{ or } d_{w1,2} = \frac{z_{1,2}}{DP_w}. \tag{2.29}$$

A necessary condition of a proper asymmetric gear engagement is equality of the drive and coast flank base pitches of the mating gears from Equations (2.7) and (2.8):

$$p_{bd1} = p_{bd2} \text{ and } p_{bc1} = p_{bc2}. \tag{2.30}$$

2.3.1 Pressure Angles

Figure 2.8 illustrates the definition of the pressure angles α_{wd} and α_{wc} in an asymmetric external, internal, and rack and pinion meshes.

The pinion and gear tooth thicknesses S_{w1} and S_{w2} at the operating pitch diameters $d_{w1,2}$ are defined by Equations (2.11) and (2.13) as follows:

- for an external and internal gearing, and rack and pinion

$$S_{w1} = \frac{d_{bd1}}{2\cos\alpha_{wd}}\left(inv(\nu_{d1}) + inv(\nu_{c1}) - inv(\alpha_{wd}) - inv(\alpha_{wc})\right), \tag{2.31}$$

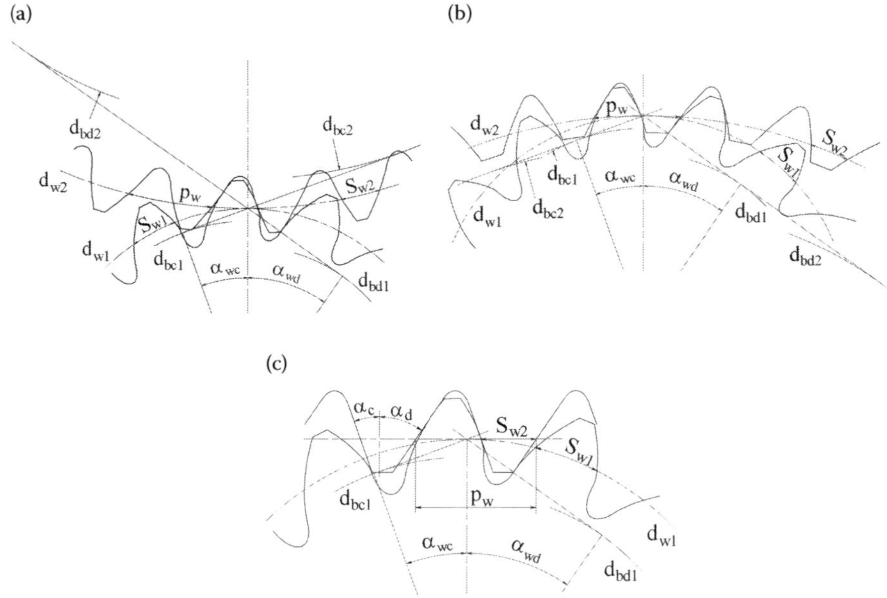

FIGURE 2.8
Pressure angles in an asymmetric gear mesh; (a) external, (b) internal, (c) rack and pinion.

- for an external gearing

$$S_{w2} = \frac{d_{bd2}}{2\cos\alpha_{wd}}\left(inv(\nu_{d2}) + inv(\nu_{c2}) - inv(\alpha_{wd}) - inv(\alpha_{wc})\right), \quad (2.32)$$

- for an internal gearing

$$S_{w2} = \frac{d_{bd2}}{2\cos\alpha_{wd}}\left(\frac{2\pi}{z_2} - inv(\nu_{d2}) - inv(\nu_{c2}) + inv(\alpha_{wd}) + inv(\alpha_{wc})\right). \quad (2.33)$$

The mating gear rack tooth thickness at the operating pitch line S_{w2} is

- for the metric system

$$S_{w2} = \pi m_w - \frac{d_{bd1}}{2\cos\alpha_{wd}}\left(inv(\nu_{d1}) + inv(\nu_{c1}) - inv(\alpha_{wd}) - inv(\alpha_{wc})\right), \quad (2.34)$$

- for the English system

$$S_{w2} = \frac{\pi}{DP_w} - \frac{d_{bd1}}{2\cos\alpha_{wd}}\left(inv(\nu_{d1}) + inv(\nu_{c1}) - inv(\alpha_{wd}) - inv(\alpha_{wc})\right). \quad (2.35)$$

The operating circular pitch p_w is

$$p_w = \frac{\pi \times d_{w1,2}}{z_{1,2}} = \frac{\pi \times d_{bd1,2}}{z_{1,2} \times \cos\alpha_{wd}} = \frac{\pi \times d_{bc1,2}}{z_{1,2} \times \cos\alpha_{wc}}. \quad (2.36)$$

It can be presented in the metric system as follows:

$$p_w = \pi m_w \quad (2.37)$$

or in the English system:

$$p_w = \frac{\pi}{DP_w}. \quad (2.38)$$

It is also

$$p_w = S_{w1} + S_{w2}. \quad (2.39)$$

The operating pressure angles for the drive flanks α_{wd} and for the coast flanks α_{wc} are defined by substitution of the S_{w1} and S_{w2} from Equations (2.31) and (2.32) or (2.33) into (2.39):

- for an external gear

$$inv(\alpha_{wd}) + inv(\alpha_{wc}) = \frac{1}{1+u}\left(inv(\nu_{d1}) + inv(\nu_{c1}) + u\left(inv(\nu_{d2}) + inv(\nu_{c2})\right) - \frac{2\pi}{z_1}\right), \quad (2.40)$$

- for an internal gear

$$inv(\alpha_{wd}) + inv(\alpha_{wc}) = \frac{1}{u-1}\left(u\left(inv(\nu_{d2}) + inv(\nu_{c2})\right) - inv(\nu_{d1}) + inv(\nu_{c1})\right), \quad (2.41)$$

where $u = z_2/z_1$ is the gear ratio.

From (2.3), the relation between pressure angles for the drive flanks α_{wd} and for the coast flanks α_{wc} is

$$\cos\alpha_{wc} = K\cos\alpha_{wd}. \quad (2.42)$$

In the rack and pinion mesh, the nominal operating pressure angles α_{wd} and α_{wc} are equal to the rack profile angles α_d and α_c:

$$\alpha_{wd} = \alpha_d \text{ and } \alpha_{wc} = \alpha_c. \quad (2.43)$$

2.3.2 Interference in Asymmetric Tooth Gears

Figure 2.9 shows the definition of the profile angles $\alpha_{pd1,2}$ and $\alpha_{pc1,2}$ at the points of the tooth contact, near the root fillet in asymmetric external, internal, and rack and pinion meshes.

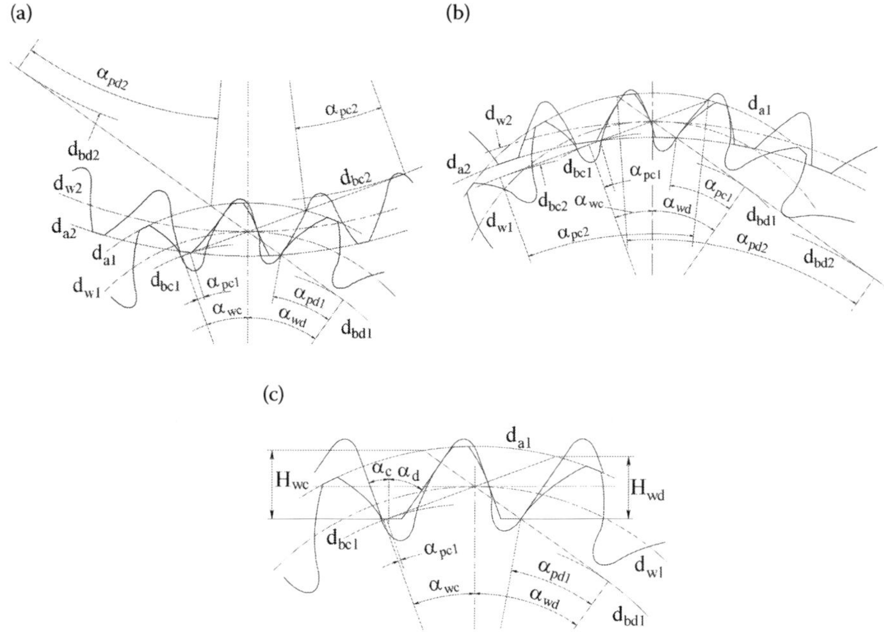

FIGURE 2.9
Profile angles $\alpha_{pd1,2}$ and $\alpha_{pc1,2}$ at the points of the tooth contact, near the root fillet in an asymmetric gear mesh; a: external, b: internal, c: rack and pinion.

The profile angles at the points of the tooth contact, near the root fillet, are as follows:

- for an external gear mesh, drive flanks

$$\alpha_{pd1} = \arctan\left((1+u)\tan\alpha_{wd} - u\tan\alpha_{ad2}\right), \tag{2.44}$$

$$\alpha_{pd2} = \arctan\left(\frac{1+u}{u}\tan\alpha_{wd} - \frac{1}{u}\tan\alpha_{ad1}\right), \tag{2.45}$$

- for an external gear mesh, coast flanks

$$\alpha_{pc1} = \arctan\left((1+u)\tan\alpha_{wc} - u\tan\alpha_{ac2}\right), \tag{2.46}$$

$$\alpha_{pc2} = \arctan\left(\frac{1+u}{u}\tan\alpha_{wc} - \frac{1}{u}\tan\alpha_{ac1}\right), \tag{2.47}$$

- for an internal gear mesh, drive flanks

$$\alpha_{pd1} = \arctan\left(u\tan\alpha_{ad2} - (u-1)\tan\alpha_{wd}\right), \tag{2.48}$$

$$\alpha_{pd2} = \arctan\left(\frac{u-1}{u}\tan\alpha_{wd} + \frac{1}{u}\tan\alpha_{ad1}\right), \tag{2.49}$$

- for an internal gear mesh, coast flanks

$$\alpha_{pc1} = \arctan\left(u\tan\alpha_{ac2} - (u-1)\tan\alpha_{wc}\right), \tag{2.50}$$

$$\alpha_{pc2} = \arctan\left(\frac{u-1}{u}\tan\alpha_{wc} + \frac{1}{u}\tan\alpha_{ac1}\right), \tag{2.51}$$

- for the rack and pinion mesh, drive flanks

$$\alpha_{pd1} = \arctan\left(\tan\alpha_d - \frac{2H_a}{d_{bd1}\sin\alpha_d}\right), \tag{2.52}$$

$$\alpha_{pd2} = \alpha_d, \tag{2.53}$$

- for a rack and pinion mesh, coast flanks

$$\alpha_{pc1} = \arctan\left(\tan\alpha_c - \frac{2H_a}{d_{bc1}\sin\alpha_c}\right), \tag{2.54}$$

$$\alpha_{pc2} = \alpha_c. \tag{2.55}$$

The rack tooth operating depths are as follows:

$$H_{wd} = \frac{d_{bd1}\sin\alpha_d}{2}\left(\tan\alpha_{ad1} - \tan\alpha_{pd1}\right), \tag{2.56}$$

$$H_{wc} = \frac{d_{bc1}\sin\alpha_c}{2}\left(\tan\alpha_{ac1} - \tan\alpha_{pc1}\right). \tag{2.57}$$

If asymmetry factor $K > 0$, interference occurs first at the coast involute flanks. If the profile angles α_{pc1} or α_{pc2} in the external mesh or angle α_{pc1} in the internal and rack and pinion meshes are less than zero, then their involute flanks close to the coast base diameters interfere with the mating gear tooth tips. This is called "tip-root" interference and it leads to the involute profile undercut near the tooth root.

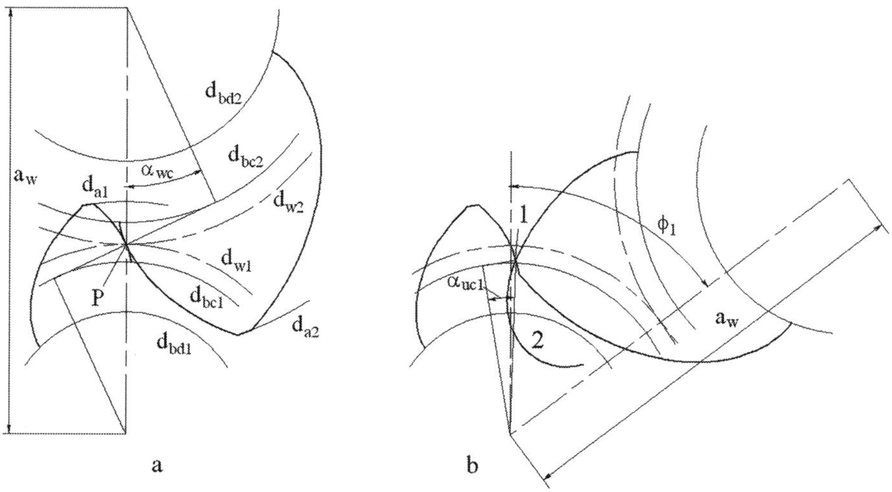

FIGURE 2.10
Definition of the undercut profile angle α_{uc1} in external mesh; a: initial tooth mesh position—gear teeth in contact at the pitch point P; b: undercut position; 1: undercut profile point, 2: trajectory of the mating gear tooth tip.

In an external mesh, the pinion profile angle α_{u1} at the undercut point is defined (Figure 2.10) by the equation system

$$\frac{\sin\left(inv(\alpha_{wc})-inv(\alpha_{uc1})\right)}{\cos\alpha_{uc1}}=(1+u)\frac{\sin\phi_1}{\cos\alpha_{wc}}-\frac{u}{\cos\alpha_{ac2}}\sin\left(\phi_1\left(1+\frac{1}{u}\right)\right.$$
$$\left.-inv(\alpha_{ac2})+inv(\alpha_{wc})\right)$$

$$\frac{\cos\left(inv(\alpha_{wc})-inv(\alpha_{u1})\right)}{\cos\alpha_{uc1}}=(1+u)\frac{\cos\phi_1}{\cos\alpha_{wc}}-\frac{u}{\cos\alpha_{ac2}}\cos\left(\phi_1\left(1+\frac{1}{u}\right)\right.$$
$$\left.-inv(\alpha_{ac2})+inv(\alpha_{wc})\right). \tag{2.58}$$

Similarly, in case of the gear profile undercut, the angle α_{u2} is defined by the system

$$\frac{u\sin\left(inv(\alpha_{w})-inv(\alpha_{uc2})\right)}{\cos\alpha_{uc2}}=(1+u)\frac{\sin\phi_2}{\cos\alpha_{wc}}-\frac{\sin\left(\phi_2(1+u)-inv(\alpha_{ac1})+inv(\alpha_{wc})\right)}{\cos\alpha_{ac1}}$$

$$\frac{u\cos\left(inv(\alpha_{w})-inv(\alpha_{u2})\right)}{\cos\alpha_{uc2}}=(1+u)\frac{\cos\phi_2}{\cos\alpha_{wc}}-\frac{\cos\left(\phi_2(1+u)-inv(\alpha_{ac1})+inv(\alpha_{wc})\right)}{\cos\alpha_{ac1}}. \tag{2.59}$$

In an internal mesh the pinion profile angle α_{u1} at the undercut point is defined (Figure 2.11) by the system

$$\frac{\sin\left(inv(\alpha_{wc}) - inv(\alpha_{uc1})\right)}{\cos\alpha_{uc1}} = \frac{u}{\cos\alpha_{ac2}}\sin\left(\phi_1\left(1-\frac{1}{u}\right) - inv(\alpha_{ac2}) + inv(\alpha_{wc})\right) - (u-1)\frac{\sin\phi_1}{\cos\alpha_{wc}}$$

$$\frac{\cos\left(inv(\alpha_{wc}) - inv(\alpha_{uc1})\right)}{\cos\alpha_{uc1}} = \frac{u}{\cos\alpha_{ac2}}\cos\left(\phi_1\left(1-\frac{1}{u}\right) - inv(\alpha_{ac2}) + inv(\alpha_{wc})\right) - (u-1)\frac{\cos\phi_1}{\cos\alpha_{wc}}. \tag{2.60}$$

For an internal tooth gear the angle α_{pc2} is always greater than zero and this kind of interference with undercut near the gear tooth fillet is not possible.

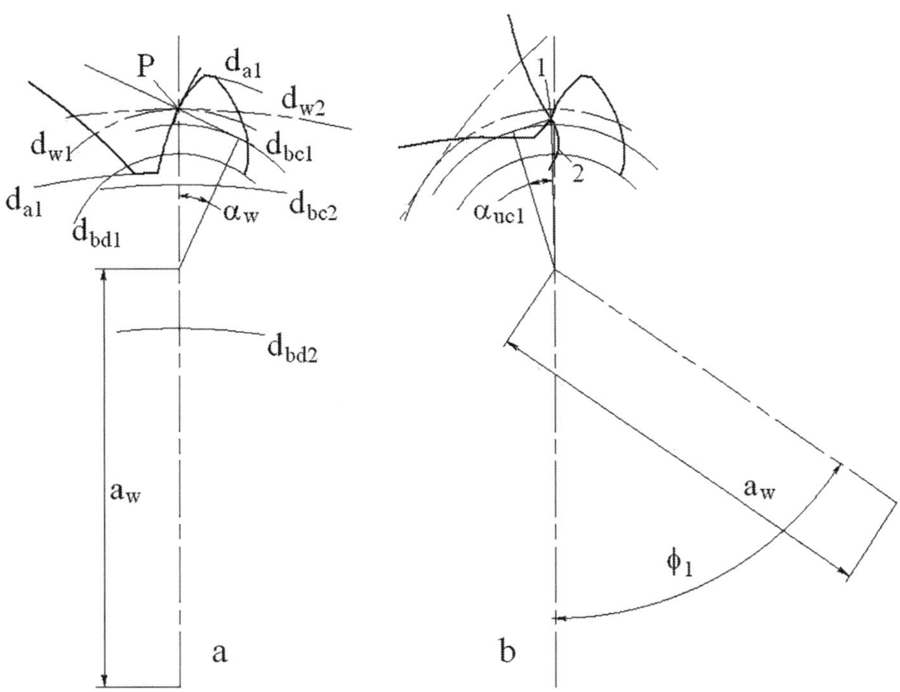

FIGURE 2.11
Definition of the undercut profile angle α_{uc1} in internal mesh; a: initial tooth mesh position—gear teeth in contact at the pitch point P; b: undercut position; 1: undercut profile point, 2: trajectory of the mating gear tooth tip.

For a rack and pinion mesh the pinion profile angle α_{u1} at the undercut point is defined (Figure 2.12) by the equation system

$$\begin{aligned}\frac{d_{bc1}\phi_1}{2\cos\alpha} - H_a\tan\alpha_c &= \frac{d_{bc1}}{2\cos\alpha_{uc1}}\sin\left(\phi_1 - inv(\alpha_c) - inv(\alpha_{uc1})\right)\\ \frac{d_{bc1}}{2\cos\alpha_c} - H_a &= \frac{d_{bc1}}{2\cos\alpha_{uc1}}\cos\left(\phi_1 - inv(\alpha_c) - inv(\alpha_{uc1})\right).\end{aligned} \tag{2.61}$$

The rack teeth do not have this kind of interference with undercut near the fillet, because $\alpha_{p2} = \alpha \geq 0$.

If an undercut occurs, the lowest contact point is coincided with the undercut point and the profile angles $\alpha_{pc1,2}$ become equal to profile angles at the undercut point $\alpha_{uc1,2}$.

There is other kind of interference, which is typical for internal gearing with a low tooth number difference $z_2 - z_1$. At certain gear geometry parameter combination, the tips of the mating gears may interfere (Figure 2.13). In the asymmetric tooth gear mesh, such "tip-tip" interference occurs first at the low pressure angle flanks. It can be avoided by increasing the coast flank pressure angle and/or reducing the tooth engagement depth.

The "tip-tip" interference does not occur if [107]

$$\Delta = \lambda_1 - u\lambda_2 \geq 0, \tag{2.62}$$

where

$$\lambda_{1,2} = \gamma_{1,2} + inv(\alpha_{ac1,2}) - inv(\alpha_{wc}), \tag{2.63}$$

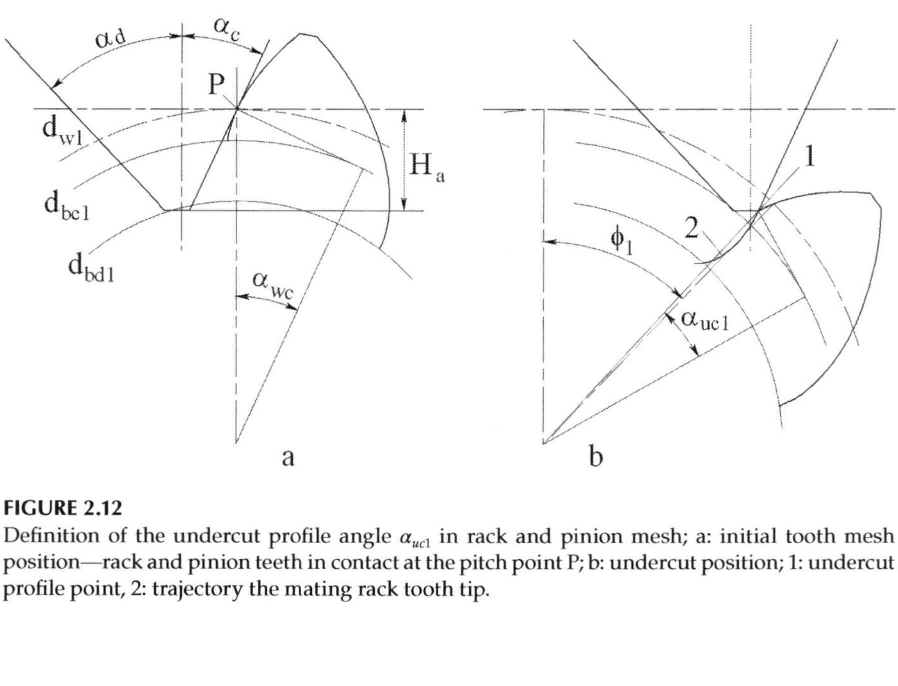

FIGURE 2.12
Definition of the undercut profile angle α_{uc1} in rack and pinion mesh; a: initial tooth mesh position—rack and pinion teeth in contact at the pitch point P; b: undercut position; 1: undercut profile point, 2: trajectory the mating rack tooth tip.

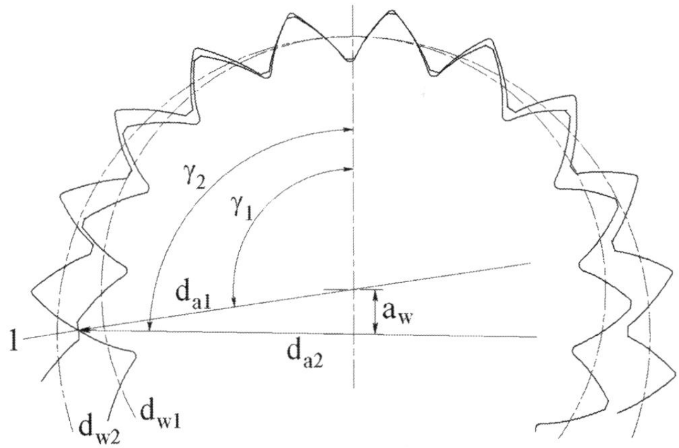

FIGURE 2.13
"Tip-tip" interference at the low pressure angle flanks in the internal mesh; 1: the interference point.

$$\gamma_1 = \pi - \arccos\left(\frac{\frac{d_{a1}^2}{4} + {a_w}^2 - \frac{d_{a2}^2}{4}}{d_{a1}a_w}\right), \tag{2.64}$$

$$\gamma_2 = \arccos\left(\frac{\frac{d_{a2}^2}{4} + {a_w}^2 - \frac{d_{a1}^2}{4}}{d_{a2}a_w}\right). \tag{2.65}$$

2.3.3 Transverse Contact Ratio

The transverse contact ratio is defined as a length of a line of contact divided by the base pitch. In asymmetric tooth gears, the transverse contact ratios are defined separately for the drive and coast tooth flanks.

Figure 2.14 illustrates the definition of the transverse contact ratios $\varepsilon_{\alpha d}$ and $\varepsilon_{\alpha c}$ in an asymmetric external, internal, and rack and pinion meshes.

The transverse contact ratios are as follows:

- for an external gear mesh, drive flanks

$$\varepsilon_{\alpha d} = \frac{z_1}{2\pi}\left(\tan\alpha_{ad1} + u\tan\alpha_{ad2} - (1+u)\tan\alpha_{wd}\right), \tag{2.66}$$

- for an external gear mesh, coast flanks

$$\varepsilon_{\alpha c} = \frac{z_1}{2\pi}\left(\tan\alpha_{ac1} + u\tan\alpha_{ac2} - (1+u)\tan\alpha_{wc}\right) \tag{2.67}$$

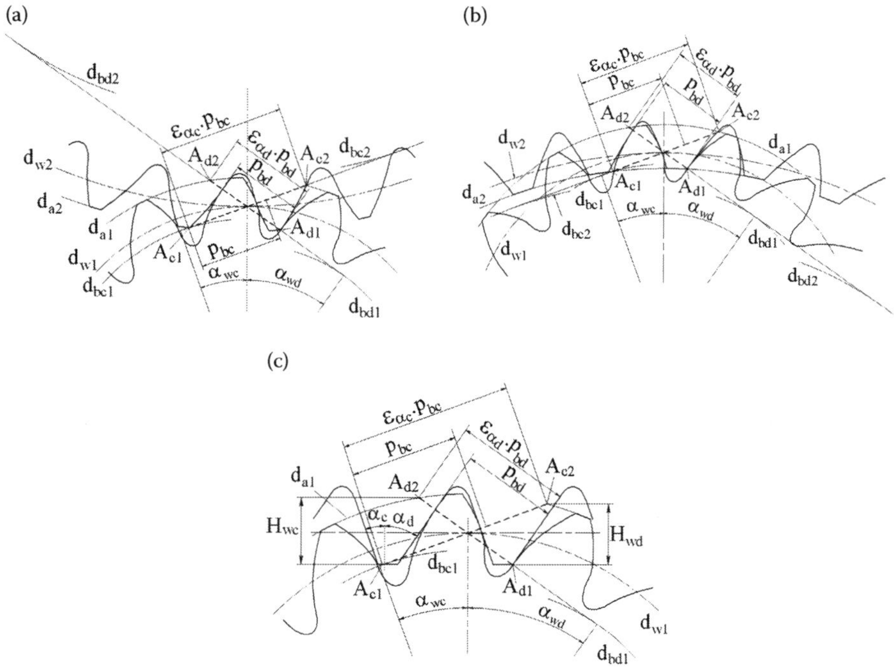

FIGURE 2.14
Transverse contact ratios in asymmetric gear mesh; A_{d1}–A_{d2}: drive flank line of contact, A_{c1}–A_{c2}: coast flank line of contact; a: external, b: internal, c: rack and pinion.

- for an internal gear mesh, drive flanks

$$\varepsilon_{\alpha d} = \frac{z_1}{2\pi}\left(\tan\alpha_{ad1} - u\tan\alpha_{ad2} + (u-1)\tan\alpha_{wd}\right) \tag{2.68}$$

- for an internal gear mesh, coast flanks

$$\varepsilon_{\alpha c} = \frac{z_1}{2\pi}\left(\tan\alpha_{ac1} - u\tan\alpha_{ac2} + (u-1)\tan\alpha_{wc}\right) \tag{2.69}$$

- for a rack and pinion mesh, drive flanks

$$\varepsilon_{\alpha d} = \frac{z_1}{2\pi}\left(\tan\alpha_{ad1} - \tan\alpha_d + \frac{2H_a}{d_{bd1}\sin\alpha_d}\right), \tag{2.70}$$

- for a rack and pinion mesh, coast flanks

$$\varepsilon_{\alpha c} = \frac{z_1}{2\pi}\left(\tan\alpha_{ac1} - \tan\alpha_c + \frac{2H_a}{d_{bc1}\sin\alpha_c}\right). \tag{2.71}$$

Alternatively, the contact ratios for external asymmetric gear meshes are defined with the following equations:

- for drive flanks

$$\varepsilon_{\alpha d} = \frac{z_1}{2\pi}\left((1+u)\tan\alpha_{wd} - \tan\alpha_{pd1} - u\tan\alpha_{pd2}\right), \tag{2.72}$$

- for coast flanks

$$\varepsilon_{\alpha c} = \frac{z_1}{2\pi}\left((1+u)\tan\alpha_{wc} - \tan\alpha_{pc1} - u\tan\alpha_{pc2}\right). \tag{2.73}$$

The contact ratios for internal asymmetric gear meshes are defined with the following equations:

- for drive flanks

$$\varepsilon_{\alpha d} = \frac{z_1}{2\pi}\left(u\tan\alpha_{pd2} - \tan\alpha_{pd1} - (u-1)\tan\alpha_{wd}\right), \tag{2.74}$$

- for coast flanks

$$\varepsilon_{\alpha c} = \frac{z_1}{2\pi}\left(u\tan\alpha_{pc2} - \tan\alpha_{pc1} - (u-1)\tan\alpha_{wc}\right). \tag{2.75}$$

The rack and pinion mesh contact ratios are defined with the following equations:

- for drive flanks

$$\varepsilon_{\alpha d} = \frac{z_1}{2\pi}\left(\tan\alpha_d - \tan\alpha_{pd1} + \frac{2(H_{wd} - H_{aw})}{d_{bd1}\sin\alpha_d}\right), \tag{2.76}$$

- for coast flanks

$$\varepsilon_{\alpha c} = \frac{z_1}{2\pi}\left(\tan\alpha_c - \tan\alpha_{pc1} + \frac{2(H_{wc} - H_{aw})}{d_{bc1}\sin\alpha_c}\right) \tag{2.77}$$

where H_{wd} and H_{wc} are effective rack tooth depths.

The transverse contact ratios of gears with asymmetric teeth can be presented using parameters for only one of mating gears:

- for drives flanks

$$\varepsilon_{\alpha d} = \frac{z_1}{2\pi}\left(\tan\alpha_{ad1} - \tan\alpha_{pd1}\right), \tag{2.78}$$

$$\varepsilon_{\alpha d} = \frac{z_2}{2\pi}\left(\pm\tan\alpha_{ad2} \mp \tan\alpha_{pd2}\right), \tag{2.79}$$

- for coast flanks

$$\varepsilon_{\alpha c} = \frac{z_1}{2\pi}\left(\tan\alpha_{ac1} - \tan\alpha_{pc1}\right), \tag{2.80}$$

$$\varepsilon_{\alpha c} = \frac{z_2}{2\pi}\left(\pm\tan\alpha_{ac2} \mp \tan\alpha_{pc2}\right). \tag{2.81}$$

In Equation (2.81) the top sign (+ or –) is for external gears and the bottom sign (+ or –) is for the internal gears.

For a rack and pinion mesh, the transverse contact ratios can also be defined in the same way:

- for drives flanks

$$\varepsilon_{\alpha d} = \frac{H_{wd}}{p_{bd}\sin\alpha_d} = \frac{2H_{wd}}{p_w\sin 2\alpha_d}, \tag{2.82}$$

- for coast flanks

$$\varepsilon_{\alpha c} = \frac{H_{wc}}{p_{bc}\sin\alpha_c} = \frac{2H_{wc}}{p_w\sin 2\alpha_c} \tag{2.83}$$

or in the metric system

- for drives flanks

$$\varepsilon_{\alpha d} = \frac{2H_{wd}}{\pi m_w \sin 2\alpha_d} \tag{2.84}$$

- for coast flanks

$$\varepsilon_{\alpha c} = \frac{2H_{wc}}{\pi m_w \sin 2\alpha_c} \tag{2.85}$$

and in the English system

- for drives flanks

$$\varepsilon_{\alpha d} = \frac{2DP \times H_{wd}}{\pi\sin 2\alpha_d} \tag{2.86}$$

- for coast flanks

$$\varepsilon_{\alpha c} = \frac{2DP \times H_{wc}}{\pi\sin 2\alpha_c}. \tag{2.87}$$

2.3.4 Helical Gearing Characteristics

A helical gear has true involute tooth flank profiles only in the transverse section perpendicular to the gear axis. However, helical gears are typically specified by parameters in the normal to a tooth line section, because in many cases they are fabricated using the generating rack tooling, for example, hob cutters. Parameters of this tooling are defined in the normal section and include a normal module or diametral pitch, normal pressure angle, and tooth proportions, like an addendum, whole depth, cutter tip radius, and addendum modification, also known as the X-shift. Since the Direct Gear Design method utilizes an established standard gear nomenclature and specifications as much as it possible, asymmetric helical gears are typically also specified with the normal tooth section parameters.

In the metric system gear, the normal module is

$$m_n = m_w \cdot \cos\beta_w, \quad (2.88)$$

In the English system gear, the normal diametral pitch is

$$DP_n = DP_w / \cos\beta_w, \quad (2.89)$$

where β_w is the helix angle at the operating pitch diameters d_{w1} and d_{w2}.

The normal pressure angles are as follows:

- for drive flanks

$$\alpha_{nd} = \arctan\left(\tan\alpha_{wd} \cdot \cos\beta_w\right), \quad (2.90)$$

- for coast flanks

$$\alpha_{nc} = \arctan\left(\tan\alpha_{wc} \cdot \cos\beta_w\right). \quad (2.91)$$

Spur gears have only a transverse contact ratio. Helical gears have also an overlap ratio or face contact ratio that in addition to a transverse contact ratio results in a total contact ratio, which is

- for drive flanks

$$\varepsilon_{\gamma d} = \varepsilon_{\alpha d} + \varepsilon_\beta, \quad (2.92)$$

- for coast flanks

$$\varepsilon_{\gamma c} = \varepsilon_{\alpha c} + \varepsilon_\beta. \quad (2.93)$$

An overlap ratio (or a face contact ratio) ε_β is defined by the angular shift ϕ of the helical gear front and back sections (see Figure 2.15):

$$\phi = \frac{2AC}{d_b} = \frac{2AB}{d_b} \tag{2.94}$$

or

$$\phi = \frac{2b_w}{d_b}\tan\beta_b, \tag{2.95}$$

where AC is the arc shift of the helical gear sections, AB is the shift of helical gear sections projected on the contact plane tangent to the base cylinder at the point A, b_w is the contact face width, a distance between helical gear front and back sections in contact, β_b is the helix angle at the base cylinder.

For helical asymmetric tooth gears, the drive and coast flank helix angles at the base cylinders are as follows:

$$\beta_{bd} = \arctan\left(\tan\beta \times \cos\alpha_{wd}\right) \tag{2.96}$$

and

$$\beta_{bc} = \arctan\left(\tan\beta \times \cos\alpha_{wc}\right). \tag{2.97}$$

For two mating gears

$$\phi_1 = u \times \phi_2. \tag{2.98}$$

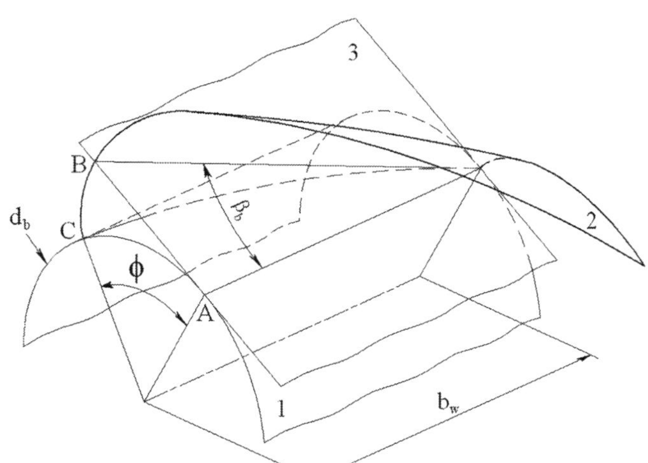

FIGURE 2.15
Angular shift ϕ of the helical gear front and back sections; 1: base cylinder, 2: helical involute surface, 3: contact plane tangent to the base cylinder.

Overlap ratio ε_β is defined by Equations (2.94), (2.95), and (2.98),

$$\varepsilon_\beta = \frac{AB}{p_b} = \frac{z_1}{2\pi}\phi_1 = \frac{z_2}{2\pi}\phi_2 = \frac{b_w}{p_b}\tan\beta_b. \tag{2.99}$$

For asymmetric gears, the overlap ratio is identical for the drive and coast flanks because according to Equations (2.7), (2.8), and (2.9)

$$p_{bc} = p_{bd}\frac{d_{bc}}{d_{bd}} = p_{bd} \times K \tag{2.100}$$

and

$$\tan\beta_{bc} = \tan\beta_{bd}\frac{d_{bc}}{d_{bd}} = \tan\beta_{bd} \times K. \tag{2.101}$$

This allows Equation (2.99) for asymmetric gears to be presented as

$$\varepsilon_\beta = \varepsilon_{\beta d} = \varepsilon_{\beta c} = \frac{b_w}{p_{bd}}\tan\beta_{bd} = \frac{b_w}{p_{bc}}\tan\beta_{bc}. \tag{2.102}$$

2.4 Pitch Factor Analysis

This paragraph presents an alternative method of involute gear geometry parameters' and mesh definition, which is called the pitch factor analysis [14,24,148].

The gear mesh operating circular pitch is

$$p_w = \frac{\pi \times d_{w1,2}}{z_{1,2}} = S_{w1} + S_{w2} + S_{bl}, \tag{2.103}$$

where S_{w1} and S_{w2} are the pinion and gear tooth thicknesses at the operating pitch diameter;

S_{bl} is the arc backlash.

The tooth thicknesses S_{w1} and S_{w2} are as follows (see Figure 2.16):

- for asymmetric gears

$$S_{w1,2} = S_{d1,2} + S_{c1,2} + S_{v1,2}, \tag{2.104}$$

where

$$S_{d1,2} = \frac{d_{w1,2}}{2}\left(\pm inv(\alpha_{ed1,2}) \mp inv(\alpha_{wd})\right) \tag{2.105}$$

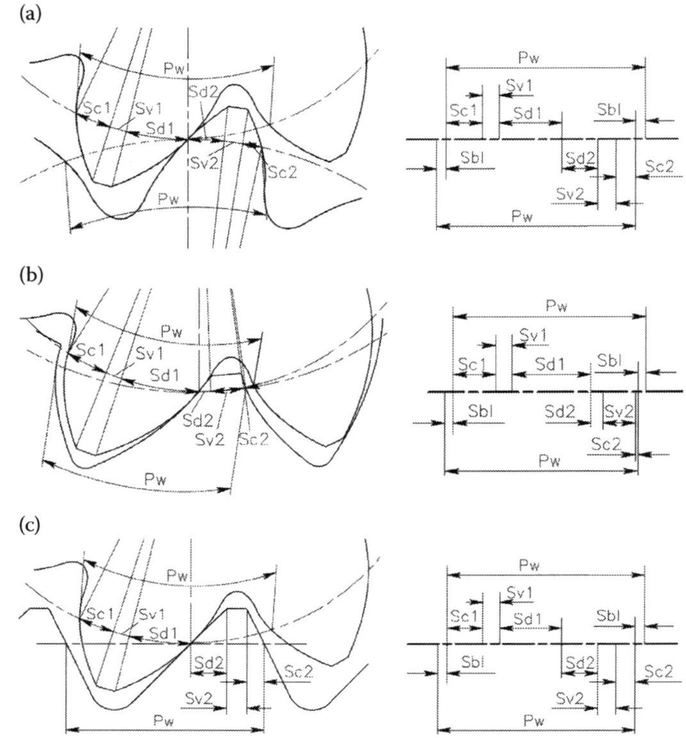

FIGURE 2.16
Asymmetric gear mesh and operating pitch components; a: external, b: internal, c: rack and pinion.

are projections of the addendum portion of the drive involute flank on the pitch circle;

$$S_{c1,2} = \frac{d_{w1,2}}{2}\left(\pm inv\left(\alpha_{ec1,2}\right) \mp inv\left(\alpha_{wc}\right)\right) \tag{2.106}$$

are projections of the addendum portion of the coast involute flank on the pitch circle; a top sign (+ or –) is for the external gears and a bottom sign (+ or –) is for the internal gears; the $S_{v1,2}$ are the pitch circle projections of the tip land and radii.

For the rack and pinion gear mesh the S_{d2}, S_{c2}, and S_{v2} are the pitch line projections. In this case a projection of the addendum portion of the drive flank on the pitch line is

$$S_{d2} = H_{aed} \tan \alpha_d. \tag{2.107}$$

A projection of the addendum portion of the coast flank on the pitch line is

$$S_{c2} = H_{aec} \tan \alpha_c. \tag{2.108}$$

Then the gear mesh operating circular pitch from (2.103) is

$$p_w = S_{d1} + S_{d2} + S_{c1} + S_{c2} + S_{v1} + S_{v2} + S_{bl}. \tag{2.109}$$

A pitch factor equation is a result of division of Equation (2.109) by operating circular pitch p_w

$$\theta_d + \theta_c + \theta_v = 1, \tag{2.110}$$

where θ_d is the drive pitch factor that is defined as follows:

$$\theta_d = \frac{S_{d1} + S_{d2}}{p_w}, \tag{2.111}$$

θ_c is the coast pitch factor that is defined as follows:

$$\theta_c = \frac{S_{c1} + S_{c2}}{p_w}, \tag{2.112}$$

and θ_v is the noncontact pitch factor that that is defined as follows:

$$\theta_v = \frac{S_{v1} + S_{v2} + S_{bl}}{p_w}. \tag{2.113}$$

The drive pitch factor is

- for the external gear mesh

$$\theta_d = \frac{z_1}{2\pi}\left(inv(\alpha_{ed1}) + u\,inv(\alpha_{ed2}) - (1+u)\,inv(\alpha_{wd})\right), \tag{2.114}$$

- for the internal gear mesh

$$\theta_d = \frac{z_1}{2\pi}\left(inv(\alpha_{ed1}) - u\,inv(\alpha_{ed2}) + (u-1)\,inv(\alpha_{wd})\right). \tag{2.115}$$

The coast pitch factor is

- for the external gear mesh

$$\theta_c = \frac{z_1}{2\pi}\left(inv(\alpha_{ec1}) + u\,inv(\alpha_{ec2}) - (1+u)\,inv(\alpha_{wc})\right), \tag{2.116}$$

- for the internal gear mesh

$$\theta_c = \frac{z_1}{2\pi}\left(inv(\alpha_{ec1}) - u\,inv(\alpha_{ec2}) + (u-1)\,inv(\alpha_{wc})\right). \tag{2.117}$$

For the rack and pinion gear mesh the drive and coast pitch factors are as follows:

$$\theta_d = \frac{z_1}{2\pi}\left(inv(\alpha_{ed1}) - inv(\alpha_d)\right) + \frac{H_{aed}}{p_w}\tan\alpha_d \tag{2.118}$$

and

$$\theta_c = \frac{z_1}{2\pi}\left(inv(\alpha_{ec1}) - inv(\alpha_c)\right) + \frac{H_{aec}}{p_w}\tan\alpha_c. \tag{2.119}$$

The drive and coast pressure angles are defined by the following equations:

- for the external gear mesh

$$inv(\alpha_{wd}) = \frac{1}{1+u}\left(inv(\alpha_{ed1}) + u\,inv(\alpha_{ed2}) - \frac{2\pi\theta_d}{z_1}\right), \tag{2.120}$$

$$inv(\alpha_{wc}) = \frac{1}{1+u}\left(inv(\alpha_{ec1}) + u\,inv(\alpha_{ec2}) - \frac{2\pi\theta_c}{z_1}\right), \tag{2.121}$$

- for the internal gear mesh

$$inv(\alpha_{wd}) = \frac{1}{u-1}\left(\frac{2\pi\theta_d}{z_1} - inv(\alpha_{ed1}) + u\,inv(\alpha_{ed2})\right), \tag{2.122}$$

$$inv(\alpha_{wc}) = \frac{1}{u-1}\left(\frac{2\pi\theta_c}{z_1} - inv(\alpha_{ec1}) + u\,inv(\alpha_{ec2})\right), \tag{2.123}$$

- for rack and pinion gear mesh

$$\alpha_{wd} = \alpha_d \tag{2.124}$$

$$\alpha_{wc} = \alpha_c \tag{2.125}$$

For gears with symmetric teeth the pitch factor θ from Equation (2.110) is

$$\theta = \theta_d = \theta_c = \frac{1}{2}\times(1-\theta_v). \tag{2.126}$$

This equation shows that for symmetric gears the pitch factor θ is always ≤ 0.5. For the standard 20° pressure angle gears $\theta = 0.25$–0.30, for the 25° pressure angle gears $\theta = 0.30$–0.35. In custom symmetric gears the pitch θ can reach values of 0.40–0.45.

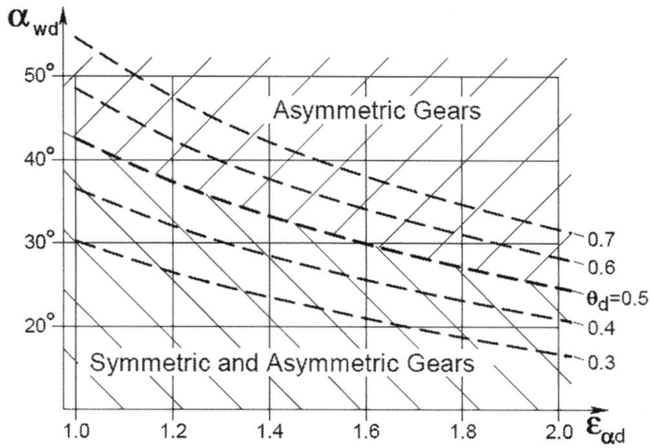

FIGURE 2.17
A sample of the α_{wd}–$\varepsilon_{\alpha d}$ chart with different values of the drive pitch factor θ_d for a gear couple with the pinion number of teeth $z_1 = 21$ and the gear number of teeth $z_2 = 37$.

For gears with asymmetric teeth the drive pitch factor θ_d from Equation (2.110) is

$$\theta_d = 1 - \theta_c - \theta_v. \tag{2.127}$$

Reduction of the coast pitch factor θ_c and the noncontact pitch factor θ_v allows the drive pitch factor θ_d to be significantly increased. Figure 2.17 presents a sample of the drive pressure angle vs. the drive contact ratio α_{wd}–$\varepsilon_{\alpha d}$ chart at different value of θ_d for a gear couple with the pinion number of teeth $z_1 = 21$ and the gear number of teeth $z_2 = 37$. The chart shows that the symmetric gear solutions lie below the curve $\theta_d = 0.5$ and the asymmetric gears are located below and above this curve. A simultaneous increase of the drive pressure angle and the drive contact ratio maximizes gear drive performance. It allows reducing the contact and bending stress increasing load capacity and power transmission density. This indicates potential advantages of the asymmetric gears over the symmetric ones for gear drives that transmit load mostly in one direction.

The pitch factor analysis is an additional Direct Gear Design analytical tool that can be used for comparison of different gear geometry solutions, helping a designer to better understand the available options and chose the optimal one.

2.5 Direct Gear Design for Different Gear Types

In the previous paragraphs, the Direct Gear Design approach to an asymmetric gearing was described for spur and helical gears. However, it is applicable to any other type of involute gears: bevel, worm, face gears, etc. Tooth macrogeometry of these gears is typically defined in a normal plane to the tooth line. This normal plane tooth profile can be considered as a tooth profile of the virtual spur gear. The formulas for calculating the number of teeth of virtual spur gears that have an identical tooth profile to the normal plane tooth profile of different types of involute types of gears are shown in Table 2.1. Virtual numbers of teeth are usually real numbers with the decimal parts. The asymmetric tooth geometry of virtual spur gears is optimized by means of Direct Gear Design (see Chapter 5). Then the optimized tooth profiles are considered as the normal section tooth profiles of the actual gears.

TABLE 2.1

Virtual Spur Gear Conversion

	Type of Gear	Number of Teeth of Virtual Spur Gears
Spur	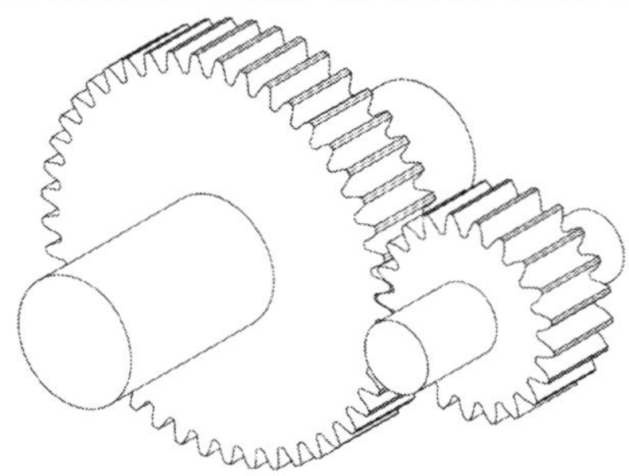	$Z_{1,2v} = Z_{1,2}$
Helical	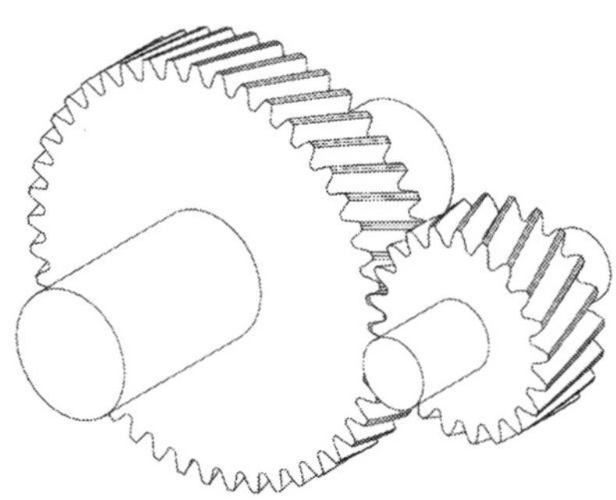	$Z_{1,2v} = \dfrac{Z_{1,2}}{\cos(\beta)^3}$
Crossed Helical	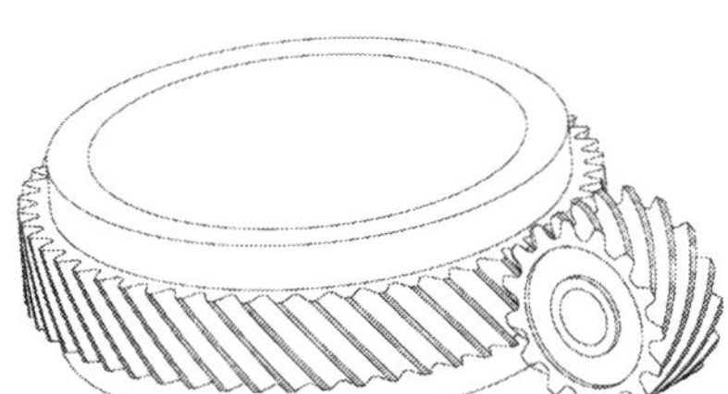	$Z_{1,2v} = \dfrac{Z_{1,2}}{\cos(\beta_{1,2})^3}$

(Continued)

TABLE 2.1 (*Continued*)

Virtual Spur Gear Conversion

Type of Gear		Number of Teeth of Virtual Spur Gears
Straight Tooth Bevel	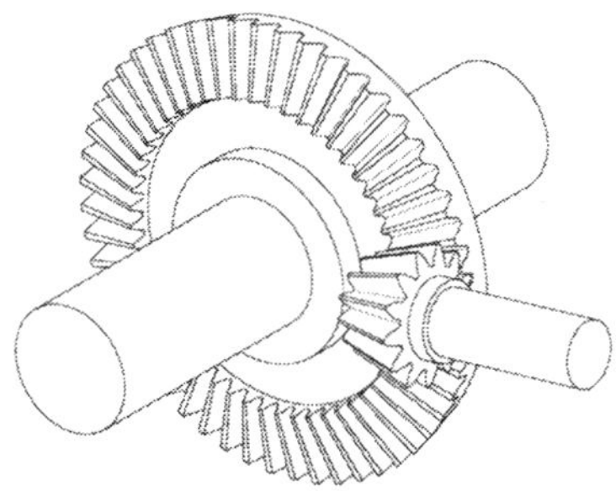	$Z_{1,2v} = \dfrac{Z_{1,2}}{\cos(\gamma_{1,2})}$
Spiral and Skewed Bevel	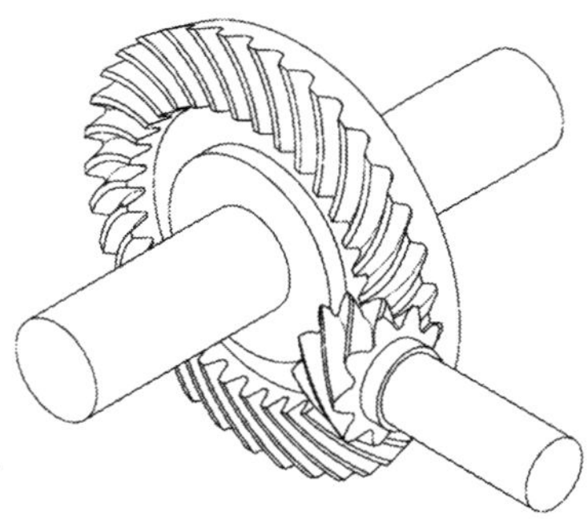 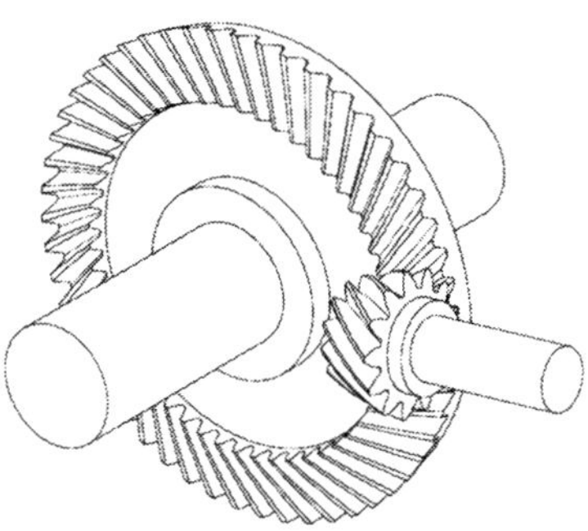	$Z_{1,2v} = \dfrac{Z_{1,2}}{\cos(\gamma_{1,2})\cos(\beta)^3}$
Hypoid	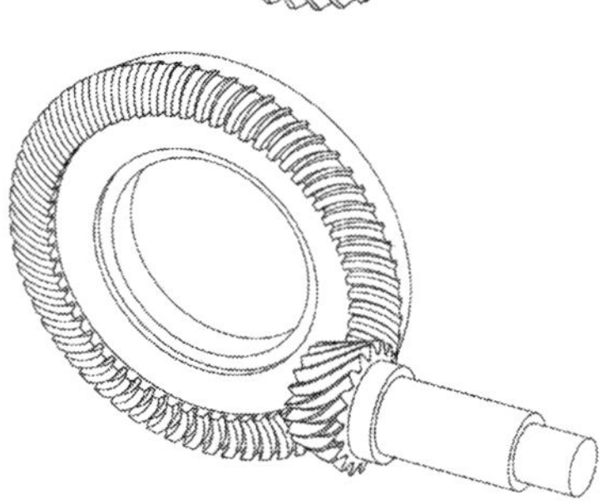	$Z_{1,2v} = \dfrac{Z_{1,2}}{\cos(\gamma_{1,2})\cos(\beta_{1,2})^3}$

(*Continued*)

TABLE 2.1 (*Continued*)

Virtual Spur Gear Conversion

	Type of Gear	Number of Teeth of Virtual Spur Gears
Worm	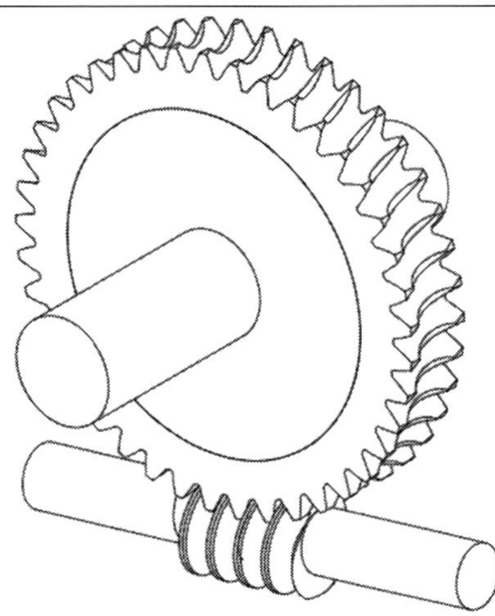	with involute worm: $Z_{wv} = Z_w / \cos(\pi/2 - \beta)^3$ $Z_{wgv} = Z_{wg} / \cos(\beta)^3$ with Archimedes worm: $Z_{wv} = \infty$ $Z_{wgv} = Z_w$
Face Spur	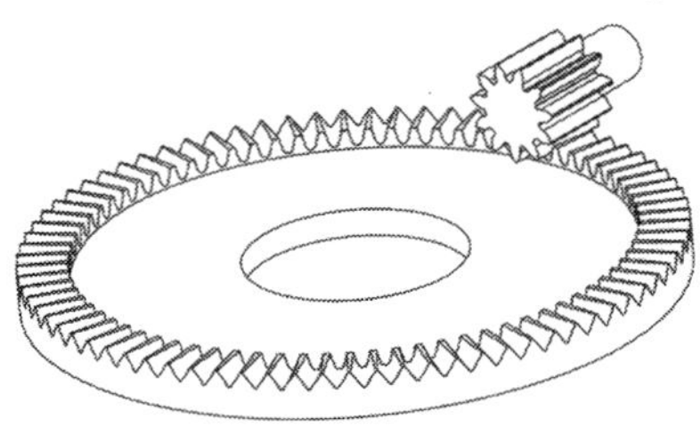	$Z_{1v} = Z_1$ $Z_{2v} = \infty$
Face Helical	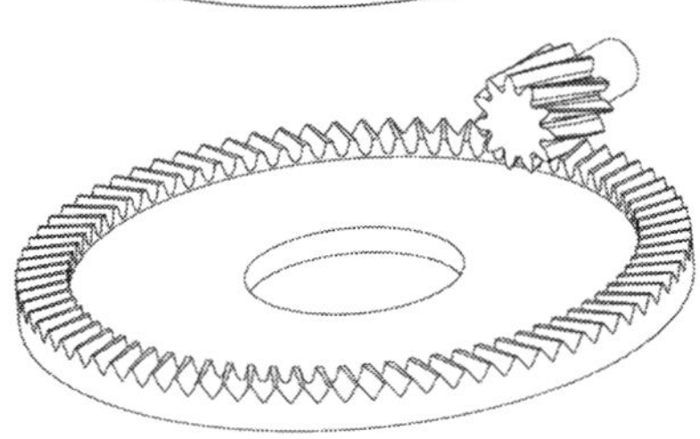	$Z_{1v} = \frac{Z_1}{\cos(\beta)^3}$ $Z_{2v} = \infty$

(*Continued*)

TABLE 2.1 (*Continued*)

Virtual Spur Gear Conversion

	Type of Gear	Number of Teeth of Virtual Spur Gears
Face Spiral	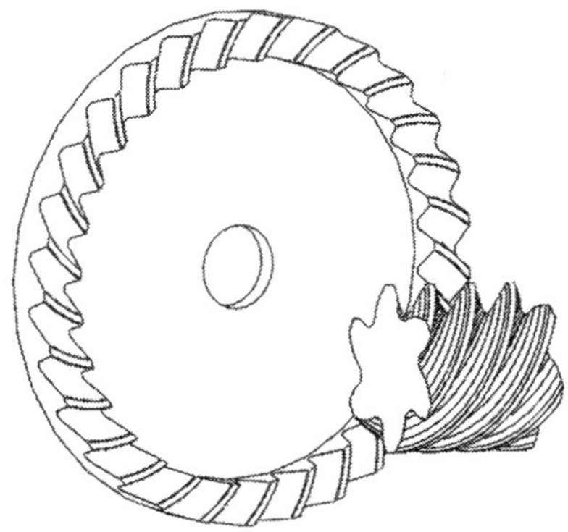	$Z_{1v} = \dfrac{Z_1}{\cos(\beta_1)^3}$ $Z_{2v} = \infty$

Note: $Z_{1,2}$: number of teeth of the real pinion and gear;
$Z_{1,2w}$: number of teeth of virtual spur pinion and gear;
Z_w *and* Z_{wg}: number of starts of real worm and number of teeth of real worm gear;
Z_{wv} *and* Z_{wgv}: number of teeth of virtual gears that replace real worm and worm gear;
β: helix angle of helical or worm gears and spiral angle of spiral bevel gears;
$\gamma_{1,2}$: pitch angle of bevel gears.

3

Area of Existence

In traditional gear design, the preselected generating rack parameters and addendum modification or X-shift factor define the nominal involute gear geometry [138,147,165,166]. The selection of addendum modifications for the mating gear pair is limited by the so-called blocking contour [167,168]. A blocking contour is constructed for a certain, preselected generating rack and given numbers of teeth for mating pinion and gear. Every point of this contour defines a gear pair with certain values of the addendum modifications or X-shift factors for the pinion and mating gear. Blocking contours are developed and explored for symmetric tooth gears, usually with the standard tooth proportions. The blocking contours of the asymmetric tooth gears were described in [17].

Direct Gear Design® does not utilize generating rack parameters to define the gear tooth geometry. However, gear pair tooth geometry parameter selection is also limited by areas of existence. Prof. E.B. Vulgakov introduced areas of existence of involute symmetric tooth gears in his Theory of Generalized Parameters [10]. Later he and his followers studied the areas of existence of asymmetric tooth gears [12,14,29,79,107].

The study of an area of existence is important for asymmetric tooth geometry understanding. It defines any gear parameters' combination and constructs the mating gear profiles within the area of existence borders that establish the gear mesh selection limits. Besides, it is a handy design tool for an initial concept development stage that makes it possible to quickly find a gear pair with required properties, visualize its tooth profiles, and preliminary calculate stress values.

Areas of existence of symmetric tooth gears were analyzed in [107]. This chapter presents detailed analysis of areas of existence of asymmetric tooth gears and defines parameter limits for a pair of spur gears or helical gears in the transverse section. In this chapter, areas of existence of an asymmetric tooth gear pair are defined and analyzed considering a backlash and all tolerances equal to zero.

The profile angles $\upsilon_{d1,2}$ at the intersection point of the drive flank involutes are used as coordinates for an area of existence of two mating asymmetric tooth gears with numbers of teeth z_1 and z_2. Other gear tooth parameters can also be used as coordinates for an area of existence, for example, the intersection point coast flank profiles angles $\upsilon_{c1,2}$, or the drive flank tooth tip profile angles $\alpha_{ad1,2}$, or the coast flank tooth tip profile angles $\alpha_{ac1,2}$.

Unlike the traditional design blocking contours, where the tooth thicknesses at the tip diameters vary depending on the X-shifts, the area of existence is constructed for selected constant relative tooth tip thicknesses. Prof. E.B. Vulgakov [10] defined the relative tooth tip thickness as a ratio of the tip diameter tooth thicknesses to the base circle diameter. However, an asymmetric tooth gear has two different drive and coast base diameters, and it is more practical to present the relative tooth tip thicknesses $m_{a1,2}$ as ratios in relation to the operating module m_w or the operating diametral pitch DP_w, because m_w or DP_w is commonly used to define a range of advisable values for the tooth tip thickness. Then the relative tooth tip thickness can be defined as follows:

$$m_{a1,2} = \frac{S_{a1,2}}{m_w} \text{ or } m_{a1,2} = S_{a1,2} \times DP_w. \tag{3.1}$$

For the external asymmetric tooth gears, it is

$$m_{a1,2} = \frac{z_{1,2} \cos \alpha_{wd}}{2 \cos \alpha_{ad1,2}} \times \left(inv\left(\nu_{d1,2}\right) + inv\left(\nu_{c1,2}\right) - inv\left(\alpha_{ad1,2}\right) - inv\left(\alpha_{ac1,2}\right) \right), \tag{3.2}$$

- for the internal tooth gears

$$m_{a2} = \frac{z_2 \cos \alpha_{wd}}{2 \cos \alpha_{ad2}} \times \left(2\pi / z_2 - inv\left(\nu_{d2}\right) - inv\left(\nu_{c2}\right) + inv\left(\alpha_{ad2}\right) + inv\left(\alpha_{ac2}\right) \right). \tag{3.3}$$

An area of existence contains a number of isograms, which present a certain constant value of gear mesh parameters or a certain constant mesh condition. The next sections describe construction of these isograms for the following gear pair parameters selected as examples:

- Pinion number of teeth $z_1 = 17$
- External or internal mating gear number of teeth $z_2 = 29$
- Pinion relative tooth tip thicknesses $m_{a1} = 0.20$
- Mating gear relative tooth tip thicknesses $m_{a1} = 0.30$
- Asymmetry factor $K = 1.1$, 1.3, and 1.5

3.1 Pressure Angle Isograms

The pressure angle $\alpha_w = \text{const}$ isograms are defined by Equation (2.40) for the external gear pair (Figure 3.1) and by Equation (2.41) for the internal gear pair (Figure 3.2). Values of the drive and coast pressure angles of a particular gear pair are bounded by Equation (2.42). Although the pressure angle isograms

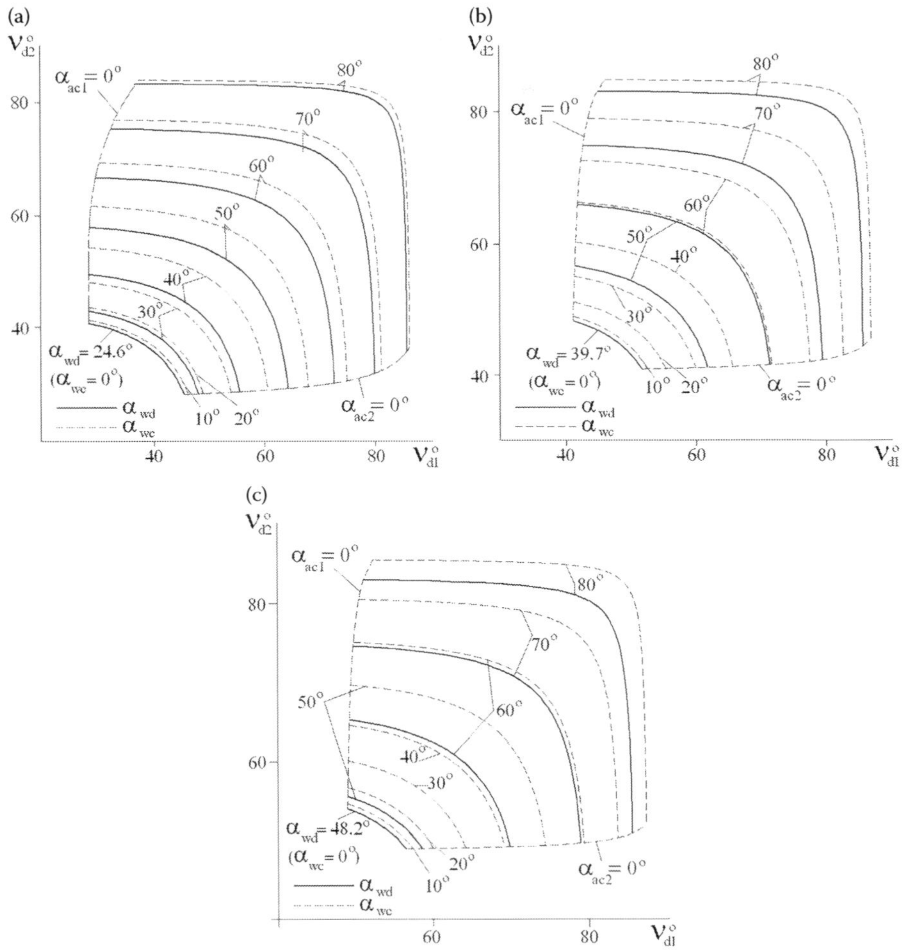

FIGURE 3.1
Pressure angle α_{wd} = const and α_{wc} = const isograms for the external gear pairs; a: asymmetry factor $K = 1.1$, b: $K = 1.3$, c: $K = 1.5$.

do not depend on the relative tooth tip thicknesses $m_{a1,2}$, they are bordered by isograms of the tooth tip profile angles $\alpha_{ac1} = 0°$ and $\alpha_{ac2} = 0°$, when the coast flanks become points on the coast base diameters d_{bc1} and d_{bc2}. When $\alpha_{ac1,2} = 0°$ Equation (3.2) for external gears becomes

$$m_{a1,2} = z_{1,2}\cos\alpha_{wd} \times \left(inv\left(\nu_{d1,2}\right) + inv\left(\nu_{c1,2}\right) - inv\left(\arccos\left(1/K\right)\right)\right). \tag{3.4}$$

When $\alpha_{ac2} = 0°$, Equation (3.3) for internal gears becomes

$$m_{a2} = z_2\cos\alpha_{wd} \times \left(2\pi/z_2 - inv\left(\nu_{d2}\right) - inv\left(\nu_{c2}\right) + inv\left(\arccos\left(1/K\right)\right)\right). \tag{3.5}$$

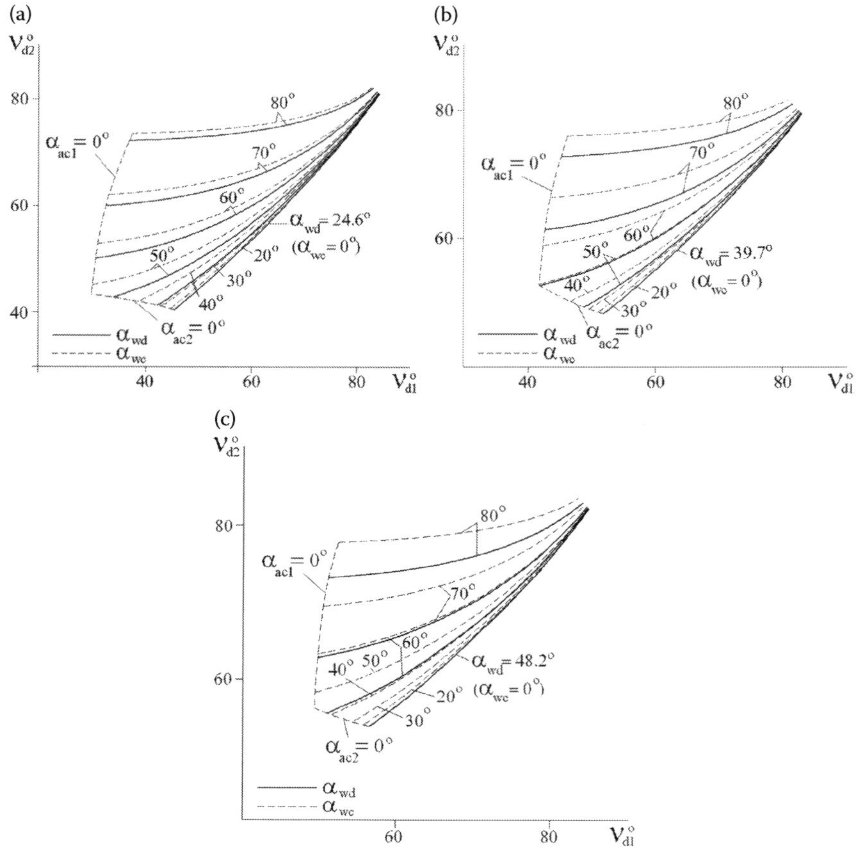

FIGURE 3.2
Pressure angle α_{wd} = const and α_{wc} = const isograms for the internal gear pairs; a: asymmetry factor $K = 1.1$, b: $K = 1.3$, c: $K = 1.5$.

Then the isograms $\alpha_{ac1} = 0°$ and $\alpha_{ac2} = 0°$ are defined for external gear pairs by Equations (2.40) and (3.4), and for internal gear pairs by Equations (2.41), (3.4), and (3.5).

Minimal values of the pressure angles $\alpha_{wc} = 0°$ and $\alpha_{wd} = \arccos(1/K)$ occur, when the gear pair pitch diameters become equal to the coast flank base diameters $d_{bc1,2}$.

The theoretical limits of the pressure angles are as follows: for the coast flanks $0° < \alpha_{wc} < 90°$ for the drive flanks $\arccos(1/K) < \alpha_{wd} < 90°$. The practical range of the pressure angles varies depending on a type of involute gears and their application. For example, some worm gears with the metal worm and polymer mating gear may have a very low pressure angle (5°–12°). A most common pressure angle range for symmetric tooth gears is 14.5°–25°. Highly stressed aerospace symmetric tooth gears may have pressure angles 25°–30°. Spur gears with asymmetric teeth can have pressure angle up to 45°

and higher. In helical gears a total contact ratio is a sum of a transverse contact ratio and an overlap ratio (Equations 2.92 and 2.93). This allows a transverse contact ratio below 1.0, realizing a significantly higher transverse pressure angles. For example, the self-locking helical gears (paragraph 7.1) can have transverse operating pressure angles up to 80° and higher.

3.2 Transverse Contact Ratio Isograms

The transverse contact ratio ε_{ad} = const and ε_{ac} = const isograms for the external gearing are shown in Figure 3.3. The drive flank contact ratio

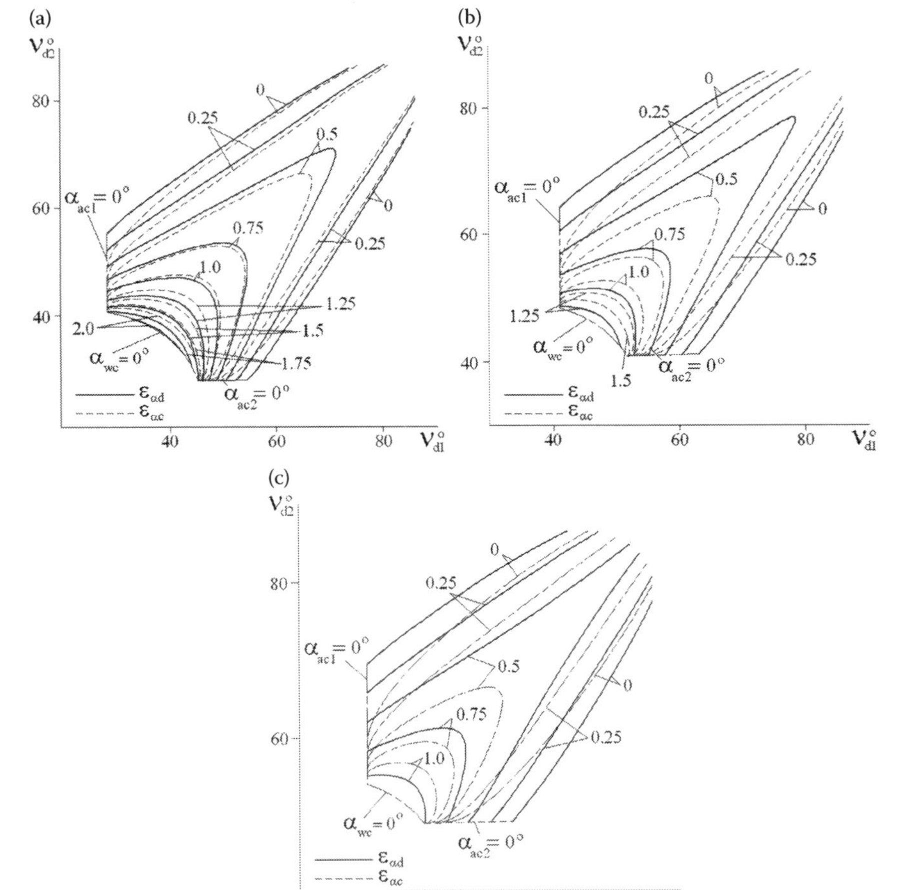

FIGURE 3.3
Transverse contact ratio ε_{ad} = const and ε_{ac} = const isograms for the external gear pairs; a: asymmetry factor $K = 1.1$, b: $K = 1.3$, c: $K = 1.5$.

isograms ε_{ad} = const in this case are described as solutions of Equations (2.40), (2.66), and (3.2). The coast flank contact ratio isograms ε_{ac} = const are solutions of Equations (2.40), (2.67), and (3.2).

The transverse contact ratio ε_{ad} = const and ε_{ac} = const isograms for the internal gearing are shown in Figure 3.4. The drive flank contact ratio isograms ε_{ad} = const in this case are described as solutions of Equations (2.41), (2.68), (3.2), and (3.3). The coast flank contact ratio isograms ε_{ac} = const are solutions of Equations (2.41), (2.69), (3.2), and (3.3).

Spur asymmetric tooth gears should have the contact ratios ε_{ad} and ε_{ac} greater or equal to zero to provide smooth mesh transition from one contacting pair of teeth to the next one. The isogram $\varepsilon_{ad} = 1.0$ limits a choice of the spur gear combinations. Helical gears have an additional overlap ratio (a face contact ratio) ε_β that allows the transverse contact ratios ε_{ad} and ε_{ac} to be within a range from 0.0 to 1.0. The total contact ratios of helical gears must

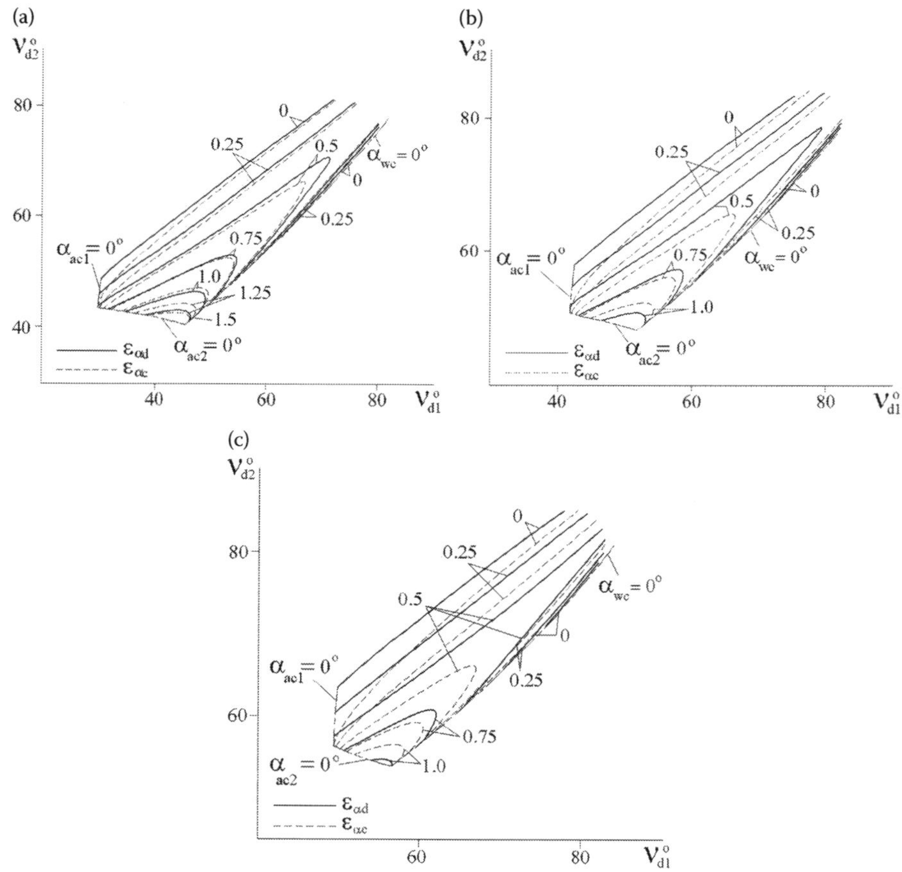

FIGURE 3.4
Transverse contact ratio $\varepsilon\alpha_d$ = const and $\varepsilon\alpha_c$ = const isograms for the internal gear pairs; a: asymmetry factor $K = 1.1$, b: $K = 1.3$, c: $K = 1.5$.

be $\varepsilon_{\gamma d} = \varepsilon_{\alpha d} + \varepsilon_{\alpha\beta} \geq 1.0$ and $\varepsilon_{\gamma c} = \varepsilon_{\alpha c} + \varepsilon_{\beta} \geq 1.0$. The isograms $\varepsilon_{\alpha c} = 0$ limit a choice of the helical gear combinations [169]. However, commonly helical symmetric tooth gears have the transverse contact ratio greater than 1.0. This should be the same for most helical asymmetric tooth gears.

3.3 Overlap Ratio Isograms

From Equations (2.2), (2.7), (2.29), (2.88), (2.96), and (2.102), an overlap ratio (face contact ratio) can be presented as follows:

$$\varepsilon_{\beta} = \frac{b_0}{\pi} \sin \beta, \tag{3.6}$$

where b_0 is a relative contact face width equal to b_w/m_n for the metric system or $b_w \times DP_n$ for the English system.

Figure 3.5 shows the overlap ratio $\varepsilon_{\beta} = \text{const}$ isograms.

In helical gearing, it is preferable to have the integer overlap ratio (ε_{β} =1, 2, 3, ...) that provides the minimal variation of a sum of the contact line lengths [138].

3.4 Interference Isograms

The interference isograms of asymmetric gears are shown in Figure 3.6. Although the interference isograms present the borders of the area of existence, it does not mean that the gear meshes do not exist beyond these

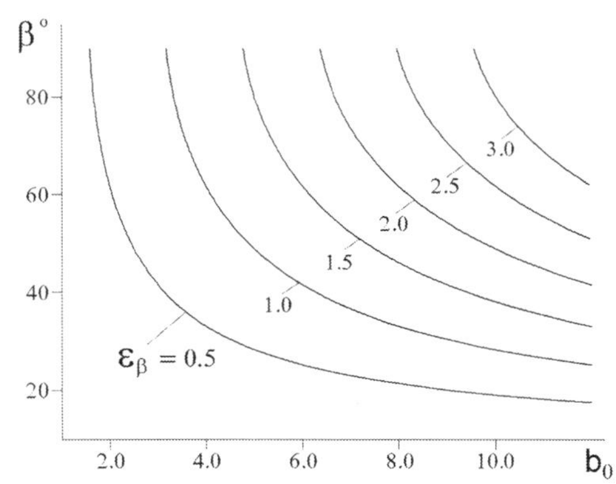

FIGURE 3.5
Overlap ratio $\varepsilon\alpha\beta = \text{const}$ isograms.

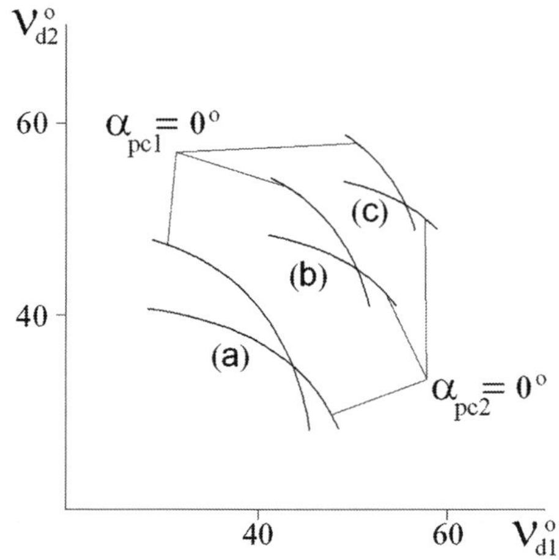

FIGURE 3.6
Root interference isograms $\alpha_{pc1} = 0^o$ and $\alpha_{pc2} = 0^o$ for the external gearings; a: asymmetry factor $K = 1.1$, b: $K = 1.3$, c: $K = 1.5$.

borders. However, those gear combinations have the tooth root undercut at least in one of mating gears and they are usually undesirable. For asymmetric gears with an asymmetry factor $K > 1.0$ the coast tooth flank has lower involute profile angles than the drive one. As a result, an interference occurs first at the coast tooth flanks. The root interference isograms $\alpha_{pc1} = 0°$ and $\alpha_{pc2} = 0°$ for the external gearing reflect the coast flank root undercut conditions. They are defined as solutions of Equations (2.40), (2.46), (2.47), and (3.2) and are shown in Figure 3.6.

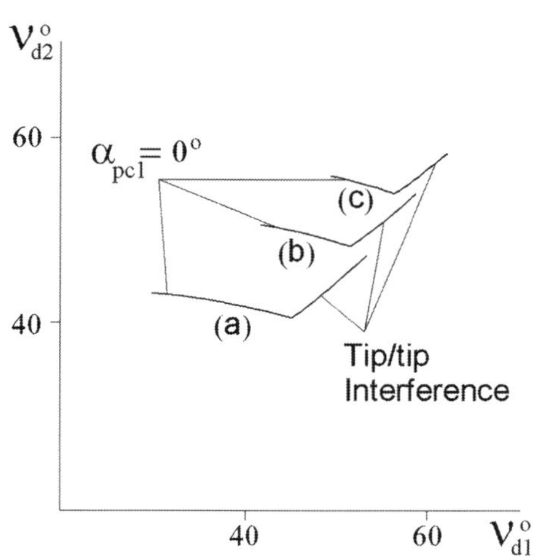

FIGURE 3.7
Root interference isograms $\alpha_{pc1} = 0^o$ and "tip/tip" interference isograms for internal gears; a: asymmetry factor $K = 1.1$, b: $K = 1.3$, c: $K = 1.5$.

The root interference $\alpha_{pc1} = 0°$ and "tip/tip" interference isograms for the internal gearing reflect the coast flank root undercut condition for the pinion and the "tip/tip" interference at the coast flanks. They are shown in Figure 3.7. The $\alpha_{pc1} = 0°$ isogram is defined as a solution of Equations (2.41), (2.50), and (3.2). The "tip/tip" interference isogram is a solution of Equations (2.41), (2.62), (3.2), and (3.3).

3.5 Pitch Point Location Isograms

Pitch point location isograms are the same for the drive and coast tooth flanks, because a pitch diameter and tooth tip diameter are the same for both flanks.

In conventional gearing the operating pitch diameter d_{w1} or d_{w2} divides the gear tooth height on the addendum and dedendum portions, and the pitch point P is located on the active part of the contact line A_1–A_2 (Figure 3.8). For external gears this means that the operating pitch diameter is larger than the form diameter and smaller than the tooth tip diameter. For internal gears this means that the operating pitch diameter is smaller than the form diameter and larger than the tooth tip diameter. Normally, a gear mesh in motion has the approach and recess actions, while the contact point moves along the contact line. An approach action is when the contact point C lies between point A_2 and pitch point P of the contact line, or the driving pinion dedendum is in contact with the driven gear addendum. A recess action is

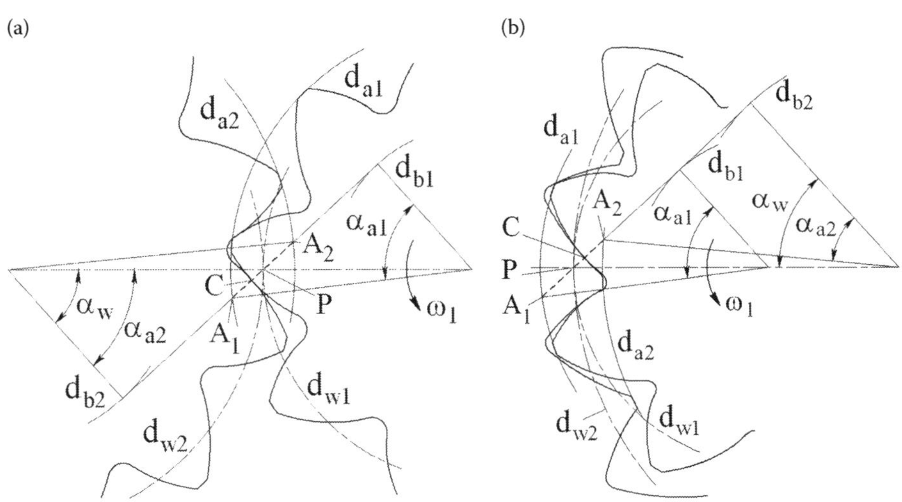

FIGURE 3.8
Conventional action gearing: a: external gears; b: internal gears.

when the contact point C lies between the pitch point P and point A_1 of the contact line, or the driving pinion addendum is in contact with the driven gear dedendum.

It is also possible to have only the approach action gearing (Figure 3.9) when the driving gear has the tip diameter $d_{a1} \le d_{w1}$ or $\alpha_{a1} \le \alpha_w$. In the approach action gearing the driving gear tooth has only the dedendum without addendum. Accordingly, the driven gear tooth has only the addendum without dedendum. The pitch point P is located outside the active part of contact line A_1–A_2 (beyond point A_1 in Figure 3.9).

The recess action gearing is shown in Figure 3.10. In the external mesh the driven gear has the tooth tip diameter $d_{a2} \le d_{w2}$ or $\alpha_{a2} \le \alpha_w$. In the internal mesh the driven gear has the tooth tip diameter $d_{a2} \ge d_{w2}$ or $\alpha_{a2} \ge \alpha_w$. In the recess action gearing the driving gear tooth has only the addendum without dedendum. Accordingly, the driven gear has only the dedendum without addendum. The pitch point P is located outside the active part of contact line A_1–A_2 (beyond point A_2 in Figure 3.10).

The isograms of the pitch point location are shown in Figure 3.11. Isogram $\alpha_{a1} = \alpha_w$ is a border between the conventional and approach action gearing areas. Considering Equations (3.2) and (2.40), this isogram equation is

$$inv(\nu_{d1}) + inv(\nu_{c1}) - inv(\nu_{d2}) - inv(\nu_{c2}) + \frac{2}{z_2} \times \left(\pi - m_{a1}(1+u)\right) = 0. \quad (3.7)$$

For the internal gearing this isogram equation derives from a common solution with (3.3) and (2.41)

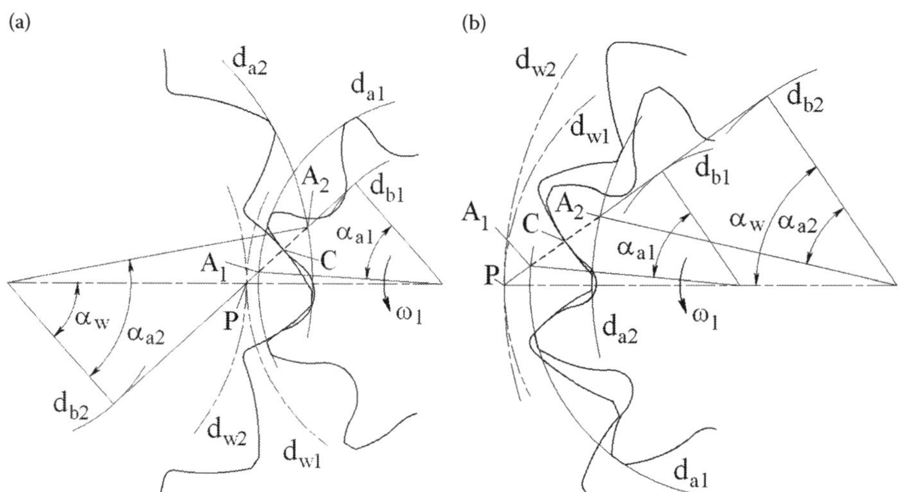

FIGURE 3.9
Approach action gearing: a: external gears; b: internal gears.

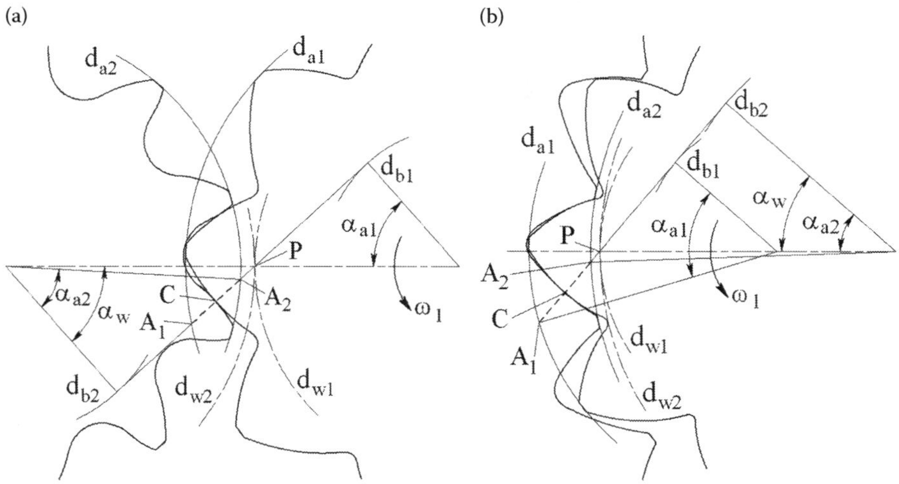

FIGURE 3.10
Recess action gearing: a: external gears; b: internal gears.

$$inv(\nu_{d1})+inv(\nu_{c1})-inv(\nu_{d2})-inv(\nu_{c2})-\frac{2}{z_2}\times m_{a1}(u-1)=0. \quad (3.8)$$

For the internal gearing the isograms $\alpha_{a1} = \alpha_w$ practically do not depend on the asymmetry factor *K*. In Figure 3.11b, these isograms for different *K* values virtually coincide.

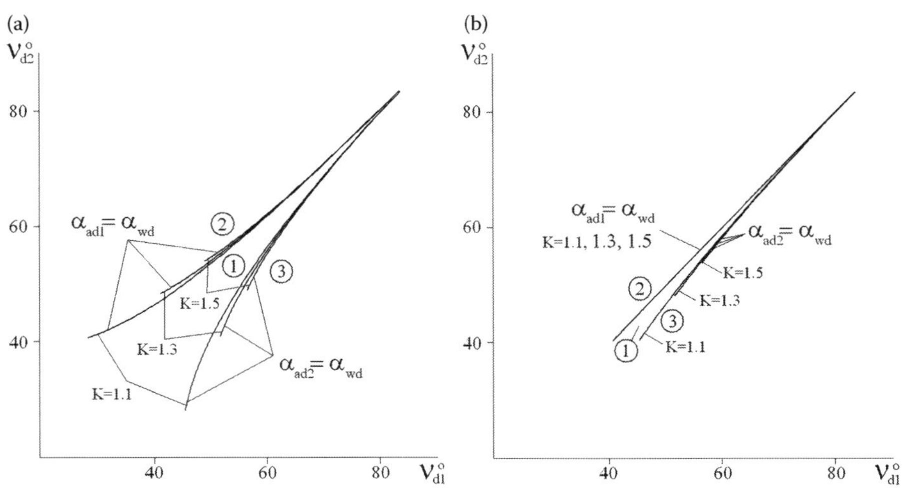

FIGURE 3.11
Pitch point location isograms: a: for external gears; b: for internal gears; 1: area of conventional action gearing; 2: area of the approach action gearing; 3: area of the recess action gearing.

Isogram $\alpha_{a2} = \alpha_w$ is a border between the conventional and recess action gearing areas. For external gearing it is a solution of a system of Equations (3.2) and (2.40):

$$inv(\nu_{d1}) + inv(\nu_{c1}) - inv(\nu_{d2}) - inv(\nu_{c2}) + \frac{2}{z_2} \times \left(m_{a2}(1+u) - \pi u\right) = 0. \quad (3.9)$$

For the internal gearing it is a solution of a system of Equations (3.3) and (2.41):

$$inv(\nu_{d1}) + inv(\nu_{c1}) - inv(\nu_{d2}) - inv(\nu_{c2}) - \frac{2}{z_2} \times m_{a2}(u-1) = 0. \quad (3.10)$$

Most gear applications use the conventional action gearing, because it provides better performance parameters, such as high mesh efficiency (minimal tooth profile sliding), tooth surface durability, bending stress balance, etc. However, the approach and recess action gearings also may have rational areas of applications. For example, the recess action gearing is used for the self-locking gears (Section 7.1).

3.6 Performance Parameters' Isograms

There are many other isograms that can be drawn in the area of existence. This section presents a few of them, which define gear pairs with certain constant performance characteristics. These isograms are constructed for the drive tooth flanks in contact.

Gear transmission power density and load capacity in many cases are defined by the tooth surface durability. Maximization of gear transmission power density requires minimization of the contact (Hertz) stress σ_H. Gear pairs with the minimal contact stress should have the maximum transverse contact ratio ε_α, if the operating pressure angle α_w is given, or the maximum operating pressure angle α_w, if the transverse contact ratio ε_α is given. These conditions occur for the drive flank of the asymmetric tooth gears when isograms α_{wd} = const and $\varepsilon_{\alpha d}$ = const are tangent (Figure 3.12). The $\sigma_{H\min}$ isograms (Figure 3.13) present the gear meshes that correspond to the tangent points of isograms α_{wd} = const and $\varepsilon_{\alpha d}$ = const. For the external asymmetric tooth gears these isograms practically do not depend on the asymmetry factor K and located approximately at 45° from the horizontal axis, where $\nu_{d1} \approx \nu_{d2}$.

Maximization of gear efficiency is critically important for many gear applications. Gear efficiency depends on gear mesh losses that are defined as follows [165]:

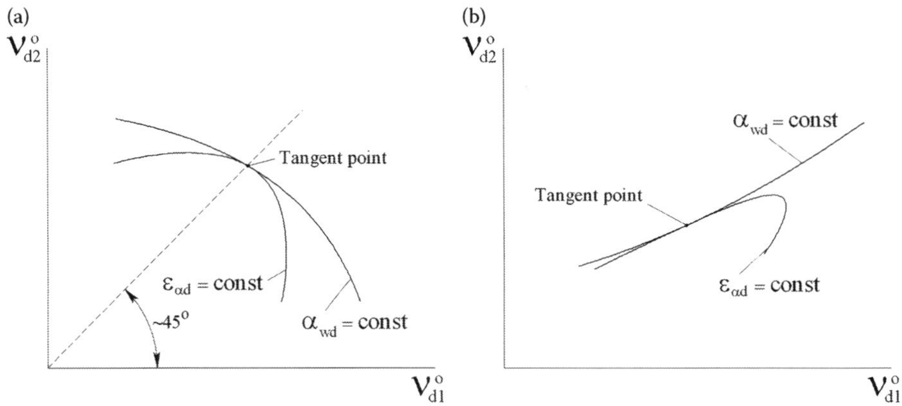

FIGURE 3.12
Tangent point of α_{wd} = const and $\varepsilon_{\alpha d}$ = const isograms: a: for external gears; b: for internal gears.

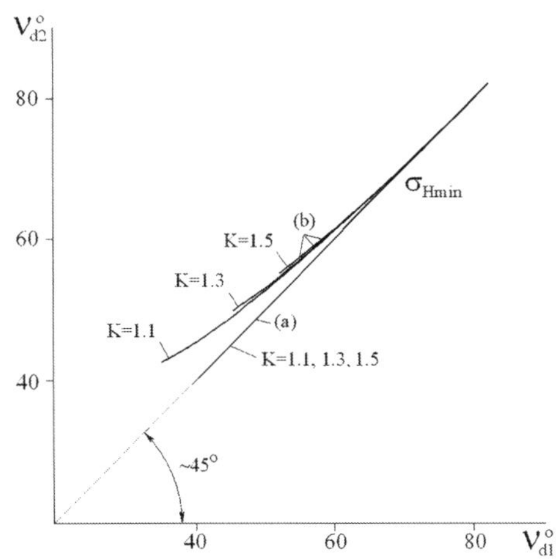

FIGURE 3.13
Minimum contact stress isograms $\sigma_{H\min}$: a: for external gears; b: for internal gears.

- for the drive flanks a pair of spur gears

$$P_t = \frac{50f}{\cos\alpha_{wd}} \times \frac{H_{sd}^2 + H_{td}^2}{H_{sd} + H_{td}}, \tag{3.11}$$

- for the drive flanks a pair of helical gears

$$P_t = \frac{50f\cos^2\beta}{\cos\alpha_{nd}} \times \frac{H_{sd}^2 + H_{td}^2}{H_{sd} + H_{td}}, \tag{3.12}$$

where f is an average friction coefficient, H_{sd} is a drive flank specific sliding velocity at the start of the approach action, and H_{td} is a drive flank specific sliding velocity at the end of the recess action.

Specific sliding velocities H_{sd} and H_{td} are ratios of the sliding velocity to the rolling velocity. They can be defined for the external and internal gears as follows:

$$H_{sd} = (u \pm 1) \times \cos\alpha_{wd} \times (\pm\tan\alpha_{ad2} \mp \tan\alpha_{wd}), \tag{3.13}$$

$$H_{td} = \frac{u \pm 1}{u} \times \cos\alpha_{wd} \times (\tan\alpha_{ad1} - \tan\alpha_{wd}), \tag{3.14}$$

where the top sign (+ or –) is for external gears and the bottom sign (+ or –) is for internal gears.

Alternatively, from Equations (3.13), (2.44), or (2.45) for external gears, and from Equations (3.14), (2.48), or (2.49) for internal gears, the specific sliding velocities are as follows:

$$H_{sd} = \frac{u \pm 1}{u} \times \cos\alpha_{wd} \times \left(\tan\alpha_{wd} - \tan\alpha_{pd1}\right), \tag{3.15}$$

$$H_{td} = (u \pm 1) \times \cos\alpha_{wd} \times \left(\pm\tan\alpha_{wd} \mp \tan\alpha_{pd2}\right). \tag{3.16}$$

From Equations (3.11) and (3.12) a minimum of the gear mesh losses happen when $H_{sd} = H_{td}$. This means that a condition for the maximum gear mesh efficiency E_{max} can be defined for the external and internal gearings from (3.13) and (3.14) as follows:

$$\tan\alpha_{ad1} \mp u\tan\alpha_{ad2} \pm (u \mp 1)\tan\alpha_{wd} = 0, \tag{3.17}$$

where the top sign (+ or –) is for external gears and the bottom sign (+ or –) is for internal gears.

Equations of the maximum gear mesh efficiency E_{max} value isograms (Figure 3.14) are defined as a solution of Equations (3.17) and (3.2) for the external gearing, and Equations (3.17) and (3.3) for the internal gearing.

Gear tooth geometry, including the tooth flanks and root fillet, affects the maximum bending stress level. In traditional gear design the tooth fillet profile is typically a trochoidal trajectory of the generating tool tooth tip. In Direct Gear Design the tooth flanks and root fillets are constructed independently, and the tooth fillet profile is optimized to minimize bending stress concentration (see Section 5.2). However, the tooth fillet profile optimization is a time-consuming process that is used for the final stage of gear design. It is not practical for browsing the area of existence analyzing many sets of gear pairs with the optimized root fillets. For preliminary construction of

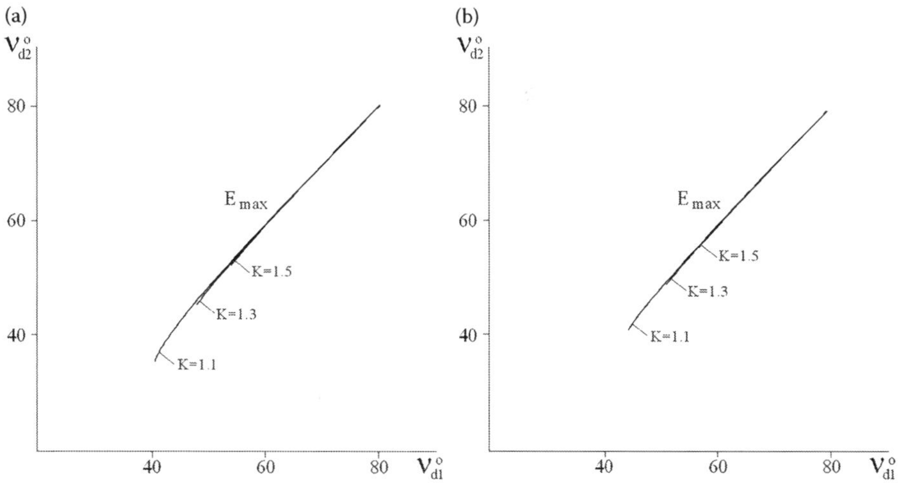

FIGURE 3.14
Maximum gear mesh efficiency isograms: a: for external gears; b: for internal gears.

the interference-free tooth root fillet profile that also provides relatively low bending stress concentration, the virtual ellipsis arc is built into the tooth tip that is tangent to the involute profiles at the tip of the tooth [79]. This makes the root fillet profile of one gear as a trajectory of the mating gear tooth tip virtual ellipsis arc (see Figure 3.15). The virtual ellipsis arc is chosen because it seamlessly fits to an asymmetric tooth tip and results in a relatively low root bending stress level. This fillet profile can be considered pre-optimized because it provides lower bending stress concentration than the full tip radius rack generated fillet profile commonly used for bending stress reduction in the traditional gear design.

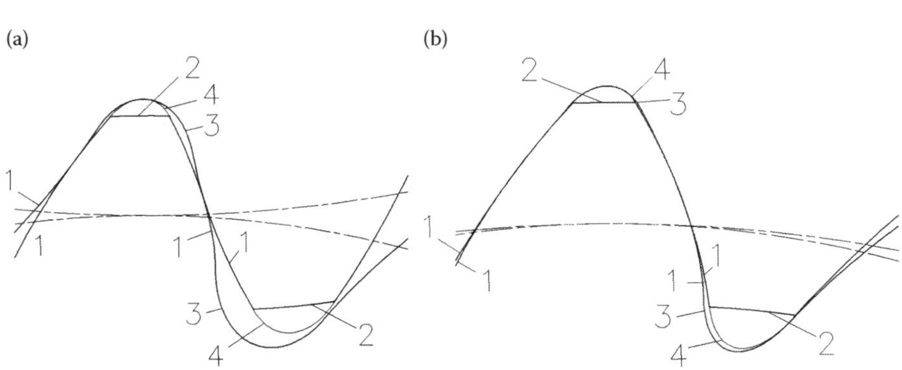

FIGURE 3.15
Tooth root fillet profile construction; a: for external gears; b: for internal gears; 1: involute profiles; 2: tooth tip lands; 3: fillet profiles; 4: ellipsis arcs.

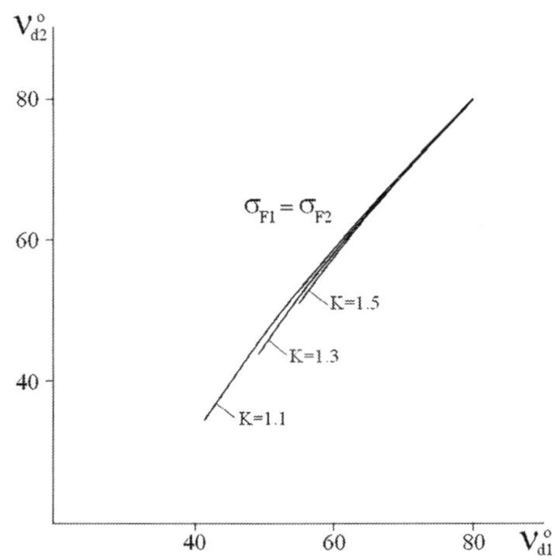

FIGURE 3.16
Equal maximum bending stress isogram $\sigma_{F1max} = \sigma_{F2max}$ for external gears.

When the gear tooth with the root fillet is defined, the bending stress is calculated utilizing the finite element analysis (FEA) method.

If mating gears are made of similar materials and have a relatively close number of load cycles, the maximum bending stresses of mating gears should be equalized. The equal maximum bending stress isogram $\sigma_{F1max} = \sigma_{F2max}$ is shown in Figure 3.16. It is defined assuming that the mating gears have identical face widths. This isogram allows a preliminary selection of a pair of gears with the equalized bending strength. Later during final gear design the mating gear face widths can be adjusted considering also a number of load cycles of each gear to achieve more accurate bending strength equalization.

The equal maximum bending stress isogram is typical for the external gearing. The internal gear tooth with the equal face width with its external mating gear usually has significantly lower bending stress, because its root tooth thickness is typically much greater. In this case, the bending stress balance can be achieved by the internal gear face width reduction.

3.7 Area of Existence and Gear Tooth Profiles

Summation of all the above-mentioned and other possible parameter or condition isograms forms the area of existence. Every point of the area of existence presents a gear pair with a certain set of parameters and gear tooth profiles. Figure 3.17 presents the areas of existence of the external and internal asymmetric gears. The shaded parts of the area of existence contain all reversible interference and tooth root undercut free gear pair options.

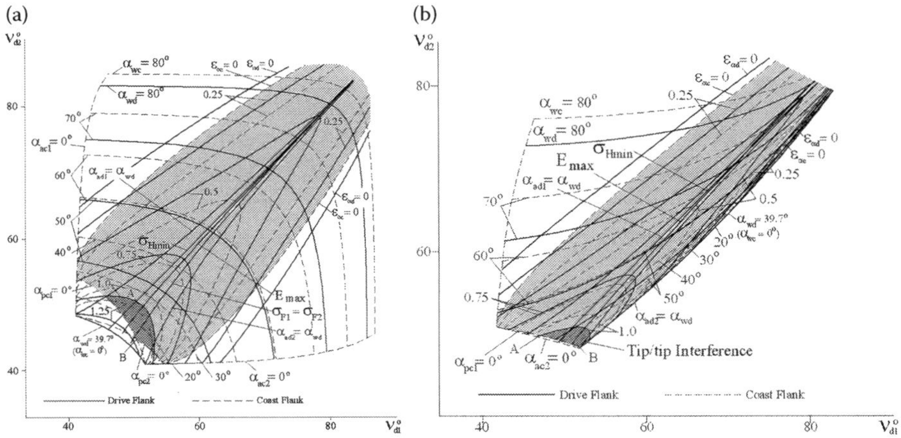

FIGURE 3.17
Area of existence of the asymmetric pinion and gear with z_1 = 17 and z_2 = 29, m_{a1} = 0.20 and m_{a2} = 0.30, K = 1.3; a: for external gears, b: for internal gears.

The dark shaded portion of the area of existence includes the spur and helical gear pairs with the drive transverse contact ratio $\varepsilon_{\alpha d} \geq 1.0$. The light shaded portion of the area of existence between the isograms $\varepsilon_{\alpha d} = 1.0$ and $\varepsilon_{\alpha c} = 0.0$ includes only helical gear pairs. Existence of helical irreversible asymmetric gear pairs is possible outside of the shaded areas between the isograms $\varepsilon_{\alpha d} = 0.0$ and $\varepsilon_{\alpha c} = 0.0$.

Figure 3.18 shows the areas of existence of asymmetric gears with the gear pair tooth profiles corresponding to certain points of those areas. More conventional or practical gear solutions are located within a dark shaded portion with the drive contact ratio $\varepsilon_{\alpha d} \geq 1.0$. Some of gear pair

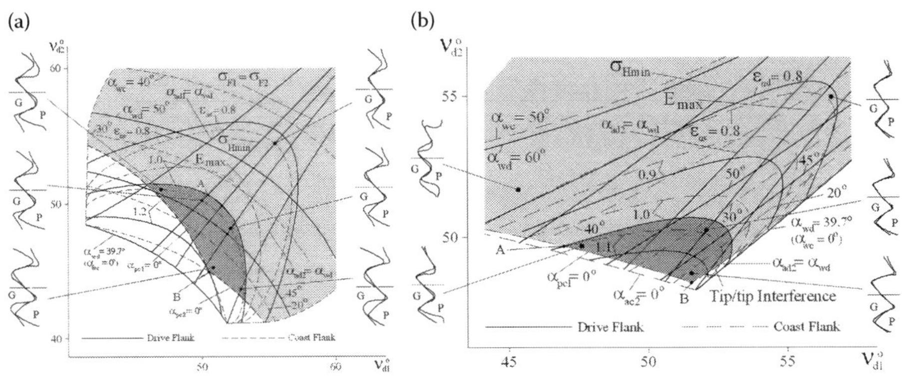

FIGURE 3.18
Area of existence and tooth profiles of the asymmetric pinion and gear with z_1 = 17 and z_2 = 29, m_{a1} = 0.20 and m_{a2} = 0.30, K = 1.3; a: for external gears, b: for internal gears; symbols P and G for the pinion and gear tooth profiles respectively.

tooth profiles have a kind of exotic shape and present rather theoretical interest. However, even they may find practical applications for some unconventional gear drives.

A maximum drive pressure angle for the spur external gearing is achieved in the point *A* (Figure 3.18a), where the drive pressure angle isogram α_{wd} = const is tangent to the drive contact ratio isogram $\varepsilon_{ad} = 1.0$ (see Figure 3.12a), except for the gear pairs with the high asymmetry factor *K* (Figure 3.20a), high relative tooth tip thicknesses m_{a1} and m_{a2} (Figure 3.24a), and a low number of teeth z_1 and z_2 (Figure 3.27a). In those cases, the point *A* is located in the intersection of the drive flank contact ratio isogram $\varepsilon_{ad} = 1.0$ and the coast flank interference isogram $\alpha_{pc1} = 0°$. The minimum drive pressure angle and maximum drive contact ratio of external spur asymmetric gears occur at the point *B* at the intersection of the interference isograms $\alpha_{pc1} = 0°$ and $\alpha_{pc2} = 0°$.

In the spur internal gearing area of existence, a point *A* (Figure 3.18b) is also present in the gear pair with the maximum for the spur gears' drive pressure angle and contact ratio $\varepsilon_{ad} = 1.0$. Its location depends on the asymmetry factor *K*, relative tooth tip thicknesses m_{a1} and m_{a2}, and pinion and gear numbers of teeth z_1 and z_2 (see Figures 3.20b, 3.24b, and 3.27b). For a high asymmetry factor *K*, high relative tooth tip thicknesses m_{a1} and m_{a2}, and a low number of teeth z_1 and z_2, the point A lies in the intersection of the drive flank contact ratio isogram $\varepsilon_{ad} = 1.0$ and the coast flank interference isogram $\alpha_{pc1} = 0°$. Otherwise, it is located at the tangent point of the drive flank contact ratio isogram $\varepsilon_{ad} = 1.0$ and the drive pressure angle isogram α_{wd} = const (see Figure 3.12b). Point *B* is at the intersection of the interference isogram $\alpha_{pc1} = 0°$ and the tip/tip interference isogram.

The next paragraphs describe how the area of existence of asymmetric gears depends on the asymmetry factor *K*, relative tooth tip thicknesses m_{a1} and m_{a2}, and pinion and gear numbers of teeth z_1 and z_2.

3.8 Areas of Existence and Asymmetry Factors

The asymmetry factor *K* not only affects the shape of an asymmetric tooth (Figure 3.19), but also the size and shape of the area of existence. Figure 3.20 presents an overlay of the areas of existence of spur asymmetric tooth gears with different values of the asymmetric factor *K*. The largest area of existence is when the asymmetry factor *K* is equal 1.0 and gear teeth are symmetric. This area is also located in a zone with relatively low flank profile angles. With growth of an asymmetry factor *K*, gear teeth become more and more asymmetric, the area of existence gets smaller, and its location changes to a zone with greater flank profile angles. At a certain maximum, for spur gears with an asymmetry factor K_{max}, the area of existence becomes a point, where the points *A* and *B* coincide. For external asymmetric gearing this is

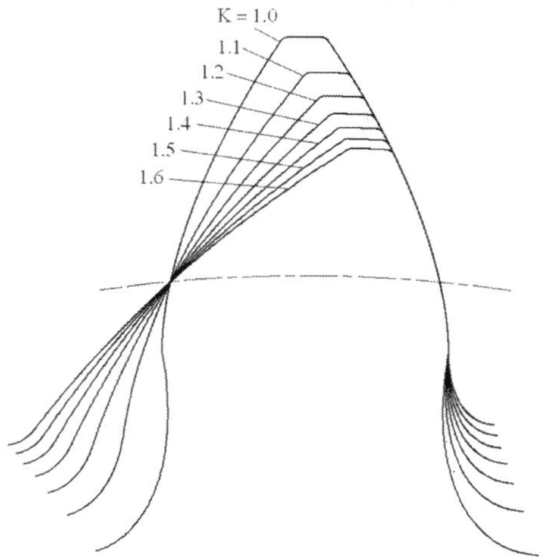

FIGURE 3.19
Tooth profiles with different asymmetry factors *K*, an identical coast flank pressure angle, and tooth thicknesses at the pitch and tip diameters.

a point of simultaneous intersection of the isogram of the drive flank contact ratio $\varepsilon_{\alpha d}$ = 1.0 and the coast flank interference isograms α_{pc1} = 0° and α_{pc2} = 0° (see Figure 3.20a). For internal asymmetric gearing this is a point of simultaneous intersection of the isogram of the drive flank contact ratio $\varepsilon_{\alpha d}$ = 1.0, the coast flank interference isogram α_{pc1} = 0°, and "tip/tip" interference isogram (see Figure 3.20b). If an asymmetry factor $K > K_{max}$, spur asymmetric gears are not possible, because in this case the transverse drive flank contact ratio $\varepsilon_{\alpha d} < 1.0$. However, this condition does not limit applications of helical asymmetric tooth gears with the total drive contact ratio $\varepsilon\gamma_d \geq 1.0$.

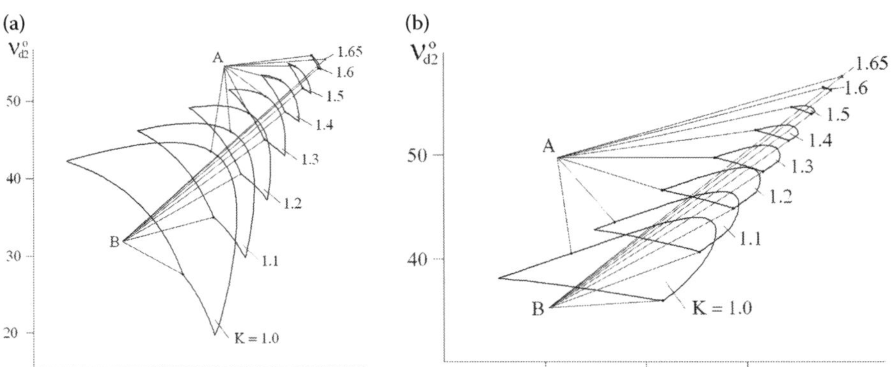

FIGURE 3.20
Areas of existence of the spur asymmetric pinion and gear with $z_1 = 17$ and $z_2 = 29$, $m_{a1,2} = 0.20$, and different values of the asymmetry factor *K*; a: for external gears, b: for internal gears.

Figures 3.21 and 3.22 show how the pressure angle and contact ratio ranges are affected by the asymmetry factor *K*. These ranges are defined by differences in parameter values between the points *A* and *B* of an area of existence. At $K = 1.0$ gear teeth are symmetric and the pressure angle and contact ratio ranges from their minimum to the maximum values are the largest. With increasing asymmetry factor *K* the pressure angle and contact ratio ranges are getting narrower. At $K = K_{max}$ these ranges are equal to zero and points *A* and *B* coincide.

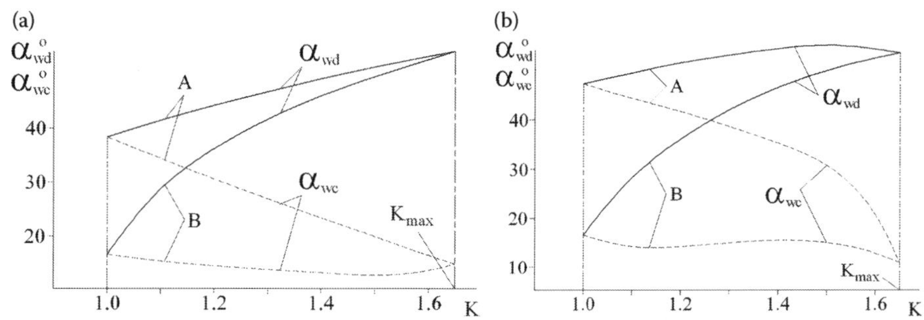

FIGURE 3.21
Drive and coast pressure angle charts of the spur asymmetric pinion and gear with $z_1 = 17$ and $z_2 = 29$, $m_{a1,2} = 0.20$, and different values of the asymmetry factor *K*; a: for external gears, b: for internal gears.

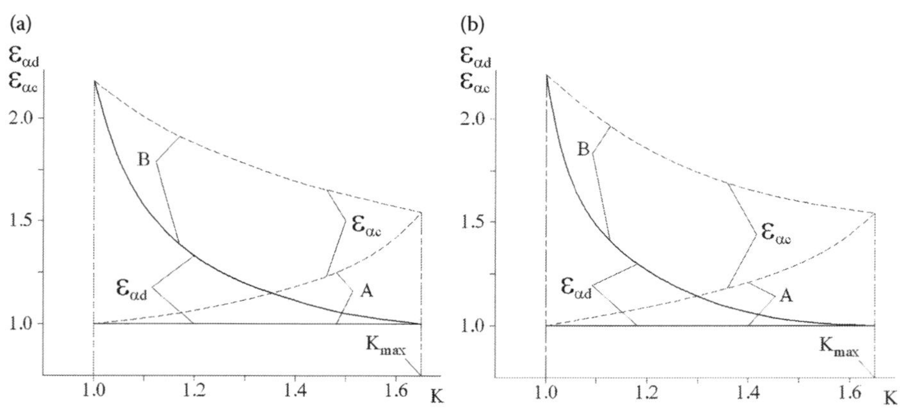

FIGURE 3.22
Drive and coast contact ratio charts of the spur asymmetric pinion and gear with $z_1 = 17$ and $z_2 = 29$, $m_{a1,2} = 0.20$, and different values of the asymmetry factor *K*; a: for external gears, b: for internal gears.

3.9 Areas of Existence and Relative Tooth Tip Thicknesses

Similar to an asymmetry factor *K*, relative tooth tip thicknesses m_{a1} and m_{a2} (Figure 3.23) affect the size and shape of the area of existence. For spur asymmetric gears, values of relative tooth tip thicknesses can vary between zero and the maximum values $m_{a1,2max}$.

Figure 3.24 presents an overlay of the areas of existence of the spur asymmetric tooth gears with different values of the relative tooth tip thicknesses m_{a1} and m_{a2}. The largest area of existence is when $m_{a1,2} = 0$ the gear teeth are pointed. This case presents only theoretical interest, because the pointed gear teeth are totally impractical. For brittle gear materials, including the case harden ones, the pointed tooth tip will be fractured under the load and for more ductile materials the pointed tooth tip will be bent.

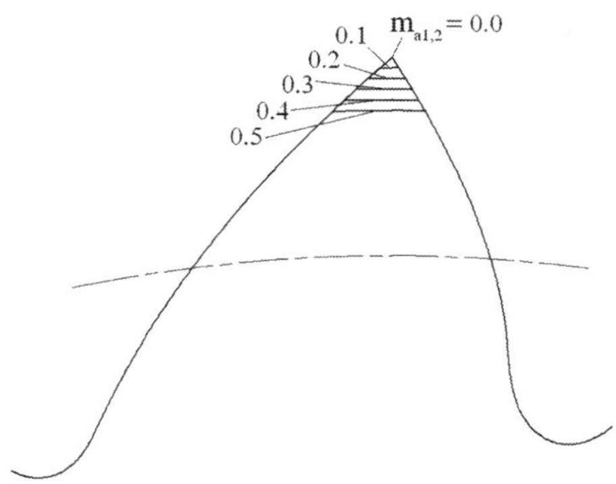

FIGURE 3.23
Different relative tooth tip thicknesses.

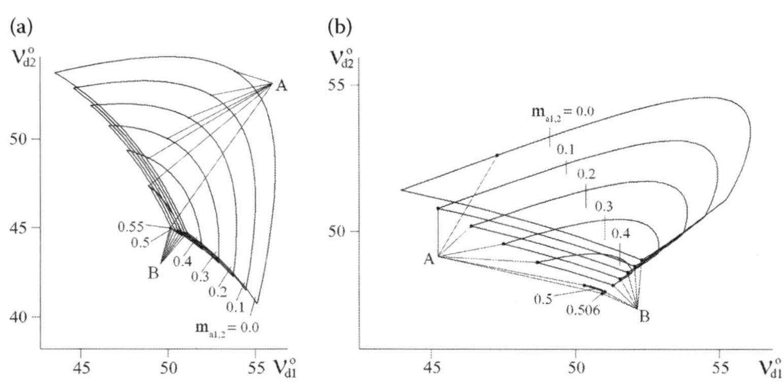

FIGURE 3.24
Areas of existence of the spur asymmetric pinion and gear with $z_1 = 17$ and $z_2 = 29$, $K = 1.2$, and different values of the relative tooth tip thicknesses $m_{a1,2}$; a: for external gears, b: for internal gears.

Both cases are unacceptable. The minimal value of the relative tooth tip thicknesses depends on gear materials, heat treatment, and applied load, and is typically equal to 0.2–0.3 for external gears and 0.3–0.4 for internal gears.

With growth of the relative tooth tip thicknesses, the area of existence is gets smaller, and its location is changing to a zone with lower flank profile angles. At a certain maximum for spur gears with relative tooth tip thicknesses $m_{a1,2\max}$ the area of existence becomes a point, where the points *A* and *B* of the area of existence coincide. For external asymmetric gearing this is a point of simultaneous intersection of the isogram of the drive flank contact ratio $\varepsilon_{\alpha d} = 1.0$ and the coast flank interference isograms $\alpha_{pc1} = 0°$ and $\alpha_{pc2} = 0°$ (see Figure 3.24a). For internal asymmetric gearing this is a point of simultaneous intersection of the isogram of the drive flank contact ratio $\varepsilon_{\alpha d} = 1.0$, the coast flank interference isogram $\alpha_{pc1} = 0°$, and the "tip/tip" interference isogram (see Figure 3.24b). If the relative tooth tip thicknesses $m_{a1,2} > m_{a1,2\max}$, spur asymmetric gears are not possible, because, in this case, the transverse drive flank contact ratio $\varepsilon_{\alpha d} < 1.0$. However, this condition does not limit applications of helical asymmetric tooth gears with the total drive contact ratio $\varepsilon_{\gamma d} \geq 1.0$.

Figures 3.25 and 3.26 show how the pressure angle and contact ratio ranges are affected by the relative tooth tip thicknesses $m_{a1,2}$. These ranges are defined by differences in the pressure angle and contact ratio values between the points *A* and *B* of an area of existence. At the $m_{a1,2} = 0.0$ the gear teeth are symmetric and the pressure angle and contact ratio ranges from their minimum to the maximum values are the largest. With increasing relative tooth tip thicknesses $m_{a1,2}$ the pressure angle and contact ratio ranges are getting narrower. At the $m_{a1,2} = m_{a1,2\max}$ these ranges are equal zero and points *A* and *B* of an area of existence coincide.

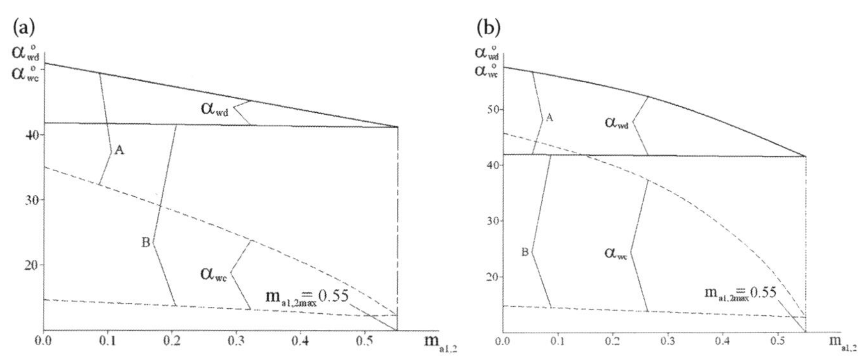

FIGURE 3.25
Drive and coast pressure angle charts of the spur asymmetric pinion and gear with $z_1 = 17$ and $z_2 = 29$, $m_{a1,2} = 0.20$, $K = 1.2$, and different values of the relative tooth tip thicknesses $m_{a1,2}$; a: for external gears, b: for internal gears.

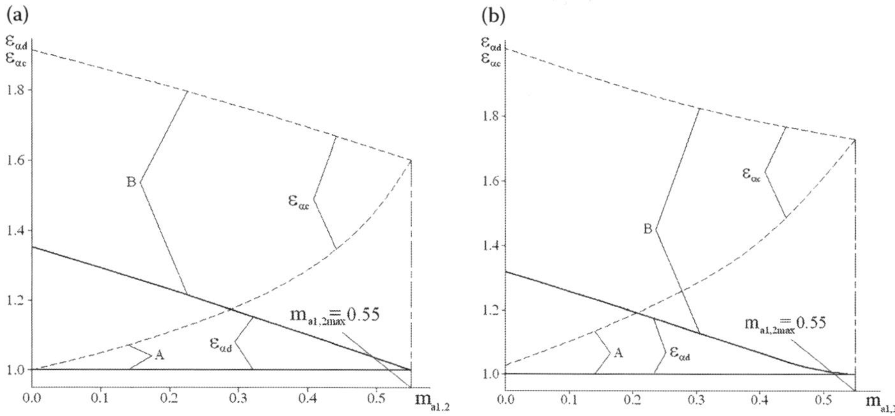

FIGURE 3.26
Drive and coast contact ratio charts of the spur asymmetric pinion and gear with $z_1 = 17$ and $z_2 = 29$, $m_{a1,2} = 0.20$, $K = 1.2$, and different values of the relative tooth tip thicknesses $m_{a1,2}$; a: for external gears, b: for internal gears.

3.10 Areas of Existence and Numbers of Teeth

Gear pairs with different numbers of teeth have areas of existence that vary in sizes and shapes. Figure 3.27 shows an overlay of areas of existence of the spur asymmetric tooth gears with different numbers of teeth and the given gear ratio, asymmetry factor K, and relative tooth tip thicknesses $m_{a1,2}$.

The greater the number of teeth the larger the area of existence. There are minimal numbers of the gear pair teeth when an area of existence of such asymmetric gears becomes very small. An additional decrease

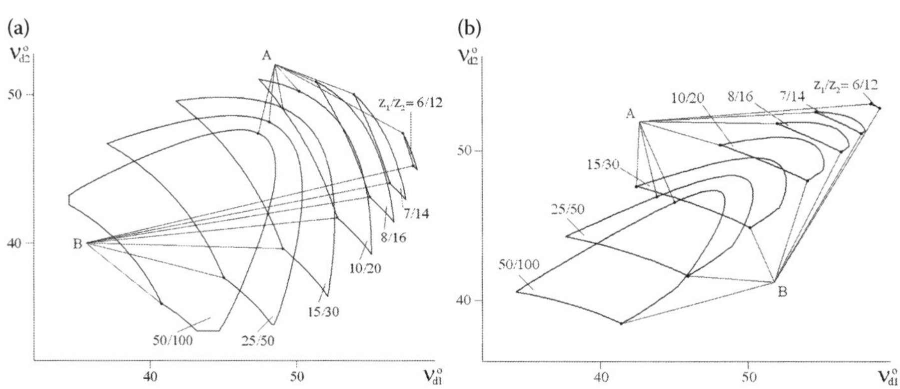

FIGURE 3.27
Areas of existence of the spur asymmetric gears with $K = 1.2$, $m_{a1,2} = 0.2$, and different numbers of teeth; a: for external gears, b: for internal gears.

in tooth numbers makes the spur gearing impossible, because the transverse drive contact ratio becomes $\varepsilon_{\alpha d} < 1.0$. However, if such reduction is necessary, an asymmetry factor K and/or relative tooth tip thicknesses $m_{a1,2}$ must be reduced.

Figures 3.28 and 3.29 show how the pressure angle and contact ratio ranges are affected by numbers of gear teeth $z_{1,2}$. These ranges are defined by

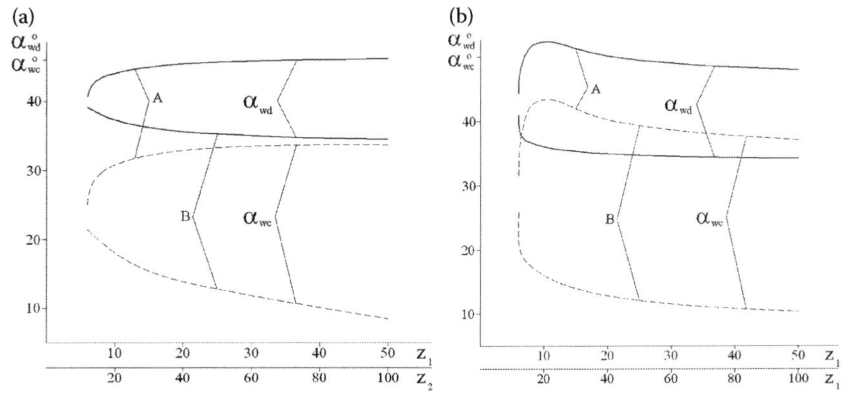

FIGURE 3.28
Drive and coast pressure angle charts of the spur asymmetric gears with $K = 1.2$, $m_{a1,2} = 0.2$, and different numbers of teeth; a: for external gears, b: for internal gears.

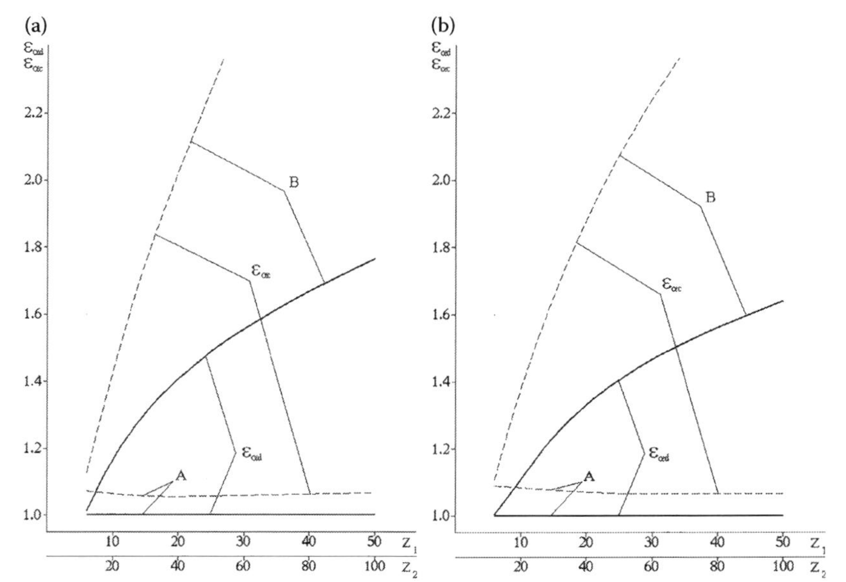

FIGURE 3.29
Drive and coast contact ratio charts of the spur asymmetric gears with $K = 1.2$, $m_{a1,2} = 0.2$, and different numbers of teeth; a: for external gears, b: for internal gears.

differences in the pressure angle and contact ratio values between the points A and B of an area of existence.

3.11 Area of Existence and Pitch Factors

In the previous sections of this chapter, areas of existence of asymmetric gears have been considered with the given constant asymmetry factor *K*, and the relative tooth tip thicknesses m_{a1} and m_{a2}. The pitch factors θ_d, θ_d, and θ_v (see Section 2.4) in this case are varying. Figure 3.30 presents the overlaid areas of existence of spur external gears with the different constant drive flank pitch factors θ_d. This type of area of existence of involute gears defines only the drive flank gear meshes. If $\theta_d \leq 0.5$, the gears can have symmetric or asymmetric teeth. If $\theta_d > 0.5$, the gears can have only asymmetric teeth. The gears with symmetric teeth are always reversible. The gears with asymmetric teeth can be reversible or irreversible, depending on the coast flank pitch factor θ_c selection.

For the drive flanks of the external gears the pressure angle isogram equation is defined from Equation (2.114) considering the tooth tip radii equal to zero. Then the effective tooth tip angles $\alpha_{ed1,2}$ should be replaced by the tooth tip angles $\alpha_{ad1,2}$:

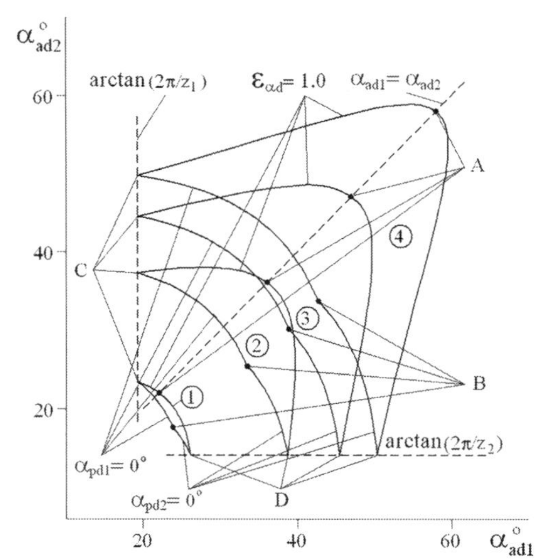

FIGURE 3.30
Areas of existence of the external spur gears with $z_1 = 18$, $z_2 = 25$ and different values of the pitch factor θ_d; 1: $\theta_d = 0.1$, 2: $\theta_d = 0.3$, 3: $\theta_d = 0.5$, 4: $\theta_d = 0.7$.

$$inv(\alpha_{ad1}) + u\,inv(\alpha_{ad2}) - (1+u)inv(\alpha_{wd}) - \frac{2\pi\theta_d}{z_1} = 0. \tag{3.18}$$

The contact ratio isogram is defined by Equation (3.18) and the equation

$$\tan\alpha_{ad1} + u\tan\alpha_{ad2} - (1+u)\tan\alpha_{wd} - \frac{2\pi\varepsilon_{\alpha d}}{z_1} = 0. \tag{3.19}$$

A result of subtracting Equation (3.18) from Equation (3.19) is

$$\alpha_{ad1} + u\alpha_{ad2} - (1+u)\alpha_{wd} - \frac{2\pi(\varepsilon_{\alpha d} - \theta_d)}{z_1} = 0. \tag{3.20}$$

The interference isograms $\alpha_{pd1} = 0°$ and $\alpha_{pd2} = 0°$ are defined by Equation (3.18) and the equations

$$u\tan\alpha_{ad2} - (1+u)\tan\alpha_{wd} = 0 \tag{3.21}$$

and

$$\tan\alpha_{ad1} - (1+u)\tan\alpha_{wd} = 0 \tag{3.22}$$

that are derived from Equations (2.44) and (2.45).

In the point A of the area of existence, where the drive flank pressure angle α_{wd} is maximum and the contact ratio $\varepsilon_{ad} = 1.0$, the pressure angle and contact ratio isograms have a common tangent point and the first derivatives of these isogram functions should be equal:

$$\left.\frac{d(\alpha_{ad2})}{d(\alpha_{ad1})}\right|_{\alpha_{wd}=const} = \left.\frac{d(\alpha_{ad2})}{d(\alpha_{ad1})}\right|_{\varepsilon_{\alpha d}=1.0} \tag{3.23}$$

or with (3.18) and (3.19),

$$\frac{(\tan\alpha_{ad2})^2}{(\tan\alpha_{ad1})^2} = \frac{(\tan\alpha_{ad2})^2 + 1}{(\tan\alpha_{ad1})^2 + 1} \tag{3.24}$$

or

$$\alpha_{ad1} = \alpha_{ad2}. \tag{3.25}$$

This means that the points A of areas of existence lie on the straight line $\alpha_{ad1} = \alpha_{ad2}$. The pressure angle equation at the point A is defined as a solution of Equations (3.18), (3.19), and (3.25):

$$\tan\alpha_{wd} + \frac{2\pi}{z_t} - \tan\left(\alpha_{wd} + \frac{2\pi(1-\theta_d)}{z_t}\right) = 0, \tag{3.26}$$

where $z_t = z_1 + z_2$ is the total number of teeth of mating gears.

Its solution is [14]

$$\alpha_{wd}^{A} = \arctan\left(\sqrt{\frac{\pi^2}{z_t^2} + \frac{2\pi}{z_t \tan\frac{2\pi(1-\theta_d)}{z_t}} - 1} - \frac{\pi}{z_t}\right). \tag{3.27}$$

Then the coordinates of the point A at the area of existence are as follows:

$$\alpha_{ad1}^{A} = \alpha_{ad2}^{A} = \arctan\left(\sqrt{\frac{\pi^2}{z_t^2} + \frac{2\pi}{z_t \tan\frac{2\pi(1-\theta_d)}{z_t}} - 1} + \frac{\pi}{z_t}\right). \tag{3.28}$$

In the point B at the intersection of the interference isograms $\alpha_{pd1} = 0°$ and $\alpha_{pd2} = 0°$ the pressure angle is minimum and the contact ratio is maximum. This maximum contact ratio is defined as a solution of Equations (3.19), (3.20), (3.21), and (3.22):

$$\arctan\left(\frac{2\pi\varepsilon_{\alpha d}}{z_1}\right) + u\arctan\left(\frac{2\pi\varepsilon_{\alpha d}}{z_2}\right) - (1+u)\arctan\left(\frac{2\pi\varepsilon_{\alpha d}}{z_t}\right) - \frac{2\pi(\varepsilon_{\alpha d} - \theta_d)}{z_1} = 0. \tag{3.29}$$

Then the coordinate angles α_{ad1} and α_{ad2}, and drive pressure angle α_{wd} at point B, are defined for drive flanks [107]:

$$\alpha_{ad1}^{B} = \arctan\left(\frac{2\pi\varepsilon_{\alpha d}}{z_1}\right), \tag{3.30}$$

$$\alpha_{ad2}^{B} = \arctan\left(\frac{2\pi\varepsilon_{\alpha d}}{z_2}\right), \tag{3.31}$$

and

$$\alpha_{wd}^{B} = \arctan\left(\frac{2\pi\varepsilon_{\alpha d}}{z_t}\right). \tag{3.32}$$

The coordinate angle α_{ad1} at the intersection point C of isograms $\varepsilon_{\alpha d} = 1.0$ and $\alpha_{pd1} = 0°$ from Equations (3.20) and (3.21) is

$$\alpha_{ad1}^{C} = \arctan(2\pi/z_1). \tag{3.33}$$

The coordinate angle α_{ad2} and pressure angle α_{wd} at the point C are defined from equations

$$\arctan\left(\frac{2\pi}{z_1}\right) + u\alpha_{ad2} - (1+u)\arctan\left(\frac{u}{1+u}\tan\alpha_{ad2}\right) - \frac{2\pi(1-\theta_d)}{z_1} = 0 \tag{3.34}$$

and

$$\arctan\left(\frac{2\pi}{z_1}\right) + u\arctan\left(\frac{1+u}{u}\tan\alpha_{wd}\right) - (1+u)\alpha_{wd} - \frac{2\pi(1-\theta_d)}{z_1} = 0. \tag{3.35}$$

The coordinate angle α_{ad2} at intersection point D of isograms $\varepsilon_{ad} = 1.0$ and $\alpha_{pd2} = 0°$ from Equations (3.20) and (3.22) is

$$\alpha_{ad2}^{D} = \arctan(2\pi/z_2). \tag{3.36}$$

The coordinate angle α_{ad1} and pressure angle α_{wd} at the point D are defined from equations

$$\alpha_{ad1} + u\arctan\left(\frac{2\pi}{z_2}\right) - (1+u)\arctan\left(\frac{1}{1+u}\tan\alpha_{ad1}\right) - \frac{2\pi(1-\theta_d)}{z_1} = 0 \tag{3.37}$$

and

$$\arctan\left((1+u)\tan\alpha_{wd}\right) + u\arctan\left(\frac{2\pi}{z_2}\right) - (1+u)\alpha_{wd} - \frac{2\pi(1-\theta_d)}{z_1} = 0. \tag{3.38}$$

The pressure angle α_{wd}, contact ratio ε_{ad} charts at the points A and B of the areas of existence from Figure 3.30 are presented in Figures 3.31 and 3.32.

Points of an area of existence with the constant drive flank pitch factor θ_d do not define complete mating gear teeth, but just their drive flanks. This allows independent selection of the tooth tip thicknesses and the coast tooth flank parameters of asymmetric gears. Considering the tooth tip radii and backlash equal to zero, the noncontact pitch factor θ_v from (2.113) is

$$\theta_v = \frac{m_{a1}\cos\alpha_{ad1} + m_{a2}\cos\alpha_{ad2}}{\pi\cos\alpha_{wd}}. \tag{3.39}$$

When some point of the area of existence with coordinates α_{ad1} and α_{ad2} is chosen, the pressure angle α_{wd} is calculated by Equation (3.18). Then after the selection of the relative tooth tip thicknesses m_{a1} and m_{a2}, the noncontact pitch factor θ_v is calculated by Equation (3.39). This allows to define the coast flank pitch factor as follows:

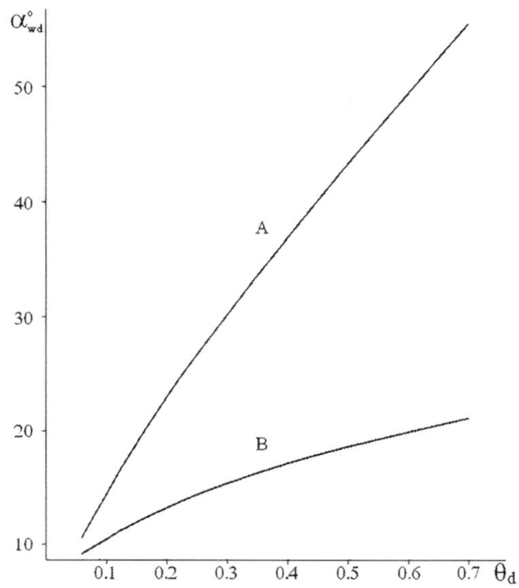

FIGURE 3.31
Drive pressure angle α_{wd} chart of the spur asymmetric gears with $z_1 = 18$, $z_2 = 25$ and different values of the drive pitch factor θ_d at the points A and B.

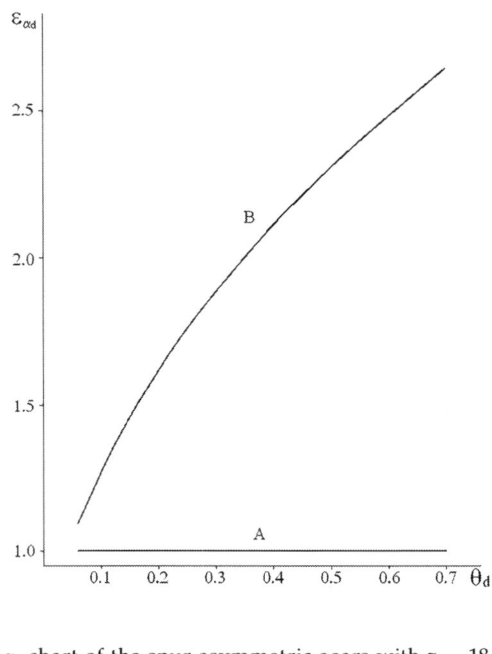

FIGURE 3.32
Drive contact ratio $\varepsilon\alpha_d$ chart of the spur asymmetric gears with $z_1 = 18$, $z_2 = 25$ and different values of the drive pitch factor θ_d at the points A and B.

$$\theta_c = 1 - \theta_d - \theta_v. \tag{3.40}$$

If the tooth tip radii are equal to zero that makes $\alpha_{ed1,2} = \alpha_{ad1,2}$ and $\alpha_{ec1,2} = \alpha_{ac1,2}$, the asymmetry factor K can be defined as a solution of Equations (2.42) and (2.116) as follows:

$$(1+u)inv\big(arc\cos(K\cos\alpha_{wd})\big) = inv\big(arc\cos(K\cos\alpha_{ad1})\big) + uinv\big(arc\cos(K\cos\alpha_{ad2})\big) - \frac{2\pi\theta_c}{z_1}. \tag{3.41}$$

Now the pressure angle and the coast flank tooth tip angles can be defined as follows:

$$\alpha_{wc} = arc\cos(K\cos\alpha_{wd}), \tag{3.42}$$

$$\alpha_{ac1,2} = arc\cos(K\cos\alpha_{ad1,2}). \tag{3.43}$$

Equation (2.67) defines the coast flank contact ratio ε_{ac}. The profile angles at the points of contact near the root fillet of the drive and coast tooth flanks are described by Equations (2.44) and (2.45), and (2.46) and (2.47), accordingly.

3.12 Application of Area of Existence

Initially, the areas of existence were developed for only research purposes to learn more about capabilities and limitations of involute macrogeometry, specifically of asymmetric tooth gears that are not so systematically explored in comparison to symmetric tooth gears. However, with development of areas of existence calculation software they became a useful practical design tool.

Figure 3.33 demonstrates an interface of the areas of existence calculation program. Input data include the following: pinion and gear numbers of teeth (for an internal gear pair the gear number of teeth should be shown a negative number), relative tooth tip thicknesses, also called tip land coefficients, asymmetry factor (or coefficient of asymmetry), and a tip radius coefficient (a ratio of the tooth tip radius to the operation module or diametral pitch) that, in this case, is identical for the pinion and gear, and the drive and coast tooth flank. The *"COORDINATES"* button opens a choice of an area of existence coordinates, including their initial minimum and maximum values, and incremental grid steps. The *"DATA FOR STRESS ANALYSIS"* button opens a window that contains a module for the metric system or diametral pitch for the English system, pinion and gear face widths, material properties

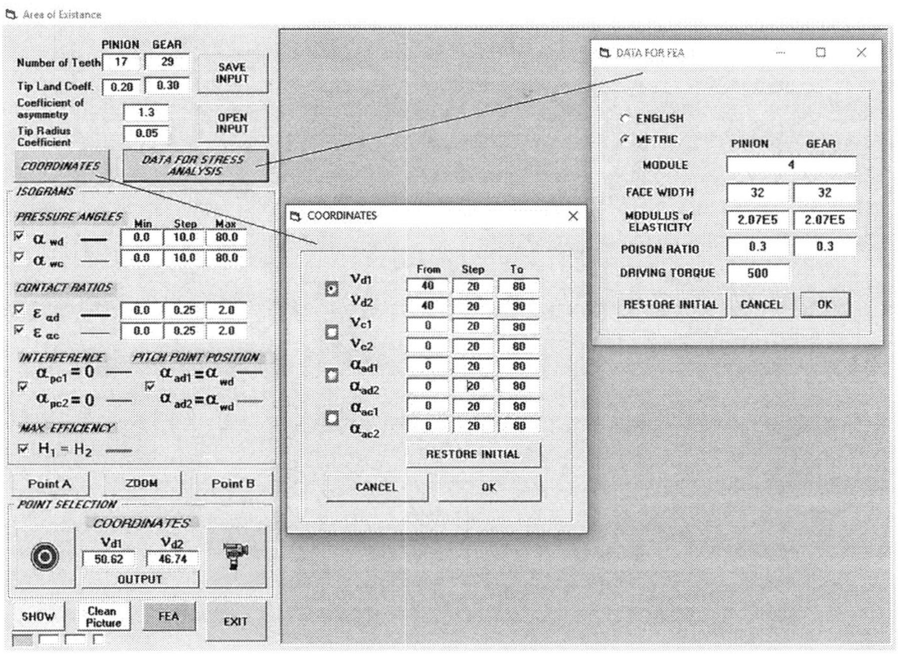

FIGURE 3.33
Area of existence program interface and input data.

including the modulus of elasticity and Poisson ratio, and pinion driving torque. The *"SAVE"* button allows to store selected data to open them later with the *"OPEN"* button.

The next step is the selection of isograms and their ranges. The *"SHOW"* button instantly shows the area of existence with selected isograms (Figure 3.34). Simultaneously, the program creates ASCII files with coordinate points of each isogram. Gear parameters of the points *A* and *B* of the area of existence are shown by pressing the corresponding button. Any zone of the area of existence can be magnified by the *"ZOOM"* button.

There are two ways to select a certain point of the area of existence: by a button with a target image or by typing the point coordinate values. A selected point of the area of existence presents a gear pair that could be animated by pushing a button with the video camera image. It starts a gear mesh animation (Figure 3.35) that can be zoomed in or out, and also speeded up or slowed down.

The *"OUTPUT"* button brings all geometry parameters of the selected gear pair to the interface screen (Figure 3.36) and creates the output file with these parameters.

Such point selection allows to quickly evaluate the number of area of existence points and choose the most suitable one. Then pressing the "FEA" button brings the second program interface to calculate the selected gear

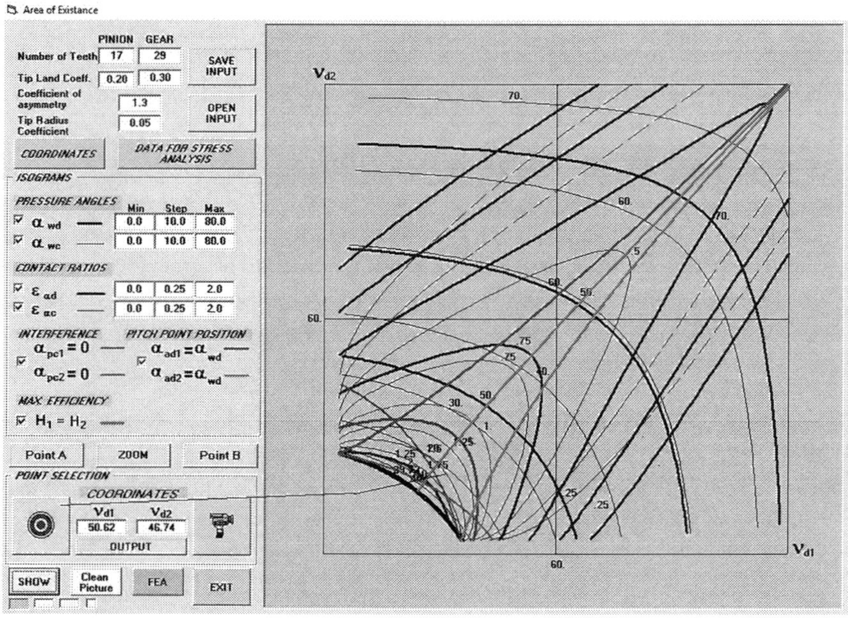

FIGURE 3.34
Area of existence isograms and section of a point presenting a required gear pair.

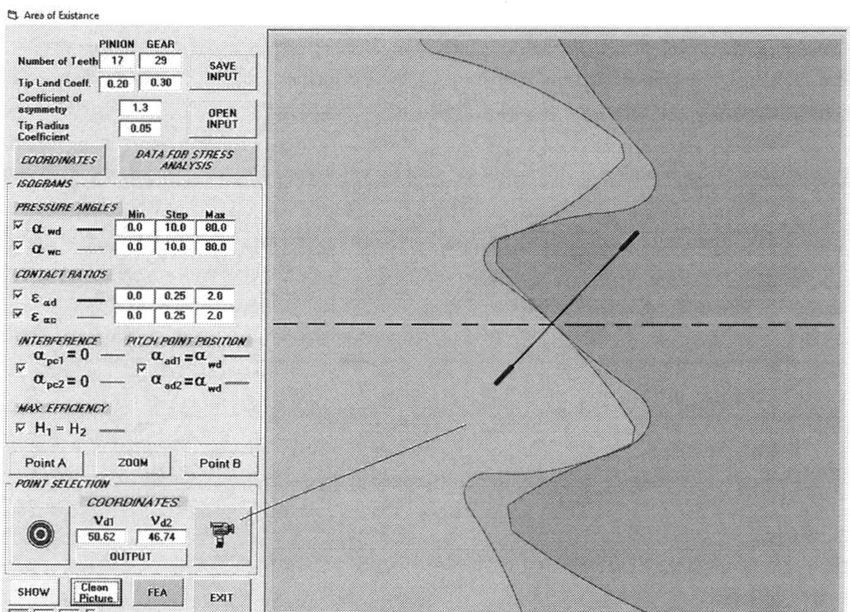

FIGURE 3.35
Animation of a gear pair, selected from area of existence.

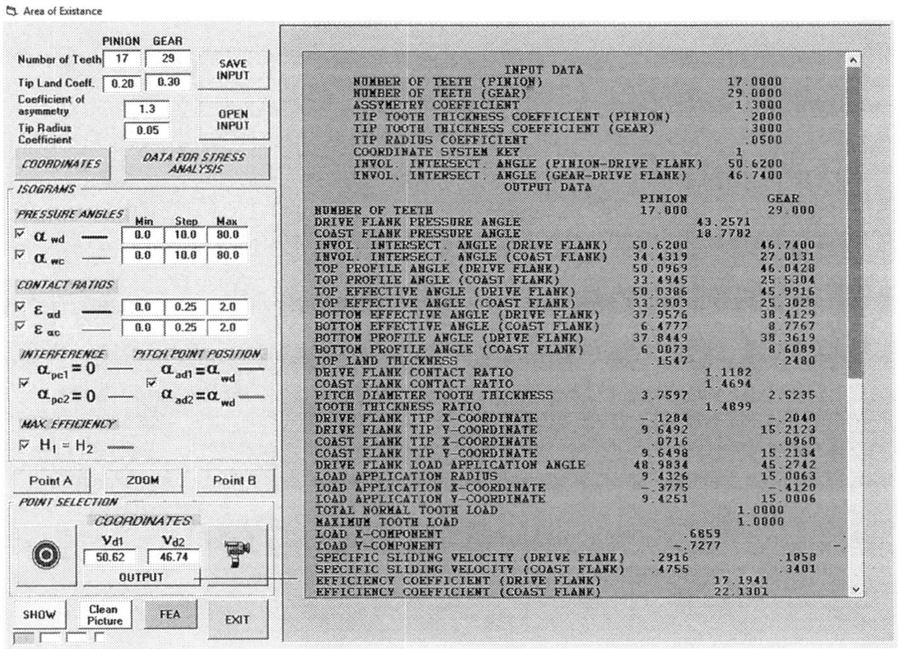

FIGURE 3.36
Output geometric parameters of a gear pair, selected from area of existence.

pair stresses (Figure 3.37). On the bottom of the screen are all input data and the area of existence coordinates of the selected gear pair. The "RUN" button initiates the calculation of the Hertzian contact stress, bending stresses defined by the FEA, gear mesh efficiency at the drive and flanks based on the chosen average friction coefficient.

The area of existence of involute gears is a research tool that can be used to find some unusual gear mesh solutions that were not known before. It also allows gear pairs with certain characteristics to be located. Its practical purpose is to define the gear pair parameters that satisfy specific performance requirements before detailed design and calculations. This involute gear research tool is incorporated into the preliminary design program with the FEA subroutine. Such a program is able not just to generate all isograms for gears with the given numbers of teeth z_1 and z_2, relative tooth tip thicknesses m_{a1} and m_{a2}, and asymmetry coefficient K, but also, with input of a module or diametral pitch, gear face widths, and material properties (the modulus of elasticity and Poisson ratio, and a friction coefficient), to promptly define the relative sliding velocities, gear mesh efficiency, bending and Hertzian contact stresses, and create the ASCII tooth profiles and gear mesh animation of any point of area of existence [79]. Such software quickly defines limits of parameter selection of involute gears, locates and animates feasible gear

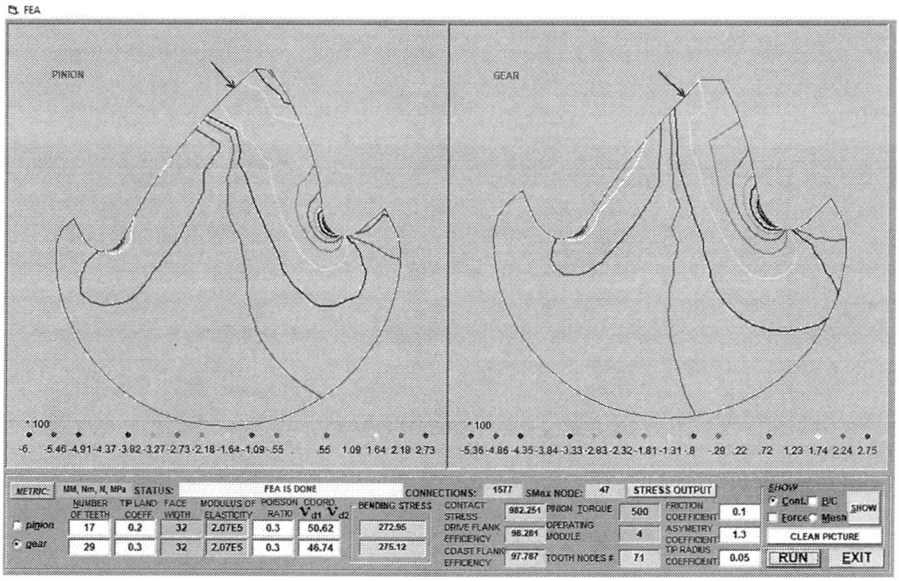

FIGURE 3.37
Stress analysis interface and calculation results.

pairs, and allows their geometry and stress levels to be reviewed. Benefits of an area of existence include the following:

- consideration of all possible gear combinations;
- instant definition of limits of the gear performance parameters;
- awareness about some uncommon gear design options;
- quick localization of gear sets suitable for particular application;
- preliminary gear design optimization.

4

Asymmetric Gearing Limits

This chapter provides the definition of the gear parameter limits for the spur and helical gears with asymmetric teeth. Some of the tooth shapes and gear mesh combinations presented below look rather unusual and may not have rational practical applications. Another goal of this chapter is to establish boundaries of the asymmetric involute gearing parameters.

4.1 Number of Teeth

Selection of the numbers of teeth of the mating gears is critically important. First of all, it provides the required gear ratio. Second, when the gear ratio and center distance are specified, the selected numbers of teeth define the gear tooth size that is described by a module in the metric system or a diametral pitch in the English system. A gear tooth size is a main parameter in the definition of root bending stress. Third, the number of gear teeth is a major factor in the definition of gear mesh efficiency. Along with other gear mesh geometry parameters (pressure angles, contact ratios, etc.), tooth number selection allows the safety factors to be balanced for bending and contact stresses, and wear resistance to optimize a gear pair design.

The maximum number of gear teeth is limited by application practicality and manufacturing technology. For example, most mechanically controlled gear hobbing machines can produce a gear with number of teeth up to 400 using one-start hobs. Usage of multi-start hobs increases this limit. Some computer numerical control (CNC) gear hobbing machines can produce gear with number of teeth up to 1,000 using one-start hobs. Other gear fabrication technologies like, for example, profile cutting or injection molding can provide gears with even greater number of teeth.

From an application point of view, the maximum number of gear teeth may also be limited by tolerance sensitivity and operating conditions. For a given gear pitch diameter, an increase in the number of teeth leads to the tooth size reduction, to a point when the size of the tooth becomes comparable with tolerances achievable by selected gear fabrication technology. Gear drive operating conditions may also result in a similar effect for fine pitch gears. For example, a wide operating temperature range and application of dissimilar gear and housing materials may lead to noticeable gear size and center distance changes that could be comparable to the tooth size of the fine pitch gears. This may lead to a proper gear engagement disruption from gear jamming to complete tooth mesh separation.

Spur reversible asymmetric gears can transmit rotational motion and load by both drive and coast tooth flanks. This means that the drive and coast contact ratios are $\varepsilon_{ad} \geq 1.0$ and $\varepsilon_{ac} \geq 1.0$. The selection of minimum numbers of teeth of the external spur reversible asymmetric gears depends on the gear ratio and is defined by the following simultaneous conditions: the coast flank profile angles at the points where involute flanks meet a root fillet $\alpha_{pc1,2} = 0°$ (point B of an area of existence) and the drive flank contact ratio $\varepsilon_{ad} = 1.0$. These minimum numbers of teeth are the same as for the external spur symmetric gears described in [170]. The maximum asymmetry factor is achieved when the relative tip thickness $m_{a1,2} = 0.0$. This means that the noncontact pitch factor $\theta_v = 0.0$ and from Equation (2.110) the sum of the drive and coast pitch factors is equal to 1.0. The external reversible asymmetric spur gears with low number of teeth are shown in Figure 4.1. Parameters of these gear pairs are presented in Table 4.1.

Minimum numbers of teeth of the internal reversible asymmetric spur gears are defined by the simultaneous conditions: the pinion coast flank profile angle at the point, where the involute flank meets the root fillet $\alpha_{pc1} = 0°$ and beginning of the "tip-tip" interference that is described by Equation (2.62) (point B of an area of existence) and the contact ratio $\varepsilon_\alpha = 1.0$. The maximum asymmetry factor is achieved when the relative tip thickness $m_{a1,2} = 0.0$. The internal spur gears with minimum numbers of teeth are shown in Figure 4.2. Its parameters are provided in Table 4.2.

Helical gears have an overlap ratio (or a face contact ratio) ε_β that compensates the lack of the transverse contact ratio ε_α (see paragraph 2.3.5), which can be reduced to zero. This makes it possible to achieve the minimum

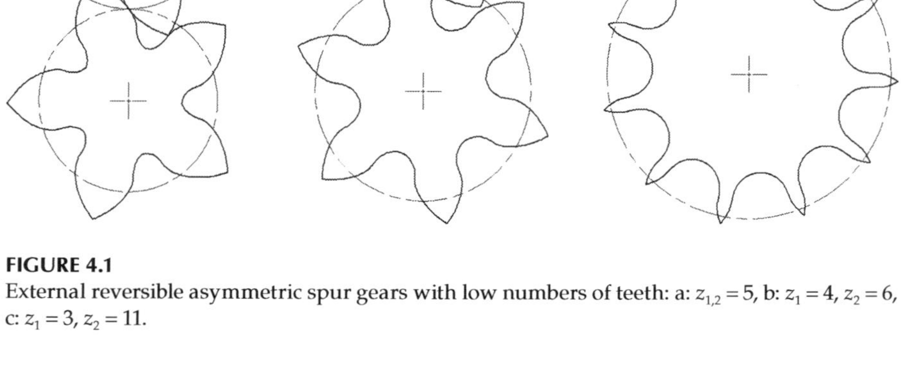

FIGURE 4.1
External reversible asymmetric spur gears with low numbers of teeth: a: $z_{1,2} = 5$, b: $z_1 = 4$, $z_2 = 6$, c: $z_1 = 3$, $z_2 = 11$.

TABLE 4.1

External Reversible Asymmetric Spur Gears with Low Number of Teeth

Pinion number of teeth, z_1	5	4	3
Gear number of teeth, z_2	5	6	11
Drive flank pitch factor, θ_d	0.533	0.514	0.502
Coast flank pitch factor, θ_c	0.467	0.486	0.498
Asymmetry factor, K	1.068	1.030	1.006
Drive pressure angle, α_{wd}°	37.62	34.80	24.94
Coast pressure angle, α_{wc}°	32.27	32.23	24.25
Drive contact ratio, $\varepsilon_{\alpha d}$	1.0	1.0	1.0
Coast contact ratio, $\varepsilon_{\alpha c}$	1.005	1.004	1.004
Pinion tooth tip drive profile angle, α_{ad1}°	54.44	58.67	64.69
Pinion tooth tip coast profile angle, α_{ac1}°	51.63	57.61	64.57
Gear tooth tip drive profile angle, α_{ad2}°	54.44	48.00	30.30
Gear tooth tip coast profile angle, α_{ac2}°	51.63	46.42	29.83

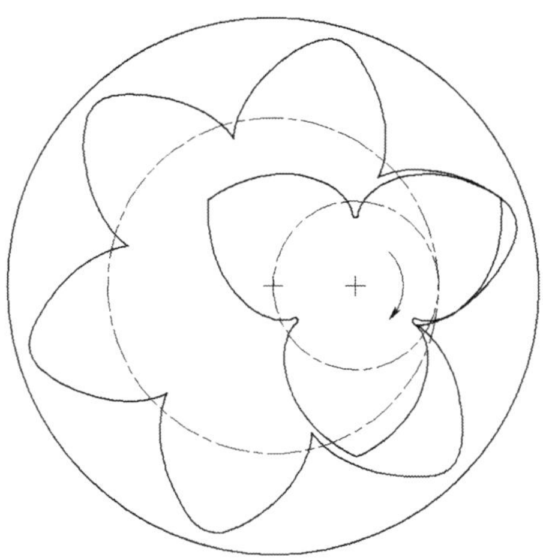

FIGURE 4.2
Internal reversible asymmetric spur gears with minimum number of teeth $z_1 = 3$, $z_2 = 6$.

number of teeth $z_1 = z_2 = 1$ [171]. Figures 4.3 and 4.4 show the external and internal helical reversible asymmetric gears with a low number of teeth. The main parameters of these gears are provided in Table 4.3.

The article [71] described the asymmetric spur gears with number of teeth 1, 2, and 3, and gear ratio $u = 1$. Those gears have the drive contact ratio $\varepsilon_{\alpha d} \geq$ 1.0, but the coast contact ratio $\varepsilon_{\alpha c} < 1.0$ that makes them irreversible. Such gears may not have the coast flank involute profiles at all. The examples of these irreversible asymmetric spur gears with number of teeth $z_{1,2} = 1$, $z_1 = 1$, and $z_2 = 2$ are shown in Figure 4.5. Their main parameters are presented in Table 4.4.

TABLE 4.2

Internal Reversible Asymmetric Spur Gears with Low Number of Teeth

Gear	External	Internal
Number of teeth, $z_{1,2}$	3	6
Drive flank pitch factor, θ_d	0.506	
Coast flank pitch factor, θ_c	0.494	
Asymmetry factor, K	1.015	
Drive pressure angle, $\alpha_{wd}{}^{\circ}$	31.13	
Coast pressure angle, $\alpha_{wc}{}^{\circ}$	29.67	
Drive contact ratio, ε_{ad}	1.0	
Coast contact ratio, ε_{ac}	1.009	
Drive tooth tip profile angle, $\alpha_{ad1,2}{}^{\circ}$	65.16	56.95
Coast tooth tip profile angle, $\alpha_{ac1,2}{}^{\circ}$	64.76	56.39

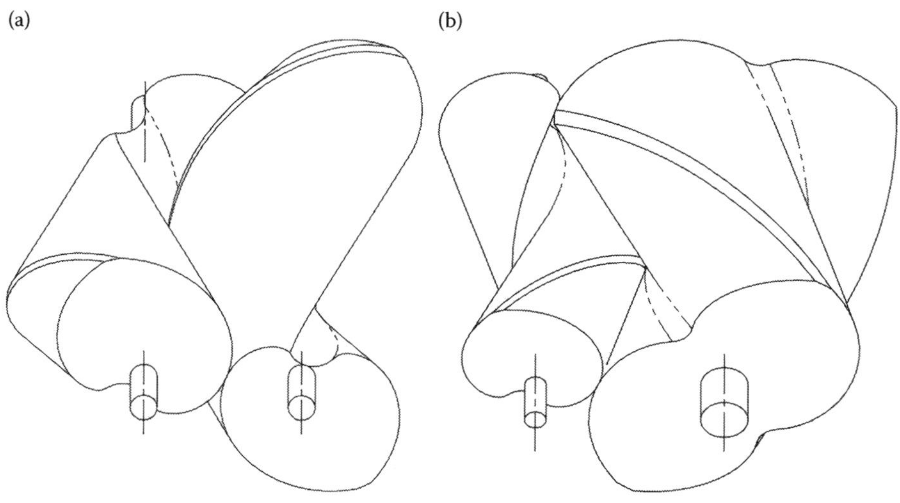

FIGURE 4.3
External helical reversible asymmetric gears with low number of teeth: a: $z_{1,2} = 1$; b: $z_1 = 1$, $z_2 = 2$.

Figure 4.6 shows the internal irreversible asymmetric spur gears with minimum number of teeth $z_1 = 2$, $z_2 = 5$. The minimal numbers of teeth of this internal gear mesh are limited by the "tip-tip" interference condition. The main parameters of this gear pair are provided in Table 4.5.

Although the gears shown in Figures 4.5 and 4.6 are irreversible, if they are assembled in two layers along the gear axes and have the mirrored tooth orientation, such a gear system will be reversible. In Figure 4.7 the front layer left driving gear operates in counterclockwise direction and the back layer gears are not engaged in motion transmission. In the reversed rotation, the back layer left driving gear operates in clockwise direction and the front layer gears are not engaged.

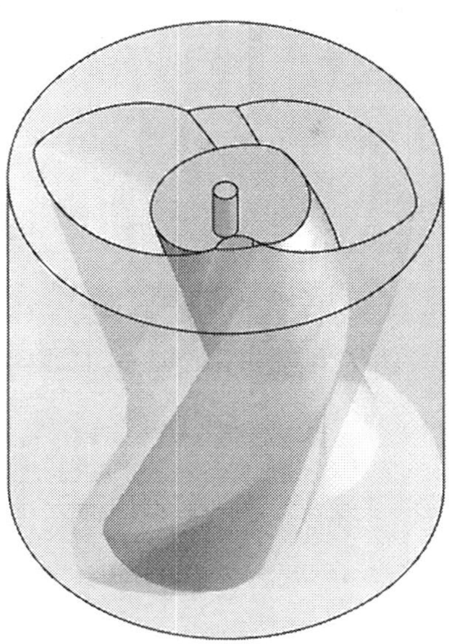

FIGURE 4.4
Internal helical reversible asymmetric gears with low number of teeth: a: $z_{1,2} = 1$; b: $z_1 = 1$, $z_2 = 2$.

TABLE 4.3
Asymmetric Reversible Helical Gears with Low Number of Teeth

Gear mesh	External Figure 4.3a	External Figure 4.3b	Internal Figure 4.4
Pinion number of teeth, z_1	1	1	1
Gear number of teeth, z_2	1	2	2
Drive flank pitch factor, θ_d	0.594	0.584	0.519
Coast flank pitch factor, θ_c	0.368	0.358	0.303
Asymmetry factor, K	1.517	1.517	1.517
Drive pressure angle, α_{wd}°	67.80	67.80	67.80
Coast pressure angle, α_{wc}°	55.02	55.02	55.02
Helix angle at pitch diameter, β_w°	36.00	36.00	36.00
Drive transverse contact ratio, $\varepsilon_{\alpha d}$	0.647	0.644	0.602
Coast transverse contact ratio, $\varepsilon_{\alpha c}$	0.455	0.458	0.455
Axial contact ratio, ε_β	0.7	0.7	0.7
Drive total contact ratio, $\varepsilon_{\gamma d}$	1.347	1.344	1.302
Coast total contact ratio, $\varepsilon_{\gamma c}$	1.155	1.158	1.155
Pinion drive tooth tip profile angle, α_{ad1}°	77.43	77.43	77.43
Gear drive tooth tip profile angle, α_{ad2}°	77.43	73.87	53.60
Pinion coast tooth tip profile angle, α_{ac1}°	54.83	54.83	54.83
Gear coast tooth tip profile angle, α_{ac2}°	54.83	65.08	35.55

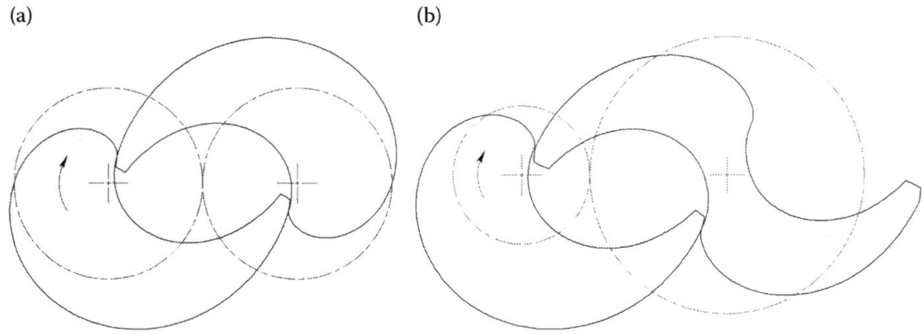

FIGURE 4.5
External spur irreversible asymmetric spur gears with low number of teeth: a: $z_{1,2} = 1$; b: $z_1 = 1$, and $z_2 = 2$.

TABLE 4.4

Irreversible External Asymmetric Spur Gears with Low Number of Teeth

Pinion number of teeth, z_1	1	1
Gear number of teeth, z_2	1	2
Drive flank pitch factor, θ_d	0.952	0.909
Drive flank pressure angle, $\alpha_{wd}{}^{\circ}$	72.34	64.50
Drive flank contact ratio, $\varepsilon_{\alpha d}$	1.0	1.0
Pinion drive flank tooth tip profile angle, $\alpha_{ad1}{}^{\circ}$	80.95	80.95
Gear drive flank tooth tip profile angle, $\alpha_{ad2}{}^{\circ}$	80.95	72.31

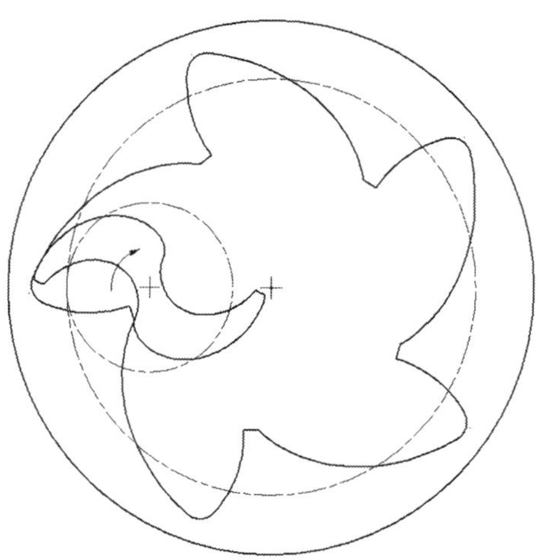

FIGURE 4.6
Internal irreversible asymmetric spur gears with minimum number of teeth $z_1 = 2$, $z_2 = 5$.

TABLE 4.5

Irreversible Internal Asymmetric Spur Gears with Low Number of Teeth

Gear	External		Internal
Number of teeth, $z_{1,2}$	2		5
Drive flank pitch factor, θ_d		0.771	
Drive pressure angle, α_{wd}°		65.04	
Drive contact ratio, ε_{ad}		1.0	
Drive tooth tip profile angle, $\alpha_{ad1,2}$°	72.31		56.95

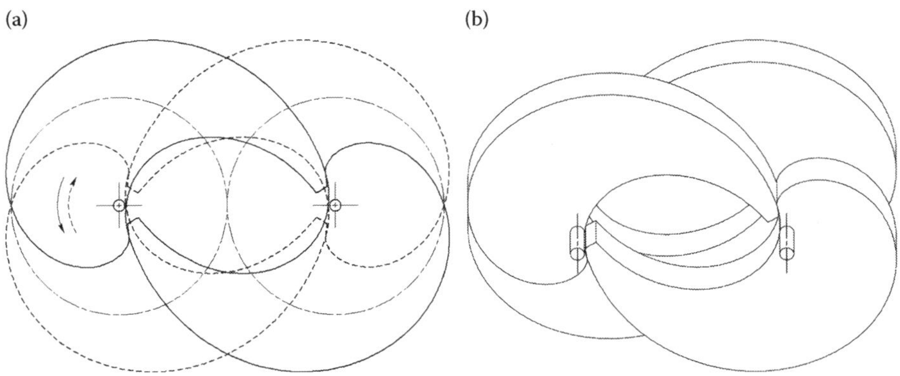

FIGURE 4.7
Two-layer assembly of reversible asymmetric spur gears with minimal number of teeth $z_{1,2} = 1$; a: gear tooth profiles, front layer gear mesh – solid lines, back layer gear mesh – dashed lines; b: isometric view.

4.2 Pressure Angles

The vast majority of gears are designed with the standard symmetric tooth proportions. One of the main tooth proportion parameters is pressure angle. Most gears are designed with the standard 20° pressure angle. The old standard pressure angle of 14½° is still in use. In some industries like, for example, aerospace, the 25° and 28° pressure angles are used [13, 172]. The term *pressure angle*, in this case, is actually related not to the gear mesh, but to the basic or generating rack that is utilized for gear design or as a cutter profile, respectively. A gear involute profile angle varies from the root form diameter to the tooth tip diameter. This section describes transverse operating pressure angles, which are defined for a pair of mating asymmetric tooth gears as follows:

$$\alpha_{wd} = \arccos\left(\frac{d_{bd1}(u \pm 1)}{2a_w}\right) \quad \text{and} \quad \alpha_{wc} = \arccos\left(\frac{d_{bc1}(u \pm 1)}{2a_w}\right). \tag{4.1}$$

For spur gears with asymmetric teeth, the minimum pressure angle is defined at the point *B* of an area of existence. For external reversible spur asymmetric gears, the point *B* is at intersection of the interference isograms $\alpha_{pc1} = 0°$ and $\alpha_{pc2} = 0°$. Then the minimum coast flank pressure angle is defined by Equations (2.46), (2.47), and (2.67) as follows:

$$\alpha_{wc}^{B} = \arctan\left(\frac{2\pi\varepsilon_{\alpha c}}{z_t}\right), \tag{4.2}$$

where $z_t = z_1 + z_2$ is the total number of teeth.

This makes the minimum drive flank pressure angle

$$\alpha_{wd}^{B} = \arccos\left(\frac{1}{K}\cos\left(\arctan\frac{2\pi\varepsilon_{\alpha c}}{z_t}\right)\right) \tag{4.3}$$

or

$$\alpha_{wd}^{B} = \arccos\left(\frac{1}{K\sqrt{4\pi\varepsilon_{\alpha c}^{2}/z_t^{2}+1}}\right). \tag{4.4}$$

If the total number of teeth is increasing and approaches to infinity ($z_t \to \infty$), the minimum theoretical limit of pressure angle for the coast tooth flanks is $\alpha_{wc}^{\min\lim} = 0°$and for the drive tooth flanks, it is

$$\alpha_{wd}^{\min\lim} = \arccos(1/K). \tag{4.5}$$

For helical gears with asymmetric teeth, the theoretical minimum pressure angles are $\alpha_{wc} = 0°$ and $\alpha_{wd} = \arccos(1/K)$.

The maximum pressure angle for external spur gears with asymmetric teeth is defined at point *A* of the area of existence, where the pressure angle isogram is tangent to the drive flank transverse contact ratio isogram $\varepsilon_\alpha = 1.0$. It depends on the number of teeth of mating gears $z_{1,2}$ and also on the relative tooth tip thicknesses $m_{a1,2}$.

A general solution for the maximum pressure angle for both symmetric and asymmetric gears is presented in Equation (3.27). When $z_t \to \infty$, the pressure angle limit for external spur gears from this equation is defined as follows [14]:

$$\alpha_{wd}^{\max\lim} = \arctan\sqrt{\frac{\theta_d}{1-\theta_d}}. \tag{4.6}$$

A chart of the maximum drive flank pressure angles $\alpha_{wd\max}$ for different total numbers of teeth and drive pitch factors is shown in Figure 4.8.

If tooth asymmetry is defined by the factor *K*, the maximum drive pressure angle is defined at the point *A* of the area of existence. Figures 4.9 and 4.10 present the ranges of the drive and coast pressure angles of external spur asymmetric reversible gears with different numbers of teeth and asymmetry factors.

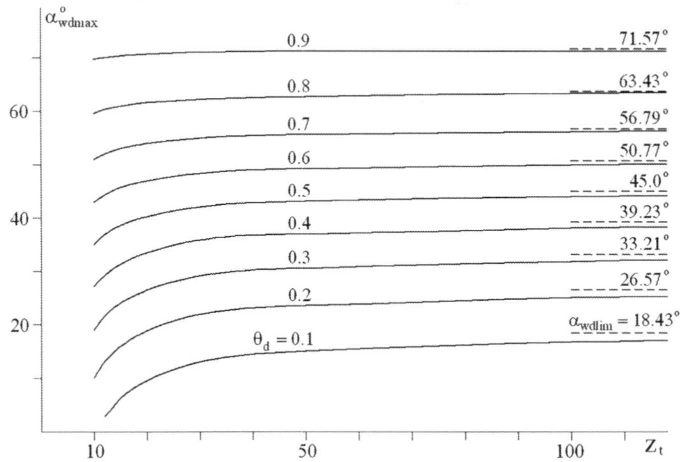

FIGURE 4.8
Maximum drive flank pressure angles $\alpha_{wd\max}$ and their limits $\alpha_{wd\lim}$ when $z_t \to \infty$.

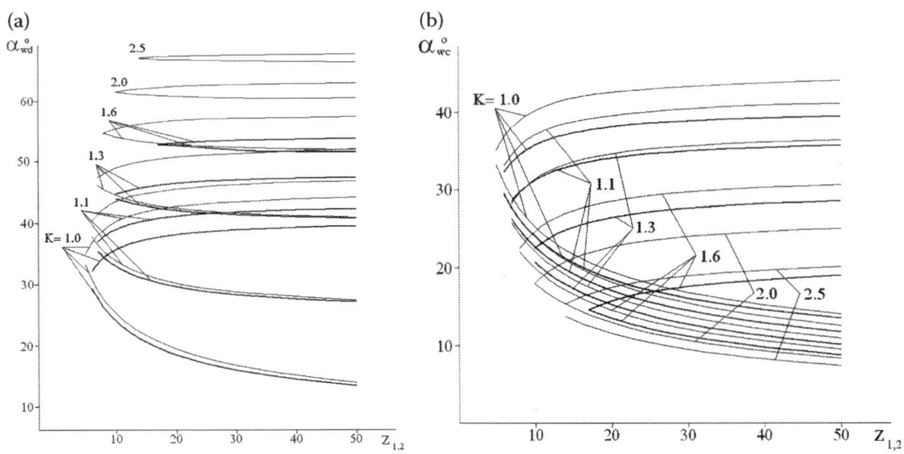

FIGURE 4.9
Minimum and maximum drive (a) and coast (b) flank pressure angles α_{wd} and α_{wc} for the external spur gears with a gear ratio $u = 1.0$, and relative tooth tip thicknesses $m_{a1,2} = 0$ (thin lines) and $m_{a1,2} = 0.25$ (thick lines).

The drive and coast pressure angle maximum limits for the external spur asymmetric reversible gears with $z_t \to \infty$ are as follows [14]:

$$\alpha_{wd}^{\max\lim} = \arctan K, \tag{4.7}$$

$$\alpha_{wc}^{\max\lim} = \arctan \frac{1}{K}. \tag{4.8}$$

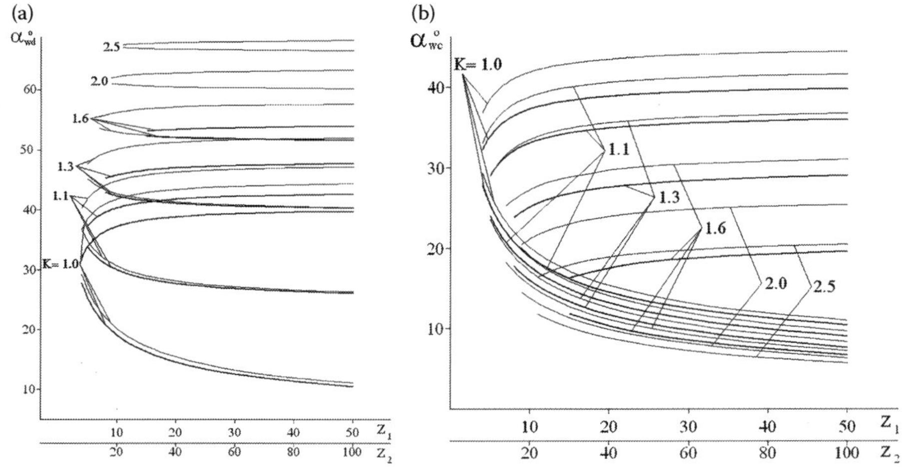

FIGURE 4.10
Minimum and maximum drive (a) and coast (b) flank pressure angles α_{wd} and α_{wc} for the external spur gears with a gear ratio $u = 2.0$, and relative tooth tip thicknesses $m_{a1,2} = 0$ (thin lines) and $m_{a1,2} = 0.25$ (thick lines).

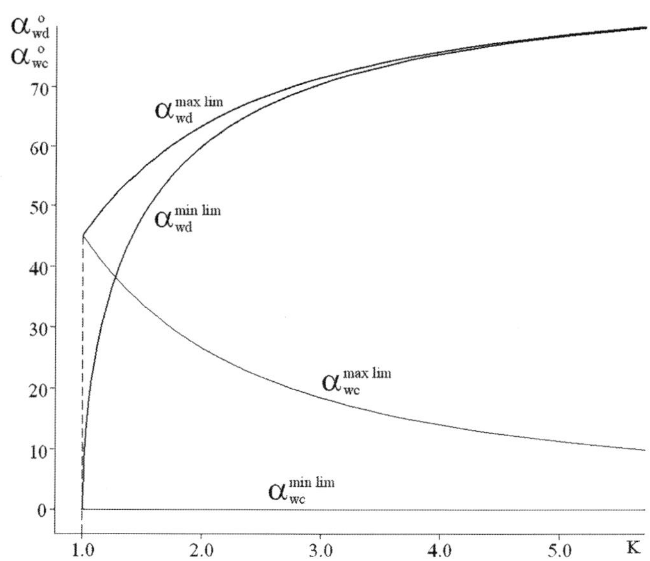

FIGURE 4.11
Drive (thick lines) and coast (thin lines) flank pressure angle limits as functions of the asymmetry factor K.

A chart of the pressure angle limits as functions of the asymmetry factor K is shown in Figure 4.11.

Helical asymmetric tooth gears can have the transverse contact ratio $0 < \varepsilon_\alpha < 1.0$. This expands a theoretical range of the transverse pressure angle to

$0° < \alpha_w < 90°$. A practical application of helical asymmetric tooth gears with very high transverse pressure angles (75° to 85°) is the self-locking gears (Section 7.1).

4.3 Contact Ratios

Reversible spur asymmetric gears require the coast flank contact ratio to be $\varepsilon_{ac} \geq 1.0$. The highest contact ratio is achieved at the point B of an area of existence at the intersection of the drive tooth flank interference isograms. Below this point, the coast flank interference occurs, resulting in involute profile undercut. Such undercut is permissible, if the coast flank contact ratio is $\varepsilon_{ac} \geq 1.0$. The charts with maximum and minimum contact ratios for the spur external reversible asymmetric gears with the relative tooth tip thicknesses $m_{a1,2} = 0$ and $m_{a1,2} = 0.25$ are shown in Figures 4.12 and 4.13.

Irreversible asymmetric gears are of more theoretical rather than practical interest, because benefits of their applications are not apparent. Table 4.6 and Figure 4.14 present such gears.

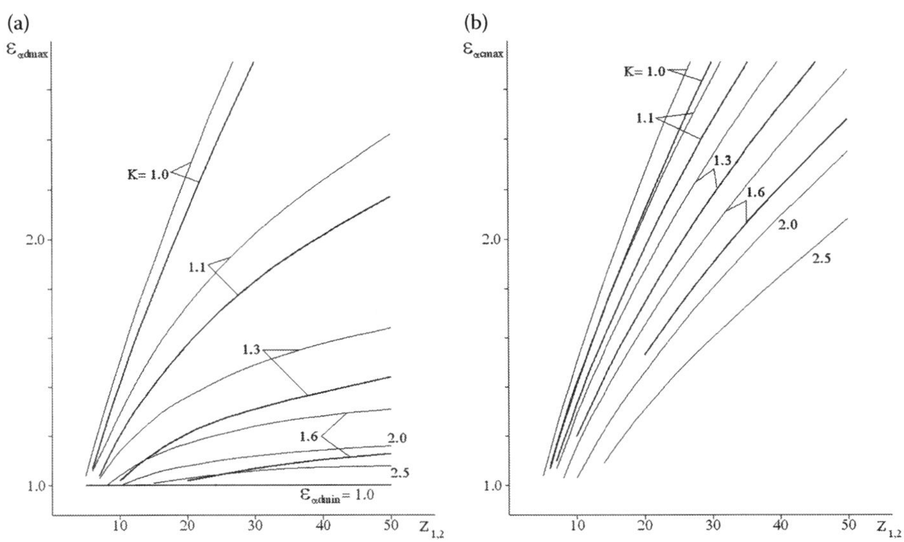

FIGURE 4.12
Minimum and maximum drive (a) and coast (b) flank contact ratios α_{wd} and α_{wc} for the external reversible asymmetric spur gears with a gear ratio $u = 1$, and relative tooth tip thicknesses $m_{a1,2} = 0$ (thin lines) and $m_{a1,2} = 0.25$ (thick lines).

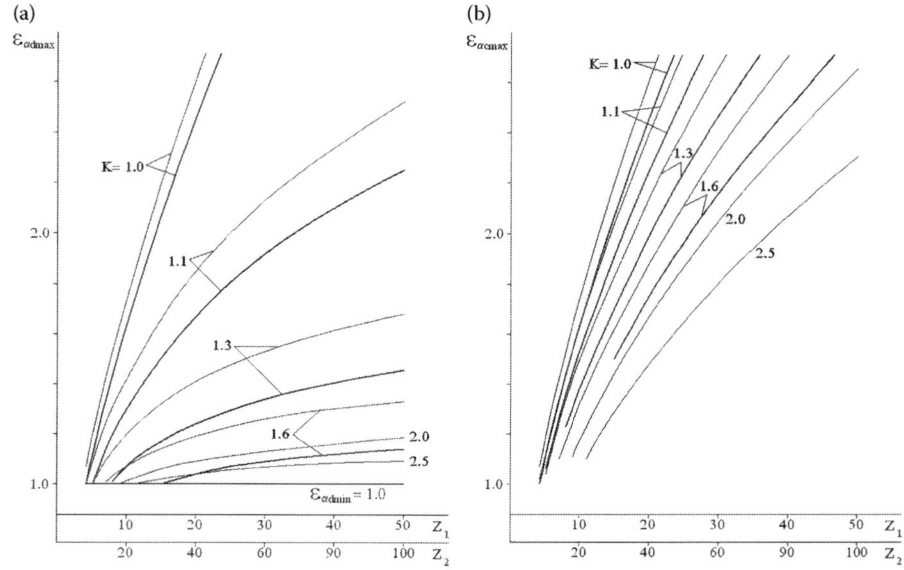

FIGURE 4.13
Minimum and maximum drive (a) and coast (b) flank contact ratios α_{wd} and α_{wc} for the external reversible asymmetric spur gears with a gear ratio $u = 2.0$, and relative tooth tip thicknesses $m_{a1,2} = 0$ (thin lines) and $m_{a1,2} = 0.25$ (thick lines).

4.4 Practical Range of Asymmetric Gear Parameters

Direct Gear Design significantly expands boundaries of involute gearing. However, in most cases gear tooth and mesh geometry parameters do not reach their theoretical limits, because of, first, specific gear application performance requirements and, second, some material and technological constraints. For example, application of gears with a very low number of teeth is limited by increased specific sliding velocities, resulting in low mesh efficiency, higher gear mesh temperature, and tooth flank scuffing probability. At the same time, this reduces tooth deflection under an operating load and flank impact absorption, resulting in higher noise and vibration. On the contrary, gears with a given pitch diameter and very high

TABLE 4.6
Maximum Transverse Contact Ratios of Irreversible Asymmetric Gears

Number of teeth, $z_{1,2}$	1	3	5	10	20	33	48
Contact ratio, $\varepsilon_{\alpha d}$	1.05	1.27	1.52	2.08	3.01	4.02	5.04
Pressure angle, α_{wd}°	73.07	53.1	43.66	33.15	25.31	20.97	18.27
Tooth tip angle, $\alpha_{ad1,2}$°	81.35	69.43	62.35	52.56	43.41	37.46	33.44
Lowest involute angle, $\alpha_{pd1,2}$	0	0	0	0	0	0	0

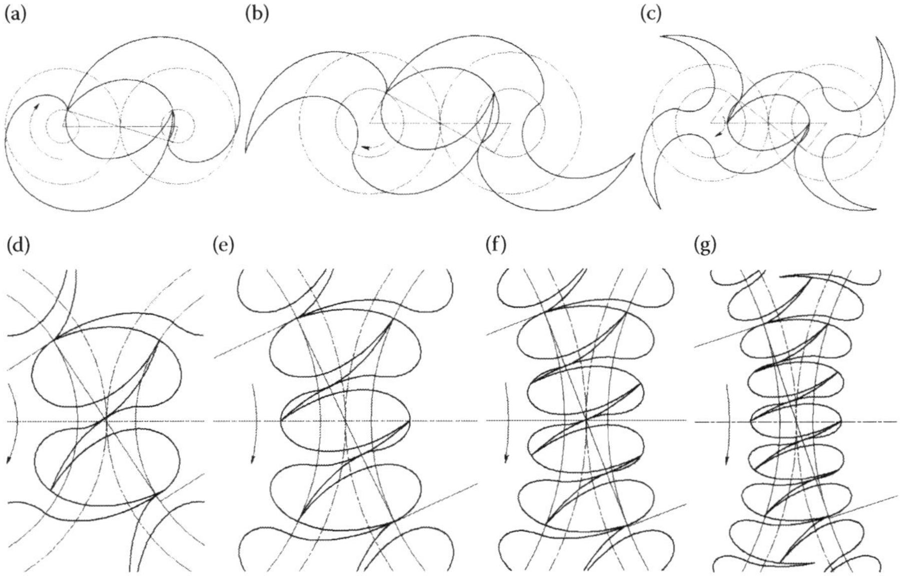

FIGURE 4.14
Irreversible external asymmetric gears with maximum contact ratio; a: numbers of teeth $z_{1,2} = 1$, b: $z_{1,2} = 3$, c: $z_{1,2} = 5$, d: $z_{1,2} = 10$, e: $z_{1,2} = 20$, f: $z_{1,2} = 33$, g: $z_{1,2} = 48$.

number of teeth have a very small tooth size. This leads to reduced bending strength and increases gear drive assembly tolerance sensitivity, when at a certain tolerance combination, the contact ratio can be reduced to $\varepsilon_\alpha < 1.0$, which also results in increased noise and vibration, and degrades gear drive performance.

The practical maximum pressure angle and transverse contact ratio are limited by the minimum tooth tip thickness. For case hardened teeth, it is sufficient to avoid the hardening through the tooth tip. For gears made out of soft metals and plastics it is sufficient to exclude tooth tip bending. The minimum relative tooth tip thickness typically is $m_{a1,2} = 0.25–0.3$. A practical minimal contact ratio for conventional spur gears is about $\varepsilon_{\alpha\min} = 1.1–1.15$. For high contact ratio (HCR) gears, it is typically about $\varepsilon_{\alpha\min} = 2.05–2.2$. These minimal contact ratio values are chosen to avoid its reduction below 1.0 for conventional spur gears and below 2.0 for HCR spur gears, considering manufacturing and assembly tolerances, and tooth tip chamfers or radii. These conditions also identify the practical maximum pressure angle. The practical minimal pressure angle for symmetric gears is defined by the beginning of the tooth involute undercut, when the involute profile angles at the lowest contact points $\alpha_{pd1,2} = 0°$, where a transverse contact ratio reaches its maximum value $\varepsilon_{\alpha\max}$. Application of gears with asymmetric teeth allows the drive flank pressure angle to be increased in comparison to gears with symmetric teeth because of the coast flank pressure angle reduction. If coast flanks are not used for load transmission and may just occasionally be engaged in contact (as a result of tooth bouncing, inertial load during gear drive deceleration, etc.), the coast flank pressure angle can be as low as $\alpha_{wc} = 10°–15°$. The practical maximum drive flank pressure angles for conventional and HCR asymmetric gears are shown in Tables 4.7 and 4.8.

The maximum drive pressure angle values in Tables 4.7 and 4.8 assume some possible small undercut of the coast flank near the root, especially for gears with a low number of teeth (15–30). However, this does not reduce the coast flank contact below $\varepsilon_{ac} = 1.0$. This undercut can be reduced or completely eliminated by using the slanted tooth tips (Figure 4.15) [14]. It increases the tooth tip land and reduces bending stress. These slanted tooth tips can be produced by the special topping gear cutter (hob) or by the secondary (after the tooth hobbing) tooth tip milling operation.

TABLE 4.7

Practical α_{wdmax} for Conventional Asymmetric Gears ($m_{a1,2} = 0.3$, $\varepsilon_{ad} = 1.1$, $\alpha_{wc} = 15°$)

	z_1						
z_2	**15**	**20**	**30**	**40**	**50**	**70**	**100**
15	43.5						
20	44.5	45.5					
30	45.9	46.4	47.3				
40	47	47.3	47.7	48.2			
50	47.6	47.8	48	48.3	48.9		
70	48	48.2	48.6	48.7	49	49.5	
100	48.7	48.9	49.2	49.5	49.1	49.6	50

TABLE 4.8

Practical α_{wdmax} for HCR Asymmetric Gears ($m_{a1,2} = 0.3$, $\varepsilon_{ad} = 2.1$, $\alpha_{wc} = 15°$)

	z_1						
z_2	**20**	**25**	**30**	**40**	**50**	**70**	**100**
20	19.3						
25	20.5	21.5					
30	21.5	22.4	23				
40	23	23.6	24.1	25			
50	24.1	24.6	25	25.6	26.1		
70	25.5	25.8	26.1	26.6	26.9	27.5	
100	26.7	27	27.2	27.5	27.7	28.1	28.5

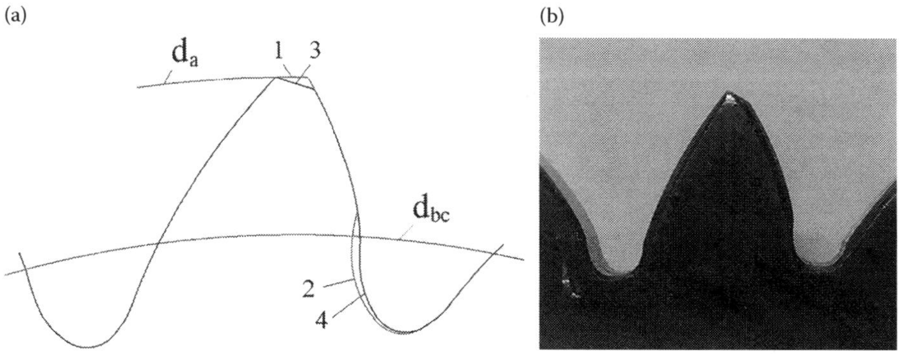

FIGURE 4.15
Asymmetric tooth with slanted tip; a: profile, b: photo (in this case a root fillet is not optimized); d_a – gear outer diameter, d_{bc} – coast flank base diameter; 1: circular tooth tip land, 2: slanted tooth tip land, 3: root fillet profile with undercut optimized from a trajectory of the mating tooth with the circular tip, 4: root fillet profile optimized from a trajectory of the mating tooth with the slanted tip.

If coast flanks are used in normal operating load transmission, as in, for example, idler or planet gears (in the epicyclic drives), the asymmetry factor K and practical range of pressure angles and contact ratios are defined based on specific application requirements (see Section 5.1.2).

5

Gear Geometry Optimization

Asymmetric tooth gears are not standard and intended for custom high performance unidirectional gear drives. This means that they require some special tooling and so are generally more expensive in production. This increase of manufacturing cost must be justified by unparalleled performance improvement. It is also clear that asymmetric tooth gears will not be considered for implementation unless they demonstrate significant advantages over the best existing symmetric tooth gears. All these factors make asymmetric gear geometry optimization, described in this chapter, an absolutely essential part of gear design.

Unlike a traditional gear design based on the generating rack parameters and its X-shift, Direct Gear Design® is not limited by any preselected tooling parameters or fabrication process requirements. This allows for significantly greater gear geometry optimization potentials in comparison to a traditional gear design method.

A starting point of gear tooth geometry optimization is to establish a set of priorities for specific gear drive applications. This "wish list" may include, for example:

- gear transmission density maximization that requires minimizing gear drive size and weight for a given transmitted load and gear ratio, or maximizing the transmitted load for a given gear drive size;
- accommodation into a required space or envelope (typically for the integrated gear drives);
- noise and vibration reduction;
- cost reduction;
- increased life and reliability;
- other performance enhancement requirements.

These application priorities dictate the Direct Gear Design tooth geometry optimization directions. As a part of a gear drive design it should be done in combination with other gear transmission component optimization. This chapter describes the macro and microgeometry optimization of asymmetric tooth gears.

5.1 Tooth Flank Profile Optimization

Flanks of asymmetric gear tooth are functionally different and practically each of their parameters could be a subject for optimization to achieve certain performance requirements. This section describes a definition of optimal values and combinations of the drive and coast pressure angles, contact ratios, and asymmetry factors.

5.1.1 Gear Pair Size Reduction

Gear pair size reduction is typically limited by the tooth flank surface endurance defined by pitting and scuffing resistance. Both types of tooth surface defects depend on the contact stress and tooth profile sliding velocity. Application of a higher operating pressure angle (α_{wd} = 30°–45° for drive flank of gears with asymmetric teeth) leads to the Hertz contact stress reduction. This allows the transmitted torque to be increased and/or the gear size (diameter, face width, or both) to be decreased, maintaining an acceptable contact stress level. This also makes gear teeth stubby, with the reduced height and increased thickness at the root area reducing the bending stress. High pressure angle gear pairs have a relatively low contact ratio of $\varepsilon_\alpha = 1.1–1.3$ and reduced profile sliding that decreases scuffing probability. Drawbacks of the application of high pressure angle gears are increased separating load taken by supporting bearings and higher gear mesh stiffness that reduces tooth engagement impact absorbing and leads to higher noise and vibration.

Another way of tooth profile optimization for gear pair size minimization is application of gears with high contact ratio (HCR). Conventional spur involute gears have a transverse contact ratio $1.0 < \varepsilon_\alpha < 2.0$ when one or two tooth pairs are in a simultaneous contact. The HCR spur gears have a transverse contact ratio $\varepsilon_\alpha \geq 2.0$ (typically 2.05–2.2) when two or three tooth pairs are in a simultaneous contact. The HCR gears provide load sharing between gear pairs in contact, and according to [13] a maximum gear pair load does not exceed 2/3 of the total transmitted load. The HCR gears have a relatively low operating pressure angle (α_{wd} = 20°–28° for drive flank of gears with asymmetric teeth, see Table 4.8), increased tooth height, and reduced thickness at the tooth root. As a result, they have an increased tooth deflection, providing a better load sharing between engaged tooth pairs and allowing reduction of contact and bending stresses, and also a noise and vibration reduction that makes them applicable for aerospace gear transmissions [13,173]. However, the HCR gears must be sufficiently accurate to have a base pitch variation lower than a tooth deflection under required operating load to provide load sharing. These gears also have some drawbacks. A minimum number of teeth of the HCR gears should be at least 20 or more to provide a transverse contact ratio $\varepsilon_\alpha \geq 2.0$. A long tooth addendum and low operating pressure angle result in higher sliding velocity that increases scuffing probability and mesh power losses. However, according to [174], "despite their higher sliding velocities high contact ratio gears can be designed to levels of efficiency comparable to those of conventional gears while retaining their advantages through proper selection of gear geometry."

Attempts to use the buttress asymmetric HRC gears in the sun/planet mesh of the planetary gear stage for noise and vibration reduction were not successful [18,26]. This could be explained by high stiffness of buttress teeth that have a low drive pressure angle and high coast pressure angle. It seems that a more rational approach for many applications of the asymmetric gears is to have the drive tooth flanks with a higher pressure angle than the coast ones. In case of epicyclic gear stages where a planet gear has both driving flanks, a higher pressure angle should be used in the sun/planet gear mesh and a lower pressure angle in the planet/ring gear mesh. Application of this approach to the asymmetric HCR gears allows the coast tooth flanks to be designed independently to reduce gear tooth stiffness for a better tooth load sharing. Table 5.1 presents the samples of the asymmetric gear tooth profile geometry for a high pressure angle and HCR gears.

5.1.2 Asymmetry Factor Selection

Asymmetric tooth profiles make it possible to increase the operating pressure angle beyond the conventional symmetric gear limits. This results in the drive flank contact stress and sliding velocity reduction, increasing the tooth surface endurance to pitting and scuffing, and providing the maximized gear transmission density.

Selection of the asymmetry factor K depends on the gear pair operating cycle that is defined by RPM and transmitted load in the main and reversed directions, and gear drive life requirements [104]. If the gear tooth is equally

TABLE 5.1

High Pressure Angle and High Contact Ratio Gear Geometry

Type		High Pressure Angle	High Contact Ratio (HCR)
Tooth form		Asymmetric	Asymmetric
Tooth profile		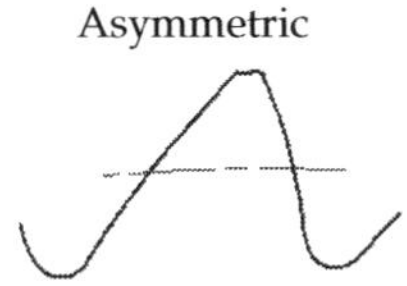	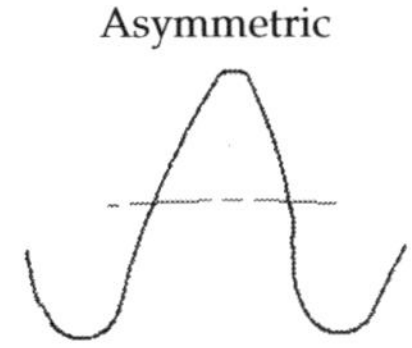
Number of teeth pinion and gear $z_{1,2}$		35	35
Pressure angle	α_{wd}	42°	26°
	α_{wc}	14°	12°
Asymmetry factor K		1.3	1.09
Transverse Contact ratio	$\varepsilon_{\alpha d}$	1.25	2.02
	$\varepsilon_{\alpha c}$	2.02	2.16

TABLE 5.2

Gears for Bidirectional and Unidirectional Load Transmission

Case #		1	2	3	4
Load transmission		Bidirectional	Mostly unidirectional	Unidirectional	Unidirectional
Loaded flanks		Both	Drive, coast at low load	Drive, coast at very low load	Drive flank only
Tooth profile		Symmetric (baseline)	Asymmetric	Asymmetric	Asymmetric
Gear mesh		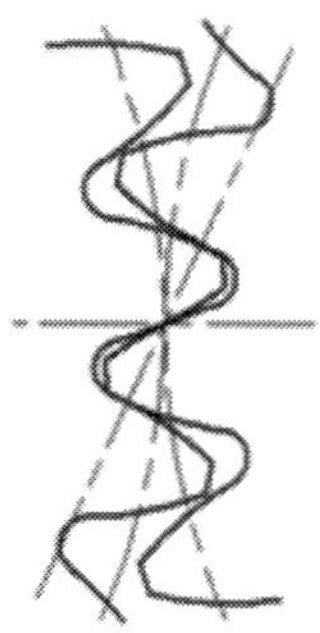	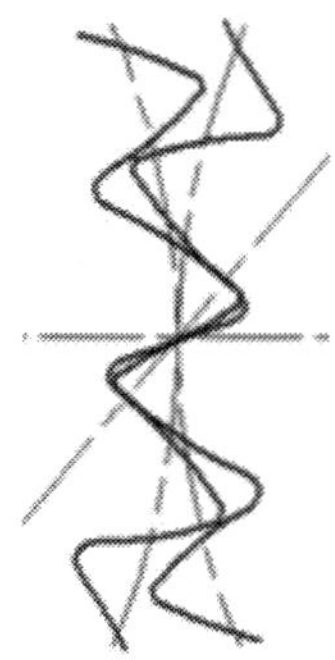	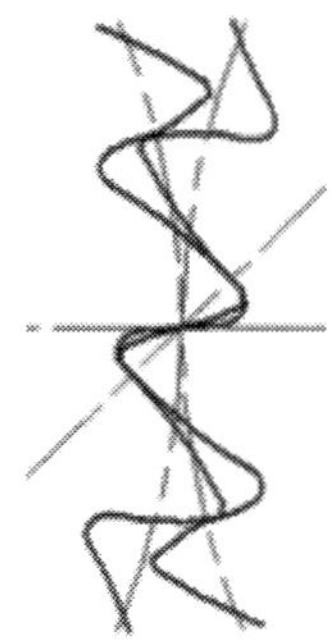	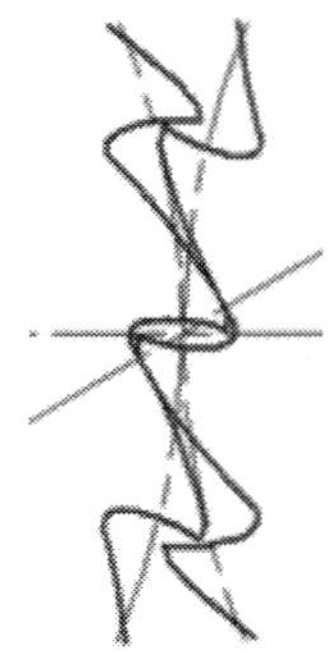
Number of teeth	pinion	24	24	24	24
	gear	24	24	24	24
Pressure angle	α_{wd}	25°	40°	46°	60°
	α_{wc}	25°	24°	10°	-*
Asymmetry coefficient K		1.0	1.19	1.42	-*
Transverse contact ratio	$\varepsilon_{\alpha d}$	1.35	1.2	1.2	1.2
	$\varepsilon_{\alpha c}$	1.35	1.44	1.0	-*
Hertz contact stress, %	drive	100	88	86	94
	coast	100	102	150	-*
Bearing load, %	drive	100	118	130	181
	coast	100	99	92	-*
Specific sliding velocity, %	drive	100	75	68	49
	coast	100	108	97	-*

*Coast flank mesh does not exist.

loaded in both the main and reversed rotation directions, asymmetric tooth profiles should not be considered. Table 5.2 illustrates the bidirectional and unidirectional load transmission cases.

CASE 1

Gear teeth are symmetric and their surface durability is identical for both drive and coast flanks. Case 1 presents a traditionally designed 25° pressure angle gear pair with a full radius fillet. This case is considered a baseline, and its Hertz contact stress, bearing load, and specific sliding velocity are assumed as 100% for comparison to other gear pairs. This type of gear profile is commonly used for aerospace gearboxes, because it provides lower bending strength and better flank surface endurance in comparison to the standard 20° pressure angle gears typical for industrial and commercial applications.

CASE 2

These asymmetric gears are for mostly unidirectional load transmission with a 40° pressure angle driving tooth flanks providing 12% contact stress and 25% sliding velocity reduction. At the same time, the contact stress and sliding velocity of the coast flanks are close to these parameters of the baseline gears and should provide tooth surface load capacity similar to that for the baseline gears. These type of gears may find applications for drives with one main load transmission direction, but it should be capable to carry a lighter load for shorter periods of time in the opposite load transmission direction.

CASE 3

These asymmetric gears have a 46° drive pressure angle that allows reduction of the contact stress by 14% and sliding velocity by 32%. The disadvantage of such gear teeth is a high (+30%) bearing load. These types of gears are only for unidirectional load transmission. Their 10° coast pressure angle flanks have insignificant load capacity. They may find applications for drives with only one load transmission direction that may occasionally have a very low load coast flank tooth contact, like in the case of tooth bouncing in the high speed transmissions.

CASE 4

These asymmetric gears have only drive tooth flanks with the extreme 60° pressure angle with no involute coast tooth flanks at all. As a result, the bearing load is significant.

There are many applications, as described in Case 2, where a gear pair transmits load in both directions, but with significantly different load magnitude and duration (Figure 5.1). In this case, the asymmetry factor K is defined by equalizing potential accumulated tooth surface damage defined by operating contact stress and number of tooth flank load cycles in each load transmission direction. In other words, a contact stress safety factor S_H should be the same for the drive and coast tooth flanks. This condition can be presented as follows:

$$S_H = \frac{\sigma_{HPd}}{\sigma_{Hd}} = \frac{\sigma_{HPc}}{\sigma_{Hc}}, \tag{5.1}$$

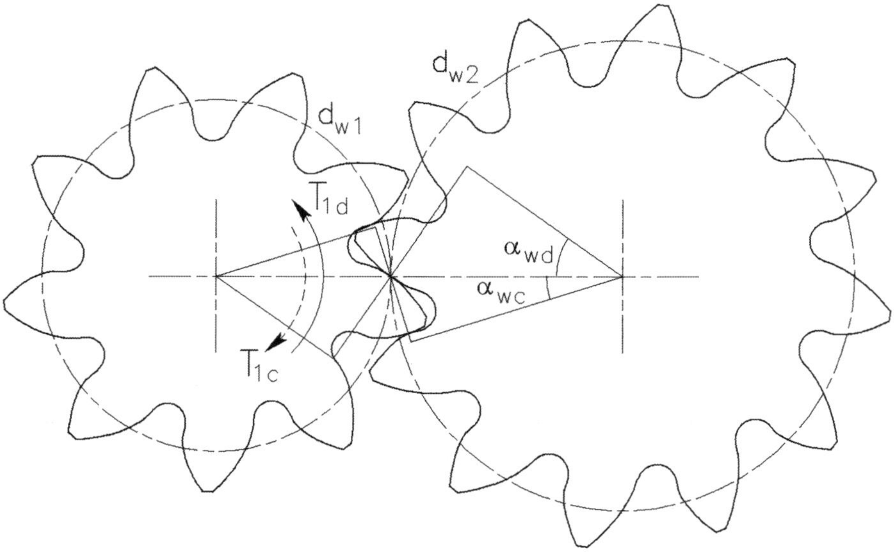

FIGURE 5.1
Asymmetric gear pair; T_{1d} and T_{1c}: pinion torque applied to the drive and coast tooth flanks.

where σ_{Hd} and σ_{Hc} are the operating contact stresses for the drive and coast tooth flanks, and σ_{HPd} and σ_{HPc} are the permissible contact stresses for the drive and coast tooth flanks that depend on a number of load cycles.

Then from (5.1)

$$\frac{\sigma_{Hd}}{\sigma_{Hc}} = \frac{\sigma_{HPd}}{\sigma_{HPc}}. \tag{5.2}$$

The contact stress at the pitch point [175] is

$$\sigma_H = z_H z_E z_\varepsilon z_\beta \sqrt{\frac{F_t}{d_{w1} b_w} \frac{u \pm 1}{u}}, \tag{5.3}$$

where

$$z_H = \sqrt{\frac{2\cos(\beta_b)\cos(\alpha_{wt})}{\cos(\alpha_t)^2 \sin(\alpha_{wt})}} \tag{5.4}$$

is the zone factor that for the directly designed spur gears is

$$z_H = \frac{2}{\sqrt{\sin(2\alpha_w)}}, \tag{5.5}$$

where Z_E is the elasticity factor that takes into account gear material properties (modulus of elasticity and Poisson's ratio); Z_ε is the contact ratio factor, its conservative value for spur gears is $Z_\varepsilon = 1.0$; Z_β is the helix factor, for spur gears $Z_\beta = 1.0$; F_t is the nominal tangent load, which at the pitch diameter d_{w1} is $F_t = 2T_1/d_{w1}$; T_1 is the pinion torque; b_w is the contact face width; and the "+" sign is for external gearing and the "−" sign for external gearing.

Then for the directly designed spur gears the contact stress at the pitch point can be presented as follows:

$$\sigma_H = z_E \frac{2}{d_{w1}} \sqrt{\frac{2T_1}{b_w \sin(2\alpha_w)} \frac{u \pm 1}{u}}. \tag{5.6}$$

Some parameters of this equation, Z_E, d_{w1}, b_w, and u, do not depend on the load transmission direction, and Equation (5.2) for the pitch point contact can be presented as follows:

$$\frac{\sin(2\alpha_{wc})}{\sin(2\alpha_{wd})} = A, \tag{5.7}$$

where a parameter A is

$$A = \frac{T_{1c}}{T_{1d}} \left(\frac{\sigma_{HPd}}{\sigma_{HPc}} \right)^2. \tag{5.8}$$

According to [175], "the permissible stress at limited service life or the safety factor in the limited life stress range is determined using life factor Z_{NT}." This allows replacement of the permissible contact stresses in Equation (5.8) for the life factors

$$A = \frac{T_{1c}}{T_{1d}} \left(\frac{Z_{NTd}}{Z_{NTc}} \right)^2. \tag{5.9}$$

When parameter A is defined and the drive pressure angle is selected, the coast pressure angle is calculated by Equation (5.7) and the asymmetry coefficient K from a common solution of (5.7) and (2.42):

$$K = \frac{\sqrt{1 + \sqrt{1 - A^2 (\sin 2\alpha_{wd})^2}}}{\sqrt{2} \cos \alpha_{wd}}. \tag{5.10}$$

Example 1

The drive pinion torque T_{1d} is two times greater than the coast pinion torque T_{1c}. The drive tooth flank has 10^9 load cycles, and the coast tooth flank has 10^6 load cycles during the life of the gear drive. From the *S-N* curve [175] for steel gears an approximate ratio of the life factors $Z_{NTd}/Z_{NTc} = 0.85$. Then the coefficient $A = 0.85^2/2 = 0.36$. Assuming that the drive pressure angle is $\alpha_{wd} = 36°$, the coast pressure angle from Equation (5.6) is $\alpha_{wc} = 10°$ and the asymmetry factor from Equation (5.9) is $K = 1.22$.

If, as a result of application of this asymmetric tooth flank optimization technique, a coast pressure angle <<10 °, its selection should be driven by some other criteria, for instance, by suitability for gear fabrication.

In many unidirectional gear drives like in, for example, propulsion system transmissions that seem irreversible, the coast tooth flanks are loaded because of the system inertia during the drive system deceleration or the tooth bouncing in the high RPM drives. This coast tooth flank load can be significant and should be taken in consideration while defining the asymmetry factor *K*.

If the gear drive is completely irreversible and the coast tooth flanks never transmit any load (Case 4 in Table 5.2), the asymmetry factor is defined only by the drive flank geometry. In this case, increase of the drive flank pressure could be limited by a minimum selected contact ratio and a separating load applied to the bearings. Application of a very high drive flank pressure angle results in the reduced coast flank pressure angle and possibly its involute profile undercut near the tooth root. Another limitation of the asymmetry factor *K* of the irreversible gear drive is growing compressive bending stress at the coast flank root. Usually for conventional symmetric gears compressive bending stress does not present a problem, because its allowable limit is significantly higher than that for the tensile bending stress. However, for asymmetric gears it may become an issue, especially for gears with thin rims.

In the unidirectional chain gear drives (Figure 5.2), the idler gear transmits the same load by both tooth flanks. This arrangement seems unsuitable for asymmetric gear application. However, in many cases, the idler's mating gears have significantly different numbers of teeth. This allows equalizing contact stresses on opposite flanks of the idler gear asymmetric teeth to achieve maximum load capacity.

Equation (5.6) is used to define the pitch point contact stress in the pinion/idler gear mesh,

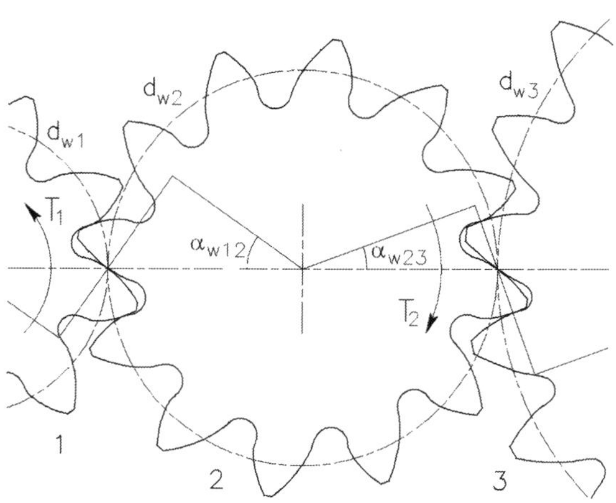

FIGURE 5.2
Chain gear arrangement; 1: input pinion; 2: idler gear; 3: output gear.

$$\sigma_{H12} = z_E \frac{2}{d_{w1}} \sqrt{\frac{2T_1}{b_{w12} \sin(2\alpha_{w12})} \frac{u_{12}+1}{u_{12}}}, \tag{5.11}$$

and in the idler/output gear mesh,

$$\sigma_{H23} = z_E \frac{2}{d_{w2}} \sqrt{\frac{2T_2}{b_{w23} \sin(2\alpha_{w23})} \frac{u_{23}+1}{u_{23}}}, \tag{5.12}$$

or, ignoring gear mesh losses,

$$\sigma_{H23} = z_E \frac{2}{u_{12} d_{w1}} \sqrt{\frac{2u_{12}T_1}{b_{w23} \sin(2\alpha_{w23})} \frac{u_{23}+1}{u_{23}}}, \tag{5.13}$$

where b_{w12} and b_{w23} are the contact face widths in the pinion/idler gear and the idler/output gear meshes, accordingly; $u_{12} = z_2/z_1$ is the gear ratio in the pinion/idler gear mesh; $u_{23} = z_3/z_2$ is the gear ratio in the idler/output gear mesh; and z_1, z_2, and z_3 are the number of teeth of the input, idler, and output gears.

Numbers of the idler gear tooth load cycles and permissible contact stresses in this case are equal in both meshes, and Equation (5.2) can be presented as $\sigma_{H12} = \sigma_{H23}$. Then assuming that all gears are made from the same material, the idler gear pressure angle ratio is defined by

$$\frac{\sin(2\alpha_{w23})}{\sin(2\alpha_{w12})} = B \tag{5.14}$$

where

$$B = \frac{b_{w12}}{b_{w23}} \frac{u_{23}+1}{u_{23}(u_{12}+1)} \tag{5.15}$$

is a parameter that reflects the gear ratios u_{12} and u_{23}, and contact face widths b_{w12} and b_{w23} in the pinion/idler gear and idler/output gear meshes, respectively.

Then considering Equation (2.42) the asymmetry factor K can be presented as follows:

$$K = \frac{\sqrt{1+\sqrt{1-B^2(\sin 2\alpha_{wd})^2}}}{\sqrt{2}\cos\alpha_{wd}}. \tag{5.16}$$

If $z_1 = z_3$ and $b_{w12} = b_{w23}$, then both the coefficient B and the asymmetry factor K are equal to 1.0 and gear teeth are symmetric. If $z_1 \neq z_3$ or $b_{w12} \neq b_{w23}$, application of asymmetric gears can be considered.

Example 2

The pinion number of teeth is $n_1 = 9$, the idler gear number of teeth is $n_2 = 12$, the output gear number of teeth is $n_3 = 20$, the contact face width ratio is $b_{w12}/b_{w23} = 1.2$. This makes parameter $B = 0.82$. Then assuming that the pinion/idler gear mesh pressure angle is α_{w12} = 35°, the idler/output pressure angle from (5.14) is 25.32° and the asymmetry factor from Equation (5.16) is $K = 1.10$.

Similarly, the contact stress equalization technique can be applied for the unidirectional epicyclic gear stage (Figure 5.3), because the planet gear can be considered as the idler gear engaged with the sun gear and ring gear. The nature of a contact in the sun/planet gear mesh is different from that in the planet/ring gear mesh. Convex flanks of the external tooth planet gear are in simultaneous contact with convex flanks of the external tooth sun gear and concave flanks of the internal tooth ring gear. This results in significant difference in Hertzian contact stresses for symmetric tooth gears. However, application of asymmetric tooth gears allows equalizing contact stresses on opposite flanks of the planet gear teeth to achieve maximum load capacity.

In this case, the asymmetry factor K is also defined by Equation (5.16), where parameter B is

$$B = \frac{b_{w12}}{b_{w23}} \frac{u_{23} - 1}{u_{23}(u_{12} + 1)}, \tag{5.17}$$

where b_{w12} and b_{w23} are the contact face widths of the sun/planet gear and planet/ring gear meshes; $u_{12} = z_2/z_1$ is the gear ratio in the sun/planet gear mesh; $u_{23} = z_3/z_2$ is the gear ratio in the planet/ring gear mesh; and z_1, z_2, and z_3 are the number of teeth of the sun, planet, and ring gears.

In a typical epicyclic gear stage $z_2 = (z_3 - z_1)/2$. This allows simplification of Equation (5.17):

$$B = \frac{b_{w12}}{u_{13} b_{w23}}, \tag{5.18}$$

where $u_{13} = z_3/z_1$ is the epicyclic stage gear ratio. This gear ratio is always greater than 1.0, which makes an epicyclic gear stage suitable for asymmetric gear application.

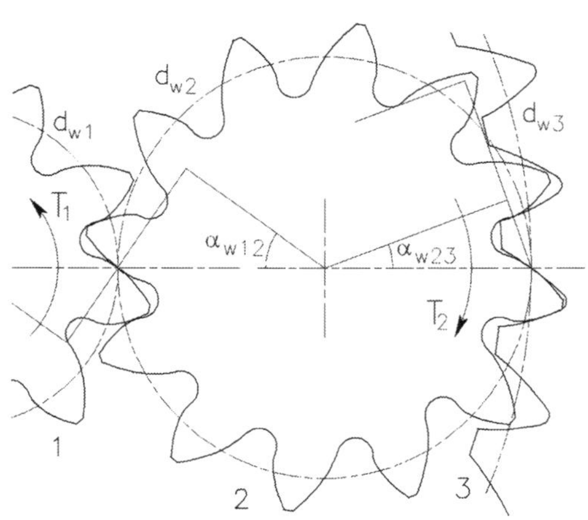

FIGURE 5.3
Planetary gear arrangement; 1: sun gear; 2: planet gear; 3: ring gear.

Example 3

The sun gear number of teeth is $n_1 = 9$, the planet gear number of teeth is $n_2 = 12$, the ring gear number of teeth is $n_3 = 33$, and the contact face width ratio is $b_{w12}/b_{w23} = 1.8$. This makes parameter $B = 0.49$. Then assuming that the sun/planet gear mesh pressure angle is $\alpha_{w12} = 40°$, the planet/ring gear mesh pressure angle from (5.14) is 14.5° and the asymmetry factor from Equation (5.16) is $K = 1.26$.

While utilizing this asymmetric tooth flank optimization technique, it is important to know that selecting of a relatively low sun/planet gear mesh pressure angle may result in an unacceptable low planet/ring gear mesh pressure angle (<<10°). In such cases, selection of a planet/ring gear mesh pressure angle should be driven by other criteria, for example, by suitability for gear fabrication.

5.1.3 Mesh Efficiency Maximization

Gear mesh power losses depend on the gear tooth geometry and friction coefficient. For spur gears, the percent of mesh losses is defined in Equation (3.11). In the optimized gear pair, maximum mesh efficiency (minimum of mesh losses) is achieved if the specific sliding velocities at the start of the approach action and at the end of the recess action are equalized: $H = H_{sd} = H_{td}$. Then Equation (3.11) can be presented as follows:

$$P_t = \frac{50 fH}{\cos\alpha_{wd}}. \tag{5.19}$$

If this equation is solved with Equations (2.72), (2.74), (3.13), and (3.14) the mesh power loss percent for spur gears can be defined as follows:

$$P_t = 50 f\pi\varepsilon_{\alpha d}\left(\frac{1}{z_1} \pm \frac{1}{z_2}\right), \tag{5.20}$$

and the maximized gear mesh efficiency as follows:

$$E_{\max} = 100 - P_t = 100 - 50 f\pi\varepsilon_{\alpha d}\left(\frac{1}{z_1} \pm \frac{1}{z_2}\right), \tag{5.21}$$

where in Equations 5.21 and 5.22 the top sign (+) is for external gears and the bottom sign (–) for internal gears.

Equation (5.21) indicates that for spur gears the optimized maximum efficiency gear geometry depends only on the type of gearing (external or internal), numbers of teeth, and transverse contact ratio $\varepsilon_{\alpha d}$. For helical gears the mesh power losses are described by Equation (3.12). For the optimized for maximum efficiency gear pair these power losses can be presented as follows:

$$P_t = 50 f\pi\varepsilon_{\alpha d}\cos^2\beta_w\left(\frac{1}{z_1} \pm \frac{1}{z_2}\right), \tag{5.22}$$

where signs ± are for external gears and internal gears, respectively.

The maximum gear mesh efficiency of the spur gears is limited by the minimum transverse contact ratio $\varepsilon_\alpha = 1.0$. Helical gears make it possible to increase the gear mesh efficiency by reducing $\varepsilon_\alpha < 1.0$ and increasing the operating helix angle β_w. Such helical gears were suggested to reduce automotive transmission power losses [184].

5.1.4 Contact Ratio Optimization

The contact ratio of spur gears is a critical parameter that affects the gear drive performance. It influences the gear drive load capacity, efficiency, and noise and vibration. Spur asymmetric tooth gears with a low and high contact ratio are described in paragraphs 4.4 and 5.1.1, where a nominal contact ratio is considered as designed without influence of operating condition factors, such as, for example, tooth deflections under the operating load. Typically, comparison of gear pairs with different contact ratios is done when numbers of teeth and tooth size (module or diametral pitch) are identical. Such a comparison might have some theoretical value, but for practical gear design it is more important to compare gear sets with identical performance parameters like, for example, mesh efficiency.

This paragraph presents an analysis of asymmetric tooth gears considering the effective contact ratio that is defined considering bending and contact tooth deflections [157]. The goal is to find an optimal solution for high performance gear drives, which would combine high load capacity and efficiency, as well as low transmission error (which affects gear noise and vibration).

A nominal drive flank contact ratio of external spur asymmetric tooth gears is defined by Equation (2.66). An effective contact ratio can be defined as the ratio of the tooth engagement angle to the angular pitch. The tooth engagement angle is a gear rotation angle from the start of the tooth engagement with the mating gear tooth to the end of the engagement. The effective contact ratio is

$$\varepsilon_{\alpha de} = \frac{\varphi_1}{360^o/z_1} = \frac{\varphi_2}{360^o/z_2}, \tag{5.23}$$

where φ_1 and φ_2 are the pinion and gear engagement angles in degrees, and $360^o/z_1$ and $360^o/z_2$ are the pinion and gear angular pitches.

Transmission error is the angular difference between the actual position of the driven gear and its ideal position if the gear pair were perfectly conjugate, projected on the line of contact and defined as [165]

$$TE = r_{bd2}\left(\theta_2 - u\theta_1\right), \tag{5.24}$$

where θ_1 and θ_2 are the driving pinion and driven gear rotation angles, r_{bd2} is the drive tooth flank base radius of the driven gear.

A typical spur gear transmission error chart is shown in Figure 5.4.

The effective contact ratio and transmission error are influenced by manufacturing tolerances and operating conditions, including deflections of gears and other gearbox components under operating load, dynamic loads and inertia, temperature, etc. Dr. Yuriy V. Shekhtman has developed a calculation procedure for definition of the effective contact ratio and transmission error considering only the bending and contact tooth deflections. According to his approach, each angular position of the driven gear relative to the driving gear is iteratively defined by equalizing a sum of the tooth contact load moments of each gear to its applied torque. The related tooth contact loads are also iteratively defined to conform to tooth bending and contact deflections, where the tooth bending deflection in each contact point is determined based on the finite element analysis (FEA) calculated flexibility and the tooth contact deflection is calculated by the Hertz equation.

The Direct Gear Design method allows for having the drive flank nominal contact ratio as one of the gear design input parameters. Maximum gear mesh efficiency is achieved when the specific sliding velocities are equalized and is defined by Equation (5.21). In order to define an optimal contact ratio, the comparable gear sets are assumed to have identical maximized mesh efficiency E, average friction coefficient f, and gear ratio u. Then Equation (5.21) for the external spur asymmetric gears can be converted to

$$\frac{\varepsilon_{\alpha d}}{z_1} = \left(1 - \frac{E}{100}\right)\frac{2u}{f\pi\left(1+u\right)} = \text{const.} \tag{5.25}$$

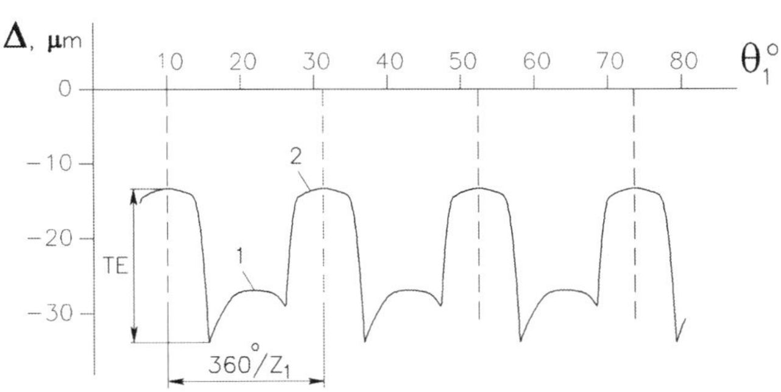

FIGURE 5.4
Typical spur gear pair transmission error chart; Δ: distance in microns between actual tooth contact point and ideal contact point; 1: single tooth pair contact, 2 – double tooth pair contact. (Kapelevich A.L. and Y.V. Shekhtman. 2017. Analysis and Optimization of Contact Ratio of Asymmetric Gears, Gear Technology. March/April, 66–71. With permission.)

This criterion is used to analyze the parameters of external spur gear sets with asymmetric teeth. Comparable asymmetric tooth gear sets have different numbers of teeth and identical center distance a_w, gear ratio u, coast flank pressure angle α_{wc}, minimal relative tooth tip thicknesses $m_{a1,2}$ that are required to avoid having the harden through tooth tips for the carburized harden gears, average friction coefficient f and gear mesh efficiency E, pinion and gear material properties, and equalized specific sliding velocities H_{sd} and H_{td}. The face widths b_1 and b_2 are defined to approximately equalize the pinion and gear tooth bending stresses considering the optimized root fillets (see Chapter 5.2). If the center distance a_w is identical for all gear sets, the operating modules are inversely proportional to a number of pinion teeth and defined from Equation (2.28) as follows:

$$m = \frac{2a_w}{z_1(1+u)}. \tag{5.26}$$

The operating pitch diameter tooth thickness ratio is

$$TTR = S_{w1}/p_w = S_{w1}/(S_{w1} + S_{w2}), \tag{5.27}$$

where S_{w1} and S_{w2} are the pinion and gear tooth thicknesses at the operating pitch diameters. p_w is the operating circular pitch.

The TTR value is defined to simultaneously equalize the specific sliding velocities H_{sd} and H_{td}, and the pinion and gear relative tooth tip thicknesses $m_{a1,2}$. The maximized drive flank pressure angle α_{wd} is defined to achieve minimal contact stress. It must also provide a nominal drive contact ratio ε_{ad} defined by Equation (2.66), the preselected values of the coast flank pressure angle α_{wc}, and pinion and gear relative tooth tip thicknesses $m_{a1,2}$.

The bearing load is

$$F = 2T_1/d_{bd1}, \tag{5.28}$$

where T_1 is the pinion operating torque, d_{bd1} is the pinion drive flank base diameter.

The load sharing factor is

$$L = F_{c\max}/F, \tag{5.29}$$

where $F_{c\max}$ is the maximum normal load in the single tooth pair contact.

If the drive flank effective contact ratio $\varepsilon_{ade} < 2.0$, the load sharing factor $L = 1.0$.

Table 5.3 presents the gear geometry and load parameters for 15 gear pairs that are defined to satisfy preselected comparison conditions, listed in the top field of this table. Gear pairs are divided into four groups based on a contact ratio type: low, medium, transitional, a HCR. Four gear sets, one from each group, are selected to define the tooth contact and bending stresses, and transmission error under variable operating loads to find a gear set with an optimal contact ratio.

The parameters of the selected gear sets are highlighted in the bold font in Table 5.3:

- Gear set 1 with a low contact ratio ($\varepsilon_{\alpha d}$ = 1.19 and $\varepsilon_{\alpha de}$ = 1.33) has a 15-tooth pinion and 30-tooth gear.
- Gear set 2 with a medium contact ratio ($\varepsilon_{\alpha d}$ = 1.51 and $\varepsilon_{\alpha de}$ = 1.69) has a 19-tooth pinion and 38-tooth gear.
- Gear set 3 with a transitional contact ratio ($\varepsilon_{\alpha d}$ = 1.83 and $\varepsilon_{\alpha de}$ = 2.04) has a 23-tooth pinion and 46-tooth gear. It is called transitional because it has a nominal contact ratio < 2.0. Such gears under a low load have one or two mating tooth pairs in contact. When a load is increased to its operating level and tooth deflections are increased respectively, the transitional contact gears have two or three mating tooth pairs simultaneously engaged in contact. These result in tooth load sharing and a single tooth pair contact load reduction.
- Gear set 4 with a HCR ($\varepsilon_{\alpha d}$ = 2.15 and $\varepsilon_{\alpha de}$ = 2.40) has a 27-tooth pinion and 54-tooth gear.

Figure 5.5 shows the selected gear set meshes in the same scale.

The main gear parameters vs. pinion number of teeth charts are shown in Figures 5.6–5.11. With increasing numbers of the pinion teeth, the module *m* is

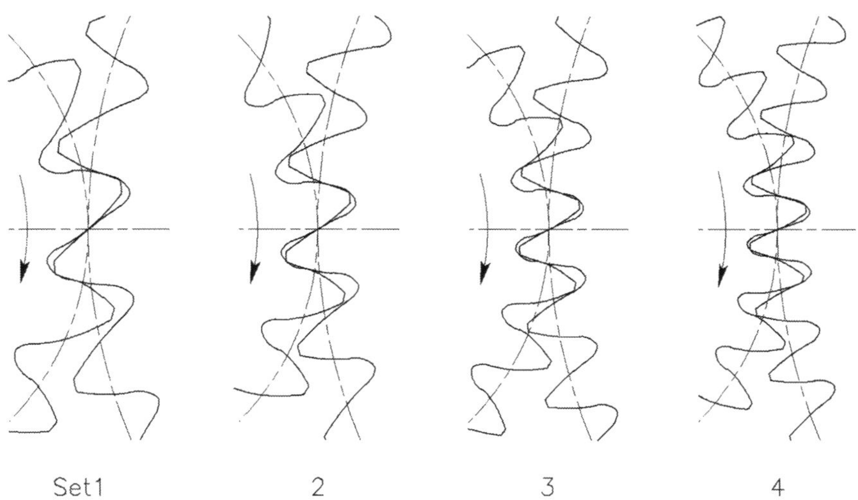

FIGURE 5.5
Selected set gear meshes at the same scale; arrows indicate the driving pinion torque direction. (Kapelevich A.L. and Y.V. Shekhtman. 2017. Analysis and Optimization of Contact Ratio of Asymmetric Gears, Gear Technology. March/April, 66–71. With permission.)

TABLE 5.3

Comparable Gear Set Parameters

Preselected gear set parameters and conditions: Center distance: 150 mm; Gear ratio: 2:1; Coast pressure angle: 15°; Pinion and gear face widths: 35 mm and 30 mm; Relative tooth tip thickness: 0.30; Average friction coefficient: 0.05; Gear mesh efficiency: 99%; Pinion and gear material properties: modulus of elasticity: 207,000 MPa, Poisson ratio: 0.3; Pinion torque: 1,500 Nm; All gears have optimized tooth root fillets.

Numbers of teeth, Pinion	14	**15**	16	17	18	**19**	20	21	22	**23**	24	25	26	**27**	28
Numbers of teeth, Gear	28	**30**	32	34	36	**38**	40	42	44	**46**	48	50	52	**54**	56
Module, mm	7.143	**6.667**	6.25	5.882	5.556	**5.263**	5.000	4.762	4.545	**4.348**	4.167	4.000	3.846	**3.704**	3.571
Tooth thickness ratio	0.518	**0.520**	0.522	0.524	0.526	**0.527**	0.529	0.531	0.532	**0.534**	0.536	0.537	0.538	**0.540**	0.541
Drive pressure angle, °	42.0	**39.1**	36.6	34.5	32.7	**31.1**	29.7	28.3	27.0	**25.8**	24.8	23.9	23.0	**22.3**	21.5
Asymmetry factor	1.300	**1.245**	1.203	1.172	1.148	**1.128**	1.112	1.097	1.084	**1.073**	1.064	1.057	1.049	**1.044**	1.038
Nominal drive contact ratio	1.11	**1.19**	1.27	1.35	1.43	**1.51**	1.59	1.67	1.75	**1.83**	1.91	1.99	2.07	**2.15**	2.23
Effective drive contact ratio	1.24	**1.33**	1.42	1.51	1.60	**1.69**	1.78	1.86	1.96	**2.04**	2.13	2.22	2.31	**2.40**	2.49
Specific sliding velocities	0.279	**0.292**	0.301	0.309	0.316	**0.321**	0.322	0.330	0.334	**0.338**	0.340	0.343	0.345	**0.347**	0.349
Bearing load, N	40,368	**38,601**	37,417	36,448	35,647	**35,037**	34,890	34,072	33,673	**33,320**	33,045	32,839	32,616	**32,422**	32,247
Load sharing factor	1.0	**1.0**	1.0	1.0	1.0	**1.0**	1.0	1.0	1.0	**0.743**	0.692	0.674	0.644	**0.630**	0.620
Contact ratio type			Low					Medium			Transitional			High	
Selected gear set number	-	1	-	-	-	2	-	-	-	3	-	-	-	4	-

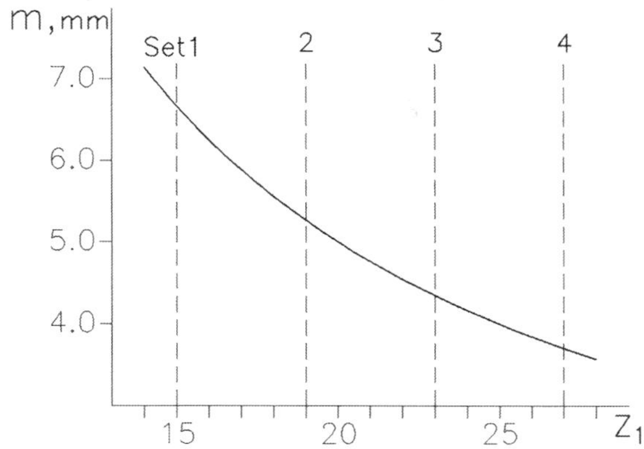

FIGURE 5.6
Gear sets' module.

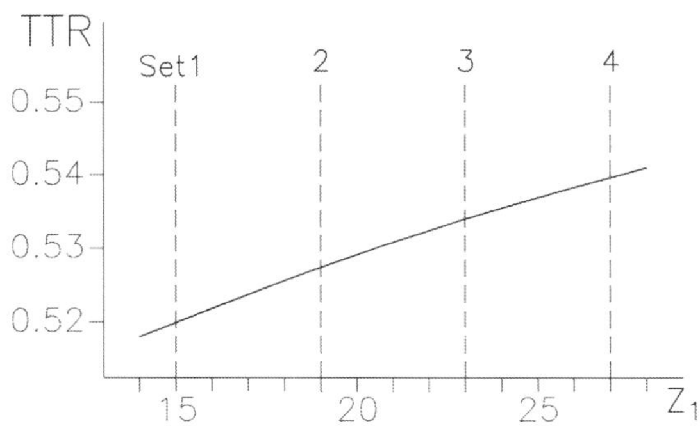

FIGURE 5.7
Tooth thickness ratio.

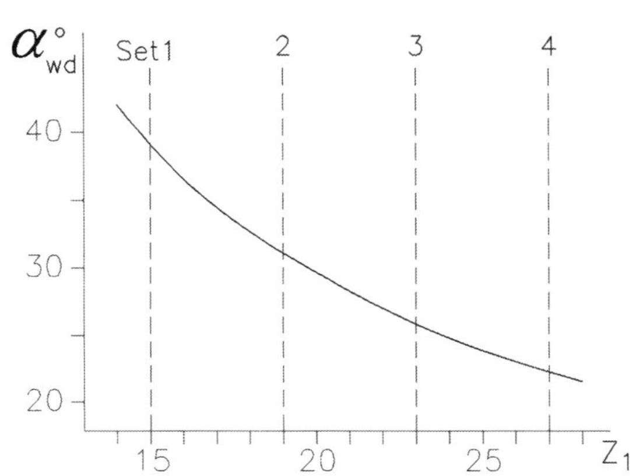

FIGURE 5.8
Drive flank pressure angle.

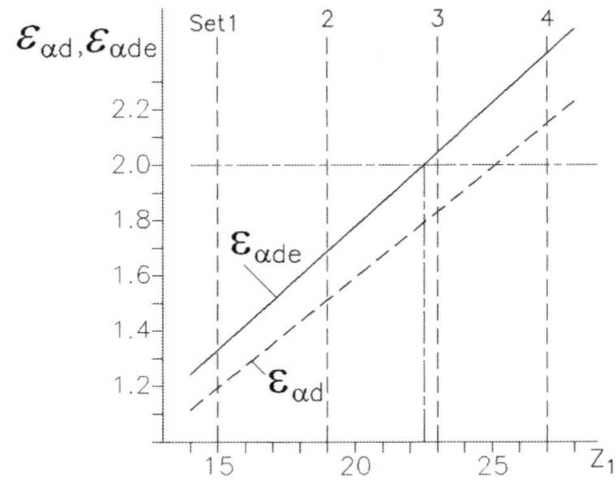

FIGURE 5.9
Nominal and effective drive contact ratios.

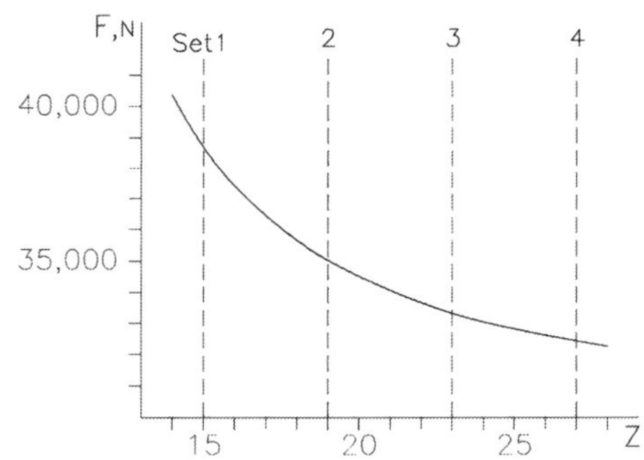

FIGURE 5.10
Bearing load.

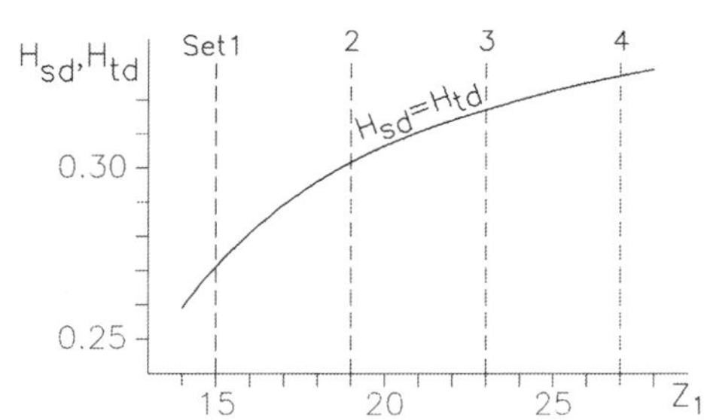

FIGURE 5.11
Drive flank specific sliding velocities.

reducing to maintain the constant center distance. The tooth thickness ratio TTR is slightly increasing, keeping equal specific sliding velocities H_{sd} and H_{td}, and the pinion and gear relative tooth tip thicknesses $m_{a1,2}$. The drive flank pressure angle α_{wd} is getting lower, and the nominal and effective contact ratios ε_{ad} and ε_{ade} are increasing, providing constant mesh efficiency. As a result of the drive flank pressure angle reduction, the bearing load F is getting lower, but the equalized specific sliding velocities H_{sd} and H_{td} are increasing because of the increased contact ratios.

The results of the selected gear sets stress analysis, and effective contact ratio, load sharing factor, transmission error calculations under the different driving torque values are shown in Table 5.4.

TABLE 5.4

Selected Gear Set Analysis Results with Different Driving Torques

Gear set 1—low contact ratio: $\varepsilon_{ad} = 1.194$, $z_1 = 15$, $z_2 = 30$, $m = 6.667$ mm, $\alpha_{wd} = 39.0°$, $\alpha_{wc} = 15.0°$								
Pinion torque, Nm	250	500	750	1,000	1,250	1,500	1,750	2,000
Effective drive contact ratio	1.25	1.28	1.29	1.31	1.32	1.33	1.34	1.35
Load sharing factor	1.0	1.0	1.0	1.0	1.0	1.0	1.0	1.0
Pinion bending stress, MPa	53.7	108	161	215	269	322	376	430
Gear bending stress, MPa	52.9	106	159	212	264	317	370	423
Contact stress, MPa	714	921	1,115	1,288	1,433	1,570	1,695	1,812
Transmission error, μm	3.1	6.1	8.8	11.3	13.7	16.1	18.4	20.8
Gear set 2—medium contact ratio: $\varepsilon_{ad} = 1.512$, $z_1 = 19$, $z_2 = 38$, $m = 5.263$ mm, $\alpha_{wd} = 31.1°$, $\alpha_{wc} = 15.0°$								
Pinion torque, Nm	250	500	750	1,000	1,250	1,500	1,750	2,000
Effective drive contact ratio	1.59	1.63	1.65	1.67	1.68	1.69	1.72	1.74
Load sharing factor	1.0	1.0	1.0	1.0	1.0	1.0	1.0	1.0
Pinion bending stress, MPa	65.0	130	195	260	325	390	455	520
Gear bending stress, MPa	66.9	134	201	267	334	401	468	535
Contact stress, MPa	717	930	1,132	1,311	1,464	1,601	1,726	1,845
Transmission error, μm	2.9	5.5	8.2	10.6	12.9	15.1	17.2	19.3
Gear set 3—transitional contact ratio: $\varepsilon_{ad} = 1.831$, $z_1 = 23$, $z_2 = 46$, $m = 4.348$ mm, $\alpha_{wd} = 25.8°$, $\alpha_{wc} = 15.0°$								
Pinion torque, Nm	250	500	750	1,000	1,250	1,500	1,750	2,000
Effective drive contact ratio	1.95	1.99	2.01	2.02	2.03	2.04	2.06	2.08
Load sharing factor	1.0	1.0	0.944	0.848	0.783	0.748	0.731	0.722
Pinion bending stress, MPa	76.8	153	216	258	301	353	412	470
Gear bending stress, MPa	81.4	162	228	271	314	359	408	458
Contact stress, MPa	740	962	1,143	1,260	1,348	1,436	1,528	1,622
Transmission error, μm	2.6	5.0	7.3	6.8	6.2	5.8	5.9	6.3
Gear set 4—high contact ratio: $\varepsilon_{ad} = 2.149$, $z_1 = 27$, $z_2 = 54$, $m = 3.704$ mm, $\alpha_{wd} = 22.3°$, $\alpha_{wc} = 15.0°$								
Pinion torque, Nm	250	500	750	1,000	1,250	1,500	1,750	2,000
Effective drive contact ratio	2.25	2.3	2.33	2.36	2.38	2.40	2.42	2.44
Load sharing factor	0.655	0.648	0.643	0.639	0.636	0.634	0.632	0.630
Pinion bending stress, MPa	70.0	140	210	280	350	420	490	560
Gear bending stress, MPa	65.2	130	196	261	326	392	457	522
Contact stress, MPa	661	820	974	1,121	1,253	1,366	1,470	1,572
Transmission error, μm	1.5	2.9	4.2	5.5	6.7	7.8	8.9	9.9

Figure 5.12 presents the charts of the effective contact ratio vs. the pinion torque. Effective contact ratios are gradually getting greater with the growing pinion torque, because of increasing bending and contact tooth deflections.

Load sharing factors of sets 1 and 2 (Figure 5.13) are equal to 1.0 because the effective contact ratios of these gears are < 2.0 within the given pinion torque range. The gear set 3 with a transitional contact also has the load sharing factor equal to 1.0 until the pinion torque exceeds a value (shown as a vertical dashed line in Figures 5.12–5.15 and 5.17) when the single/double tooth pair contact when the effective contact ratio < 2.0 becomes the double/triple tooth pair contact when the effective contact ratio ≥ 2.0. Then the load sharing factor reduces to below 1.0.

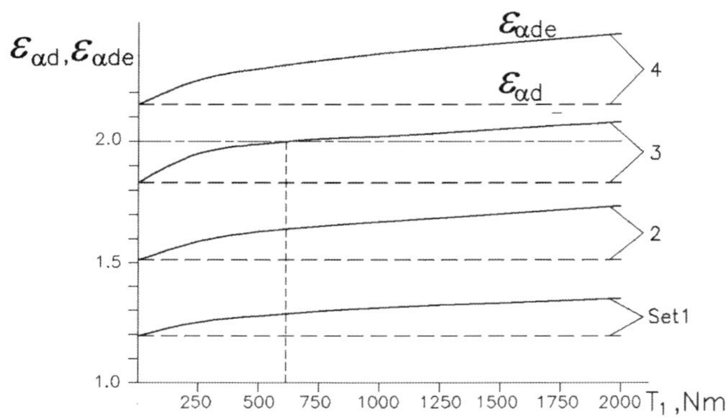

FIGURE 5.12
Effective contact ratio vs. pinion torque charts; dashed lines indicate the nominal contact ratio values. (Kapelevich A.L. and Y.V. Shekhtman. 2017. Analysis and Optimization of Contact Ratio of Asymmetric Gears, Gear Technology. March/April, 66–71. With permission.)

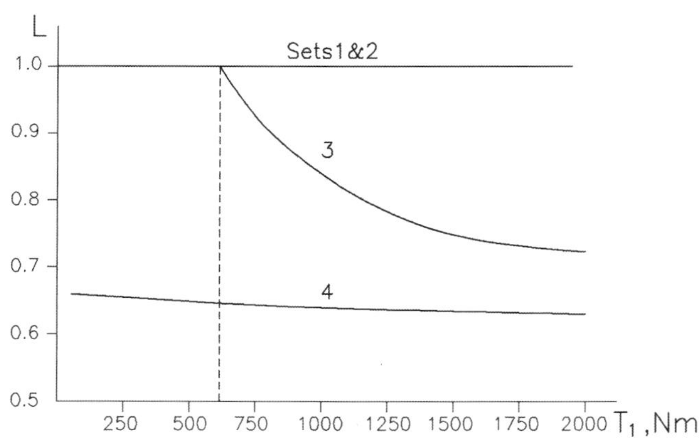

FIGURE 5.13
Load sharing factor vs. pinion torque charts. (Kapelevich A.L. and Y.V. Shekhtman. 2017. Analysis and Optimization of Contact Ratio of Asymmetric Gears, Gear Technology. March/April, 66–71. With permission.)

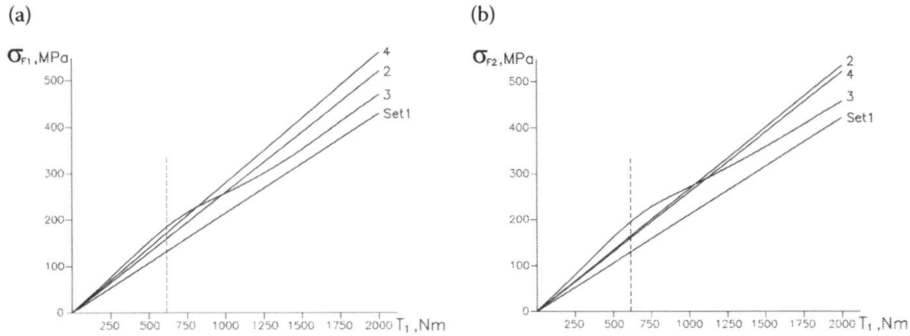

FIGURE 5.14
Root bending stress charts vs. pinion torque; (a) pinion, (b) gear. (Kapelevich A.L. and Y.V. Shekhtman. 2017. Analysis and Optimization of Contact Ratio of Asymmetric Gears, Gear Technology. March/April, 66–71. With permission.)

The tooth root bending stresses of the pinion and gears (see Figure 5.14a,b) of sets 1, 2, and 4 are increasing in direct proportion to the pinion torque. The gear set 3 tooth root bending stresses behave the same until the increasing pinion torque converts the single/double tooth pair contact into the double/triple tooth pair contact. Then the bending stress growth noticeably slows down.

The tooth flank contact stresses (see Figure 5.15) of sets 1, 2, and 4 are growing proportionally to the square root of the pinion torque. But the gear set 3 tooth contact stress growth is slowing down when the increasing pinion torque makes the effective contact ratio $\varepsilon_{\alpha de} \geq 2.0$ and the tooth load is shared in the double/triple tooth pair contact.

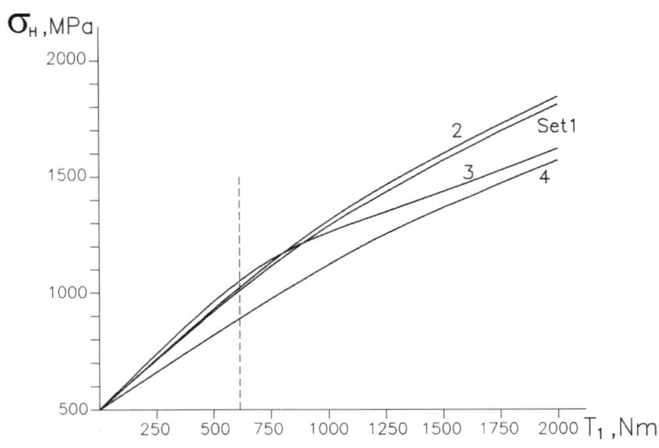

FIGURE 5.15
Contact stress vs. pinion torque charts. (Kapelevich A.L. and Y.V. Shekhtman. 2017. Analysis and Optimization of Contact Ratio of Asymmetric Gears, Gear Technology. March/April, 66–71. With permission.)

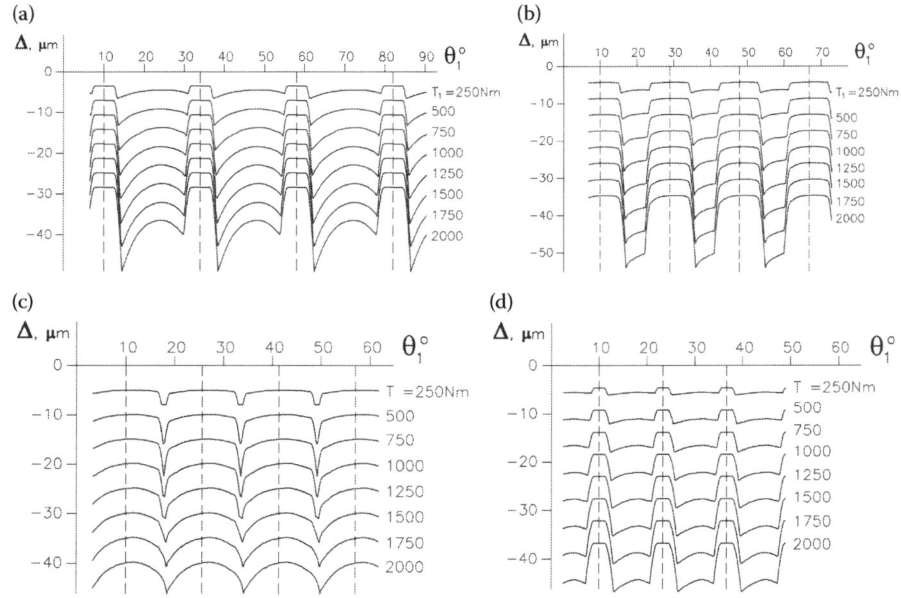

FIGURE 5.16
Transmission error charts; (a) gear set 1, (b) gear set 2, (c) gear set 3, (d) gear set 4. (Kapelevich A.L. and Y.V. Shekhtman. 2017. Analysis and Optimization of Contact Ratio of Asymmetric Gears, Gear Technology. March/April, 66–71. With permission.)

Figure 5.16 presents the transmission errors charts vs. the pinion rotation angle. A transmission error of sets 1, 2, and 4 is increasing with the growing pinion torque, but its shape remains basically the same, although the set 3 transmission error chart behaves differently. At the beginning until the pinion torque is relatively low, its growing proportionally to the torque, but when the effective contact ratio becomes ≥ 2.0 and the double/triple tooth pair contact occurs, the transmission error is getting low till reaching its minimum, then it is growing again. Its shape is changing during transition from the single/double tooth pair contact to the double/triple tooth pair contact.

Figure 5.17 presents the charts of the transmission error vs. pinion torque for all four gear sets. Set 4 with the HCR gears naturally has a transmission error lower than sets 1 and 2 with low and medium contact ratio gears. However, set 3 with a transitional contact ratio has a transmission error minimum that is lower than for the gear set 4 at the same pinion torque. The transitional contact ratio gears can be an optimal solution for low noise and vibration gear drives, if the gear parameters are defined to have a transmission error minimum in the middle of operating torque range (Figure 5.17).

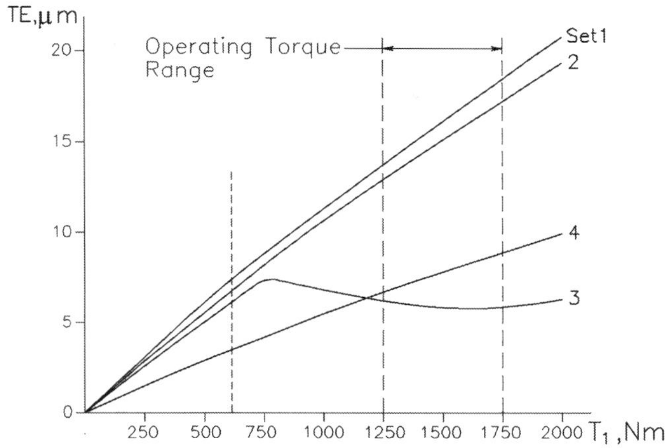

FIGURE 5.17
Transmission error vs. pinion torque charts. (Kapelevich A.L. and Y.V. Shekhtman. 2017. Analysis and Optimization of Contact Ratio of Asymmetric Gears, Gear Technology. March/April, 66–71. With permission.)

5.2 Tooth Root Fillet Optimization

Historically, gear tooth geometry improvement efforts were concentrated on the involute flanks. Although the gear tooth root fillet is an area of maximum bending stress concentration, its profile and accuracy are marginally defined on the gear drawing by a typically very generous root diameter tolerance and, in some cases, by the minimum fillet radius, which is not easy to accurately inspect. In fact, tooth bending strength is usually provided by material and heat treatment improvement, and root surface enhancement operation, like the shot peening, rather than root geometry optimization. The gear tooth root fillet profile is typically a trochoidal curve determined by the generating tool (gear hob or shaper cutter) tooth tip trajectory.

Unlike the contact Hertz stress, the bending stress typically does not define the major gear pair dimensions such as pitch diameters or a center distance. If the calculated bending stress is unacceptably high, the coarser diametral pitch (larger module) can be applied. A number of gear teeth, in this case, is proportionally reduced to keep required pitch diameters and center distance, and the same (or close) gear ratio. This makes the gear tooth larger reducing bending stress to an acceptable level, although this increases tooth stiffness and specific sliding velocity, decreasing the gear mesh efficiency.

There are two general approaches for reducing the tooth root bending stress. One of them is to alter the generating cutter tooth tip profile [162,176,177]. The most common application of such an approach is the tooling rack with the full tip radius. Another approach is to alter the gear tooth root fillet profile [161,178–180]. The most common application of such an approach is to use the circular arc root fillet profile. Further development of both of these approaches is based on a mathematical function fitting technique where the cutter tip radius or the gear tooth root trochoid fillet profile is replaced by a parabola, ellipsis [93], chain curve, "bionic tooth root contour" [181], or other curve, reducing the bending stress. Bending stress reduction achieved by these root fillet profile improvement methods varies greatly depending on the cutter and/or gear mesh parameters. The resulting tooth fillet profile must be checked for interference with the mating gear tooth tip. In the mentioned publications authors applied FEA for stress calculation. In [182] the boundary element method (BEM) is used for stress analysis of the optimized tooth root fillet.

5.2.1 Optimization Approach

Direct Gear Design optimizes a tooth root fillet after the involute flank parameters are already completely defined. The goal is to achieve a minimum of stress concentration on the tooth root fillet profile. It could be done by increasing the root fillet curvature radius in the maximum stress area. This leads to increased stress in other parts of the root fillet, where stresses are typically lower. As a result, the maximum bending stress becomes evenly distributed along the large portion of the root fillet.

Prior to optimization, the initial root fillet profile is generated as a trajectory of the mating gear tooth tip in the tight (zero-backlash) engagement (see Figure 5.18). This shape of the initial root fillet allows the interference with the mating gear tooth tip to be excluded completely.

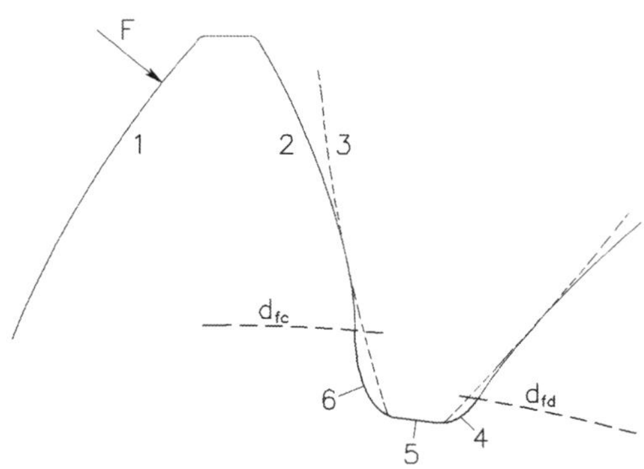

FIGURE 5.18
Generation of the initial root fillet profile as a trajectory of the mating gear tooth tip; 1: drive involute flank, 2: coast involute flank, 3: mating gear tooth in the tight (zero-backlash) engagement, 4: portion of the initial root fillet formed by the drive flank tip radius of the mating gear tooth, 5: portion of the initial root fillet formed by the tip land of the mating gear tooth, 6: portion of the initial root fillet formed by the coast flank tip radius of the mating gear tooth, d_{fd}: drive flank form circle diameter, d_{fc}: coast flank form circle diameter.

The Direct Gear Design tooth root fillet optimization method was developed by Dr. Yuriy V. Shekhtman [44,68]. It utilizes the following calculation processes:

- Definition of a set of mathematical functions that are used to describe the optimized fillet profile: This set contains a combination of the trigonometric, polynomial, and exponential functions.
- Parameters of these functions are defined during the optimization process.
- The 2D FEA with the triangle linear elements is used to calculate stress. This kind of finite elements allows the achievement of satisfactory optimization results within reasonable calculation time.
- A random search method [183] and modified combination of the gradient descent and ravine methods are used to define the next step in the multiparametric iteration process of the root fillet profile optimization.

At the beginning of the root fillet optimization process, the finite element (FE) nodes are evenly distributed on the initial fillet profile. Previous publications [44,68] described the optimization process when the initial root fillet (FE) nodes were moved along the straight lines (beams) that connected the nodes with the initial root fillet center, which was defined as the center of the best-fitted circular arc. Such an approach provides good optimization results mostly for symmetric tooth gears, although it was used for both symmetric and asymmetric gears. Figure 5.19 presents a different approach to the definition of the FE node movement paths that is more suitable for

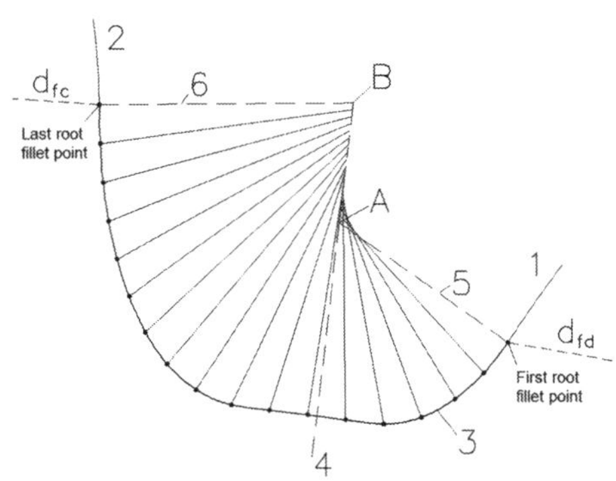

FIGURE 5.19
Root fillet FE node movement paths definition; 1: drive involute flank, 2: coast involute flank, 3: initial root fillet profile, 4: line connecting a center of the gear with a center of the initial root fillet portion that is formed by the tip land of the mating gear tooth, 5: line perpendicular to the drive involute flank at the first point of the initial root fillet, 6: line perpendicular to the coast involute flank at the last point of the initial root fillet.

asymmetric tooth root fillet optimization. The points A and B are defined as intersection points of the line connecting the center of the gear with the center of the initial root fillet portion that is formed by the tip land of the mating gear tooth (see Figure 5.18) and the lines perpendicular to the involute flanks at the first and last points of the initial root fillet. Then the line AB is evenly divided by a number of points equal to the number of the root fillet FE nodes and each of these points is connected to its FE node as shown in Figure 5.19. These straight lines (beams) are used as the root fillet FE node movement paths.

The first and last finite element nodes of the initial fillet profile located on the form diameter circle cannot be moved during the root optimization process. The rest of the initial fillet FE nodes are moved along their movement paths. The bending stresses are calculated for every fillet profile configuration iteration. Variable parameters of mathematical functions that describe the fillet profile for the next iteration are defined depending on stress calculation results of the previous iterations. Both successful (resulting with stress reduction) and unsuccessful (resulting with stress increase) iterations are taken into account for definition of the next iteration parameters.

After the specified number of unsuccessful iterations, indicating that further stress reduction cannot be achieved, the optimization process stops, resulting in the optimized fillet profile.

Figure 5.20a shows the final positions of the root fillet FE nodes that form its optimized profile. Figure 5.20b presents a photo of the optimized root fillet of the asymmetric tooth gear.

As a result of the asymmetric tooth root fillet optimization, the tensile (Figure 5.21a) and compressive (Figure 5.21b) stresses are evenly distributed along large portions of the root fillet profile.

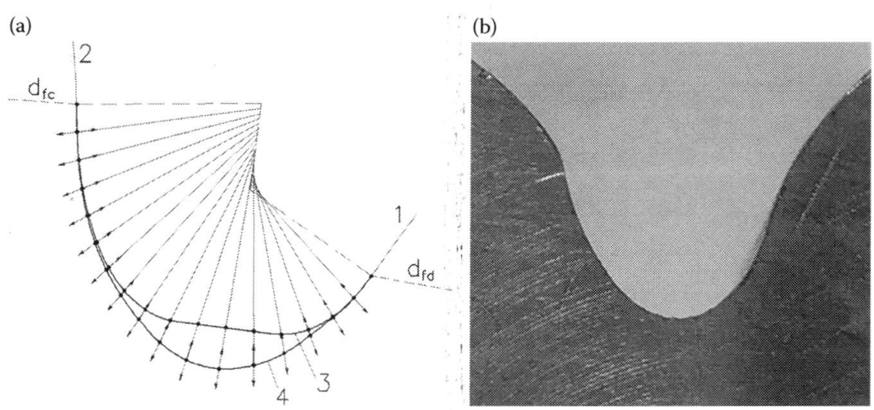

FIGURE 5.20
Optimized root fillet of the asymmetric tooth gear; (a) a final position of the root fillet FE nodes; (b) a photo of the optimized root fillet; 1: drive involute flank, 2: coast involute flank, 3: initial root fillet profile, 4: optimized root fillet profile.

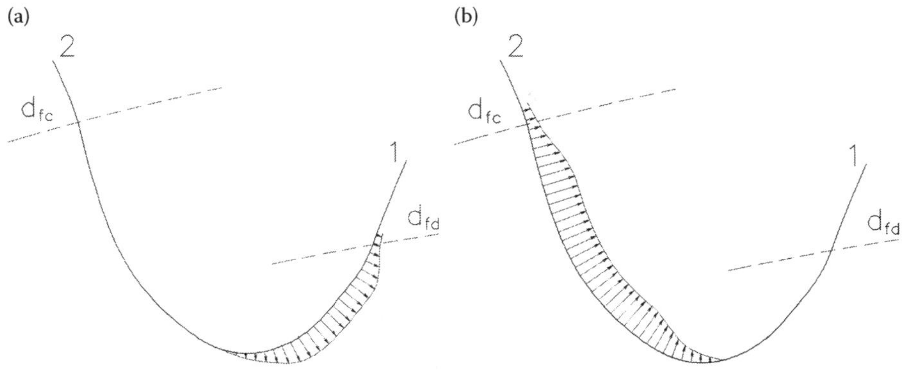

FIGURE 5.21
Tensile (a) and compressive (b) root bending stress charts along the optimized root fillet profile; 1: drive involute flank, 2: coast involute flank.

The FEA mesh of the asymmetric gear tooth with the optimized root fillet is shown in Figure 5.22. The FE nodes at the bottom of the mesh model, shown by the black dots, are constrained in all directions and present the boundary conditions of the solid body gear. The thin rim gear tooth root optimization is considered in the next paragraph. During the optimization process, a tooth force *F* is applied perpendicularly to the drive tooth flank at the highest point of single tooth contact (HPSTC) for the low and medium contact ratio spur gears or at the highest point of double tooth contact for the HCR spur gears. Typically, a tooth force *F* application point does not coincide with a FE node and located between two nodes. Then it is replaced by two forces applied to those nodes. A sum of these two forces is equal to the initial single force *F* and their separate values are defined in an inverse proportion to the distances from the initial single force *F* application point to the loaded nodes.

The typical tooth stress isograms and stress distribution along the tooth profile after the root fillet optimization are shown in Figure 5.23. They illustrate how bending tensile and compressive stresses are evenly distributed along large portions of the tooth root fillet.

The more the number of finite element nodes placed on the root fillet profile, the more accurate the stress calculation results, but this requires more iterations and the root fillet profile optimization takes more time. During the optimization process the fillet nodes cannot be moved inside the initial fillet profile because this may cause interference with the mating gear tooth tip. This makes unnecessary a checkup of the optimized root fillet for interference with the mating gear tooth tip. This is a main constraint for the root fillet optimization. Another very important constrain is to have a positive second derivative at any point of the optimized root fillet curve to exclude its waviness.

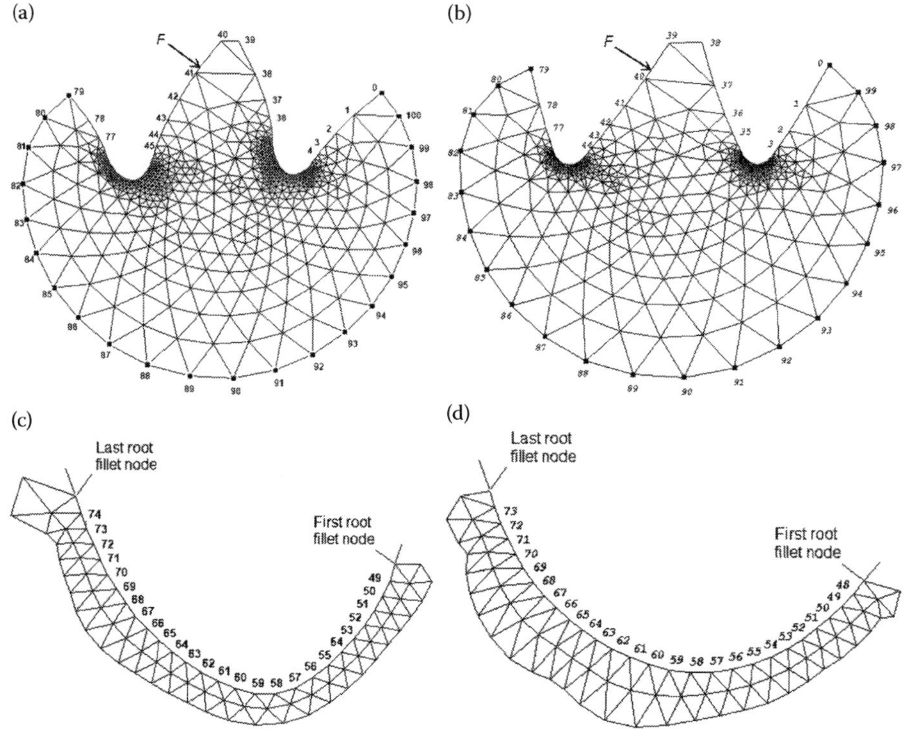

FIGURE 5.22
FEA mesh model of the asymmetric gear tooth (a) external gear and (b) internal gear and magnified optimized root fillet (c) external gear and (d) internal gear; *F:* applied tooth load; black dots—boundary FE nodes constrained in all directions.

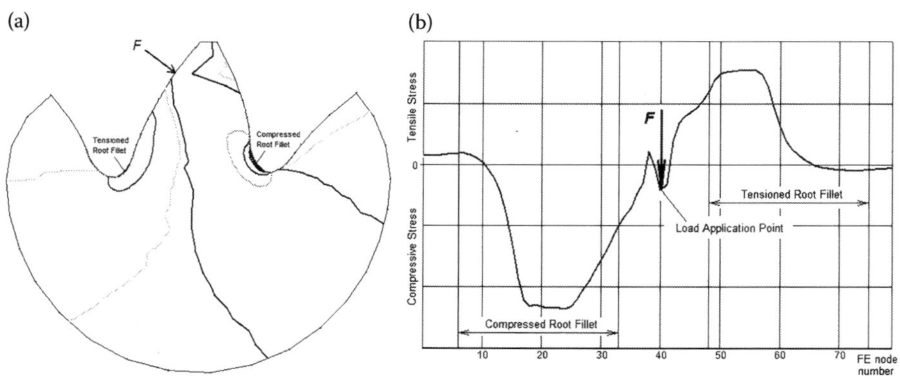

FIGURE 5.23
Asymmetric stress isograms (a) and a stress distribution along the tooth profile (b); *F:* applied tooth load.

There are other possible additional constraints. One of them is a minimum radial clearance. The optimized fillet typically results in a lower radial clearance than that in conventional standard gears. The concern here is that the radial clearance can be so small that the lubricant would be trapped in the root fillet space, resulting in additional hydraulic power losses and gear efficiency reduction. In this case, the required root diameter providing the acceptable radial clearance should be established as the root fillet optimization process constrain and the optimized root fillet profile must be tangent to this root diameter (see Figure 5.24).

Another root fillet optimization constrain is related to manufacturability of the gears with the optimized fillet. This is typical for external gears with a low (<20) number of teeth. The root fillet optimization process tries to reduce curvature (increase radius) at the maximum stress root fillet area to minimize its concentration. For gears with a low number of teeth this may result in a small fillet radius near the form diameter creating the undercut. Unlike the undercut that occurs in conventional gears, this one is made for purpose, and it does not affect the active involute flank profile. However, producing such an optimized fillet profile could be difficult or even impossible by some gear fabrication methods, such as, for example, form cutting, hobbing, etc. In order to make the root fillet profile manufacturable using these fabrication methods, the root fillet profile undercut should be limited or completely eliminated. This constrains the undercut angles φ_d and φ_c (Figure 5.24) between tangents to the initial and optimized fillet profiles at their first and last points. These additional fillet optimization constraints may compromise root stress concentration reduction in comparison to the optimized fillet constrained only by the initial fillet profile.

The presented optimization process, intended to minimize the maximum root bending stress, is trying to shape the root fillet to evenly redistribute tensile and compressive stresses along its profile (Figure 5.21). This stress redistribution may increase a total stress magnitude (a difference between

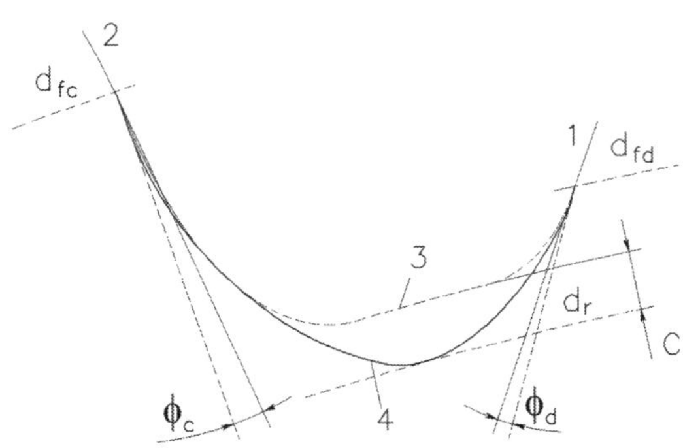

FIGURE 5.24 Tooth root fillet optimization constrains (dashed lines); 1: drive involute flank, 2: coast involute flank, 3: initial root fillet profile, 4: optimized root fillet profile, d_{fd} and d_{fc}: drive and coast flank form circle diameters, d_r: required root diameter, C: radial clearance, φ_d and φ_c: drive and coast flank undercut angles.

the maximum tensile and compressive values) at a certain point of the root fillet. Typically, this occurs at the very bottom of the root fillet near the root diameter. If this happens, a total stress magnitude would become an additional root fillet optimization constrain that must be taken into consideration. In such cases, the root fillet optimization should simultaneously combine the bending stress and total stress magnitude minimization. Figure 5.25 presents the comparison charts of the bending stress and total stress magnitude along the root fillet profile before and after optimization. The percentage of the tensile σ_{Ft} and σ_{Fc} compressive stress reduction after the root fillet

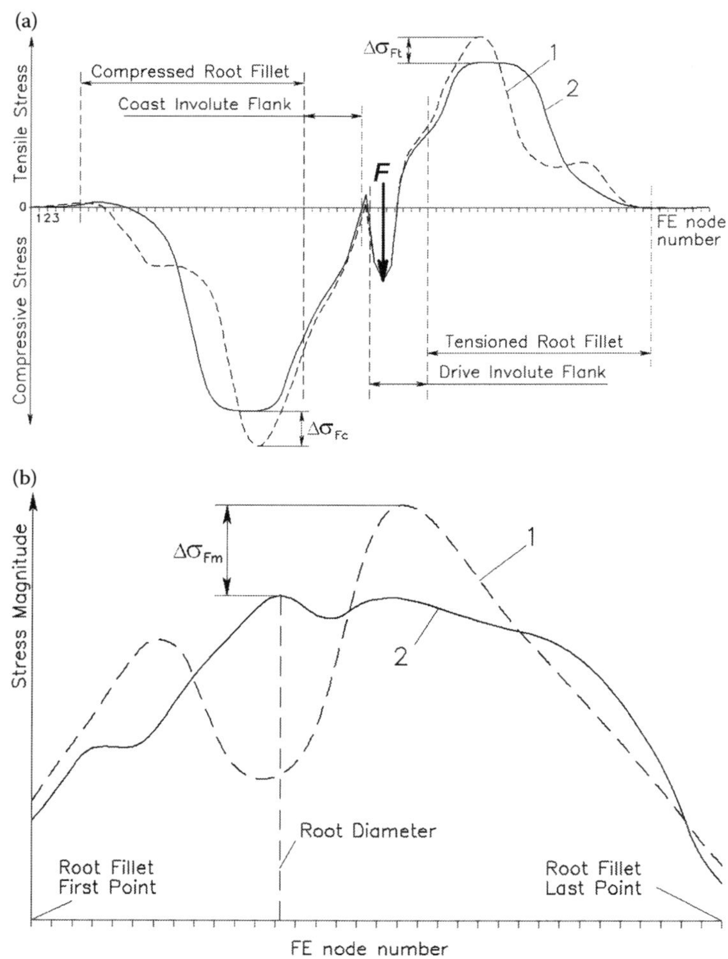

FIGURE 5.25
Bending stress (a) and total stress magnitude (b) along the root fillet; 1: before root fillet optimization, 2: after root fillet optimization; $\Delta\sigma_{Ft}$: tensile stress reduction, $\Delta\sigma_{Fc}$: compressive stress reduction, $\Delta\sigma_{Fm}$: total stress magnitude reduction.

optimization greatly depends on the shape of the initial root fillet that is a trajectory of the mating gear tooth tip in the tight (zero-backlash) engagement. If the mating gear number of teeth is low and its tip tooth thickness is small (close to the pointed tooth tip), the stress reduction after optimization is relative low 10%–15% in comparison to the initial root fillet. If the mating gear has many teeth or it has internal teeth and the tooth tip is wide, the stress reduction after optimization could be as high as 30%–40%. There are no sufficient data about maximum possible reduction of a total stress magnitude as a result of the root fillet optimization, Although about 20% of the total stress magnitude reduction in comparison to the initial root fillet is achievable.

Table 5.5 presents the comparison of the asymmetric gear pairs designed traditionally by the rack generating method and by Direct Gear Design. All parameters and dimensions of these gear pairs are identical, except the root fillet profiles. The traditionally designed asymmetric gear pair is generated by the asymmetric rack with a full tip radius that produces the trochoidal tooth root fillet with minimal bending stresses, achievable by this gear design method. The asymmetric gear pair defined by Direct Gear Design has an optimized root fillet, which demonstrates significant (~18%) bending stress reduction compared to the best trochoidal root fillet of the traditionally designed asymmetric gear pair.

TABLE 5.5

Root Bending Stress Comparison

Tooth Geometry Definition		Rack Generation		Direct Gear Design	
Gear		Pinion	Gear	Pinion	Gear
Number of teeth		19	29	19	29
Generating rack coefficients	Addendum	1.000		–	
	Dedendum	1.174		–	
	Tip Radius	0.264 (full radius)		–	
X-shift coefficient		0.0	0.0	–	
Pressure angles (drive/coast)		35°/20°		35°/20°	
Module, mm		4.00	4.00	4.00	4.00
Pitch diameter (PD), mm		76.000	116.000	76.000	116.000
Tooth tip diameter, mm		84.000	124.000	84.000	124.000
Root diameter, mm		66.609	106.608	66.811	106.964
Tooth thickness at PD, mm		6.283	6.283	6.283	6.283
Face width, mm		30.0	28.0	30.0	28.0
Torque, Nm		700	1,068	700	1,068
Max. tensile stress, MPa		441	440	363	359
Max. compressive stress, MPa		–676	–691	–556	–565
Tensile stress reduction		–	–	17.7%	18.4%
Compressive stress reduction		–	–	17.8%	18.2%

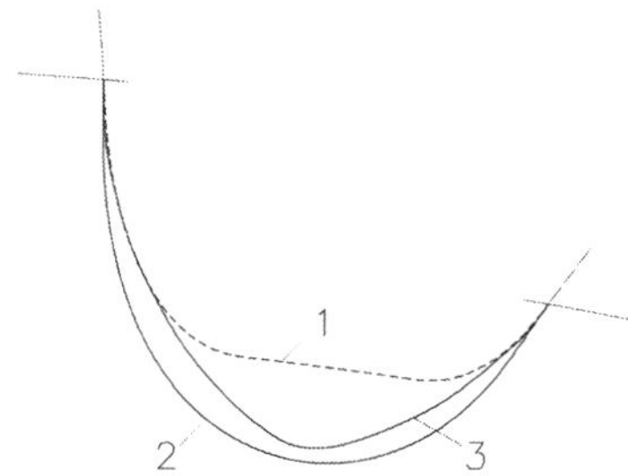

FIGURE 5.26
Pinion root fillet profile comparison; 1: trajectory of the mating gear tooth tip in the zero-backlash engagement, 2: traditionally designed trochoidal root fillet, 3: root fillet optimized by Direct Gear Design.

An overlay of the pinion root fillet profiles of the traditionally and directly designed gear pair are shown in Figure 5.26. Figure 5.27 presents stress distribution along the pinion tooth profile for both comparable asymmetric gear pairs.

5.2.2 Thin Rim Gear Root Fillet Optimization

Thin rim gears naturally have higher bending root stress compared to the solid body gears. This makes essential to reduce this stress by the root fillet optimization. In this case, the optimized root fillet profile and maximum root stress depend on the rim thickness and type of the gear rim design. There are several different rim design options that can be considered.

The first one is a "free" rim support with its possible radial deflection. Its FEA mesh model is shown in Figure 5.28a. This is typical for gears with the spokes connecting the rim with the hob.

The second option of rim design is typical for gears that have the sliding fit on the internal rim surface with a very small clearance. This restrains the rim radial deflections. Its FEA mesh model is shown in Figure 5.28b.

The third type of the gear thin rim design is when a gear is connected with a shaft by the interference fit that could be a press or shrink fit. It imposes additional hoop stress depending on press or shrink fit interference. Its FEA mesh model is shown in Figure 5.28c. If this interference is significant, resulting in high hoop stress or if a rim thickness is low, the root fillet optimization may not be possible.

As it is shown in paragraph 5.1.2, application of asymmetric tooth gears in an epicyclic stage allows the contact stresses to be balanced by choosing a greater pressure angle in the sun-planet gear mesh than in the planet-ring gear mesh in order to maximize power density, reducing a gear drive size. This means that a reduction of the center distance and all gear dimensions, including the planet gear root diameter. In many cases, a planet gear is supported by a roller bearing incorporated inside the gear, using its inner

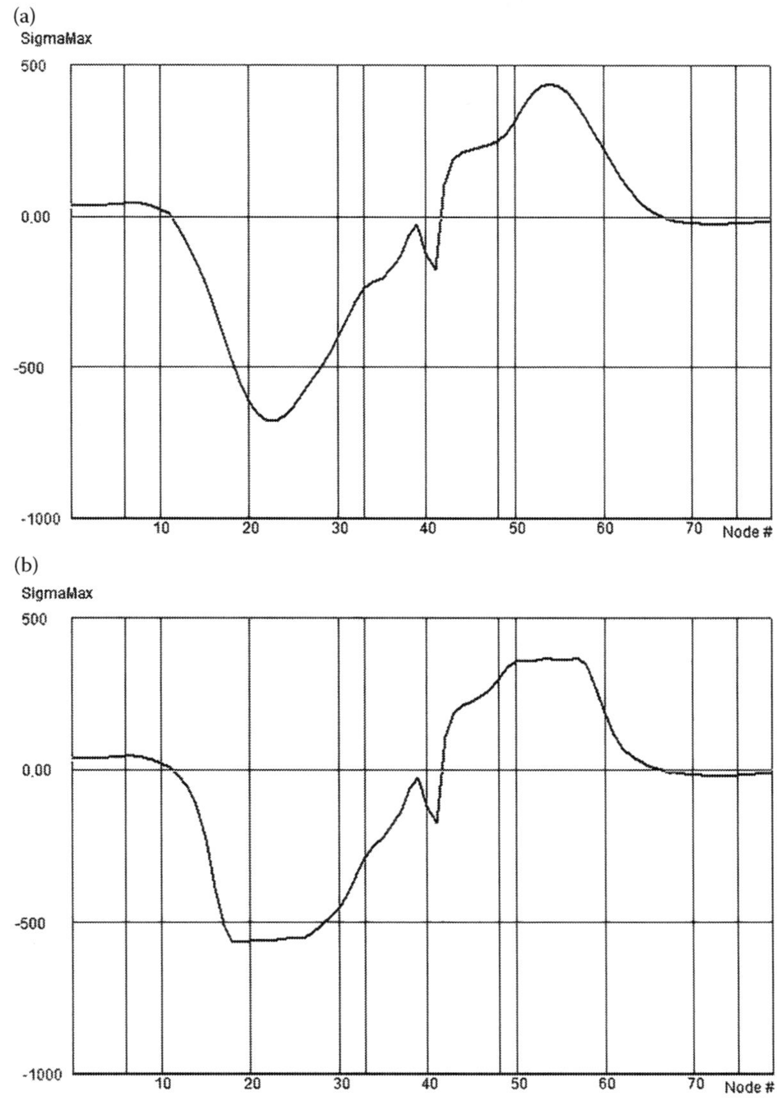

FIGURE 5.27
Stress distribution along the pinion tooth profile; (a) with the traditionally designed trochoidal root fillet, (b) with the root fillet optimized by Direct Gear Design.

diameter as a roller bearing outer race surface (see Figure 5.29). Reduction of the center distance, while the transmitted torque remains constant, increases the planet gear roller bearing load. Its leads to the necessity of increasing the diameter of the roller bearing to maintain its required load capacity and life. A simultaneous reduction of the planet gear root diameter and increase of its inner diameter significantly reduce the planet gear rim thickness, increasing

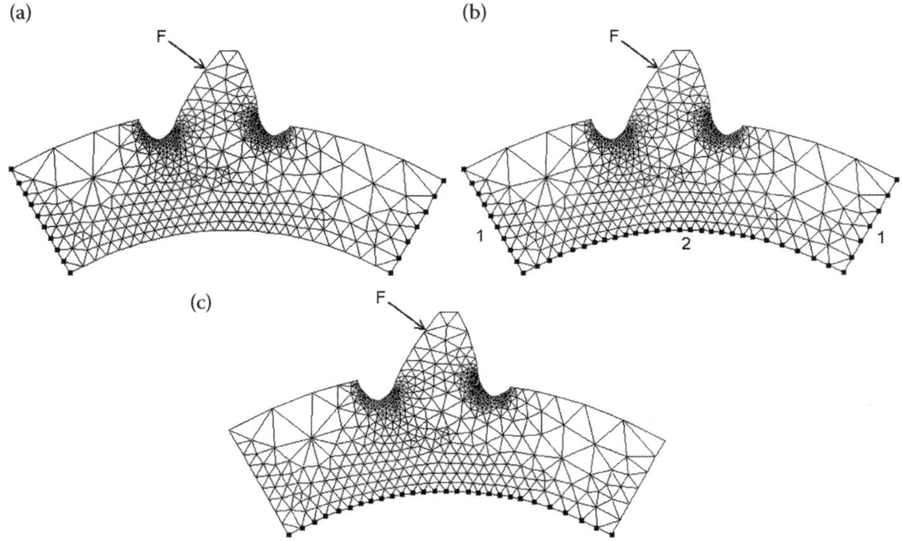

FIGURE 5.28
Thin rim gear FEA mesh models: (a) a "free" rim support, the radially located FE nodes are constrained in all directions; (b) a slip fit rim support, 1- the radially located FE nodes are constrained in the circular direction and 2: the FE nodes located on the inner radius are constrained in the radial direction; (c) an interference fit rim support, the FE nodes located on the inner radius are constrained in the circular direction and are moved in the radial direction out of a center on the distance equal a half of the press fit interference.

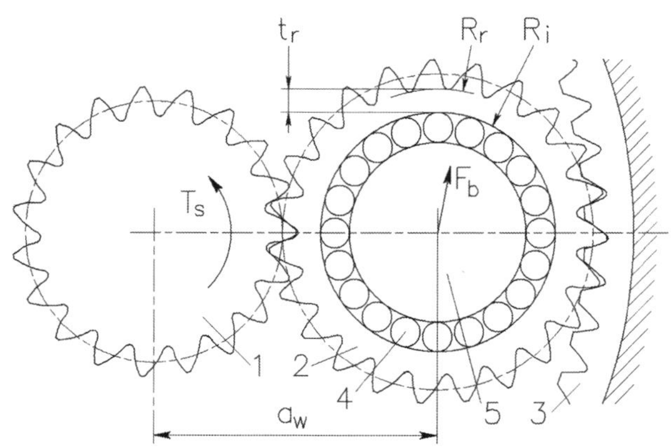

FIGURE 5.29
Epicyclic stage with the thin rim planet gear; 1: sun gear, 2: planet gear, 3: ring gear, 4: bearing rollers, 5: planet gear shaft, T_s: sun gear torque, a_w: center distance, F_b: planet gear bearing load, R_r: planet gear root radius, R_i: planet gear inner radius, t_r: rim thickness. (Kapelevich A.L. and Y.V. Shekhtman. 2018. Root Fillet Optimization of Thin Rim Planet Gears with Asymmetric Teeth. Gear Solutions, March, 52–55. With permission.)

the tooth root bending stress. In such cases, tooth root fillet optimization is essential to keep root stresses within acceptable level [160].

Figure 5.30 shows the planet gear tooth loading. Teeth of the planet gear are loaded at both flanks by forces applied from the sun and ring gears. Although the tangent components of these forces are identical for a simple asymmetric epicyclic stage (Figure 5.29), the normal force F_s of the sun/planet gear contact is greater than the normal force F_r of the planet/ring gear contact by a ratio equal to the asymmetric factor *K*. Besides, because of the tooth asymmetry and different contact ratios at the opposite tooth flanks, the HPSTC of the sun-planet gear contact is located significantly closer to the tooth tip compared to the planet/ring gear contact. For these reasons, the sun/planet gear contact force results in greater root bending stress and this loading condition (Figure 5.30a) is selected for the root fillet optimization.

A relative angular position between the planet gear teeth and supporting bearing roller is constantly changing during the epicyclic stage operation. Two of such relative angular positions are selected for the planet gear root fillet optimization. The first one (Figure 5.31a) is when the bearing roller is in line with the middle of the loaded planet gear tooth thickness and the second one (Figure 5.31b) is when the loaded planet gear tooth is located between two supporting bearing rollers.

Figure 5.32 represents the thin rim planet gear FEA mesh models of two relative angular positions of the loaded gear tooth and supporting bearing rollers. The FE nodes at the inner gear diameter are constrained in the circular direction, simulating the bearing roller contact points.

Figures 5.33 and 5.34 show the planet gear tooth stress isograms and stress distribution along the tooth profile. Since the load from the sun gear was selected as the load application condition for the root fillet optimization, it results in even stress distribution along large portions of the root fillet. But when a load from the ring gear is applied to the opposite tooth flank, a stress distribution along the root fillet is not even. Nevertheless,

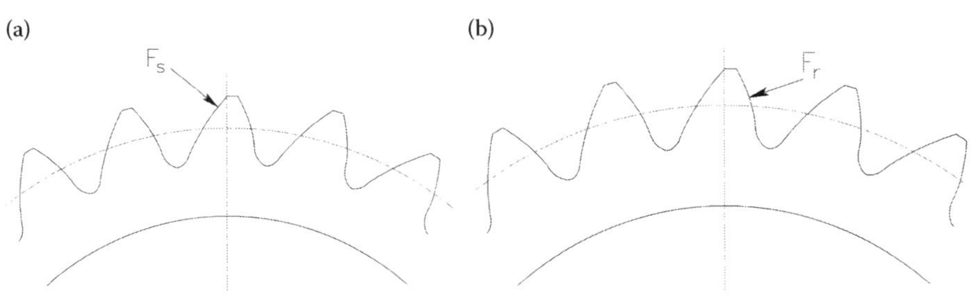

FIGURE 5.30
Planet gear tooth loading; (a) by a force F_s applied from the sun gear, (b) by a force F_r applied from the ring gear from ring gear. (Kapelevich A.L. and Y.V. Shekhtman. 2018. Root Fillet Optimization of Thin Rim Planet Gears with Asymmetric Teeth. Gear Solutions, March, 52–55. With permission.)

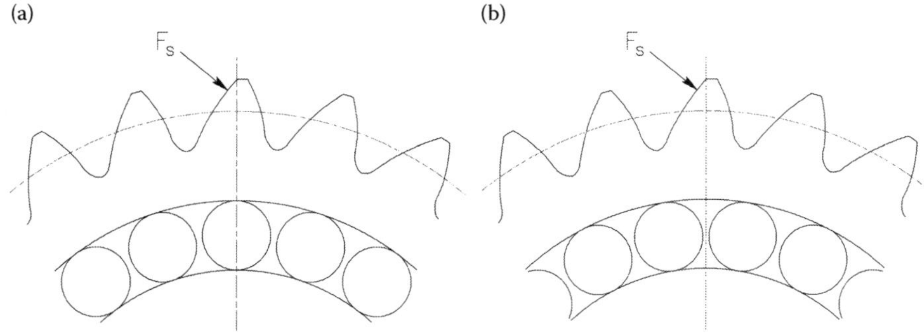

FIGURE 5.31
Loaded planet gear tooth and bearing roller positions: (a) the tooth is in line with a roller, (b) the tooth is between two rollers. (Kapelevich A.L. and Y.V. Shekhtman. 2018. Root Fillet Optimization of Thin Rim Planet Gears with Asymmetric Teeth. Gear Solutions, March, 52–55. With permission.)

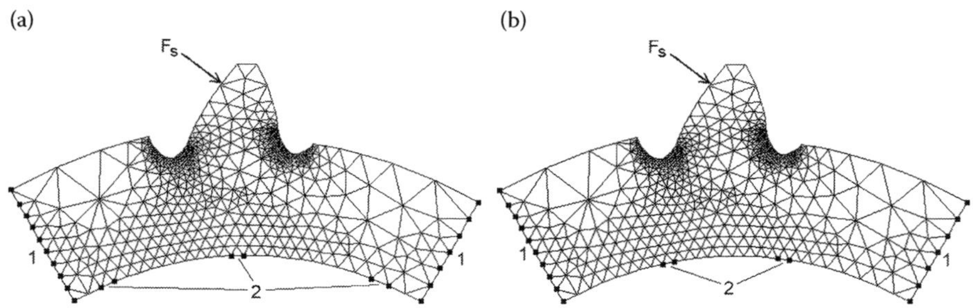

FIGURE 5.32
Thin rim planet gear FEA mesh models with the bearing roller positions: (a) a tooth is in line with a bearing roller, (b) a tooth is between two bearing rollers; 1: FE nodes constrained in the circular direction, 2: radially constrained FE nodes. (Kapelevich A.L. and Y.V. Shekhtman. 2018. Root Fillet Optimization of Thin Rim Planet Gears with Asymmetric Teeth. Gear Solutions, March, 52–55. With permission.)

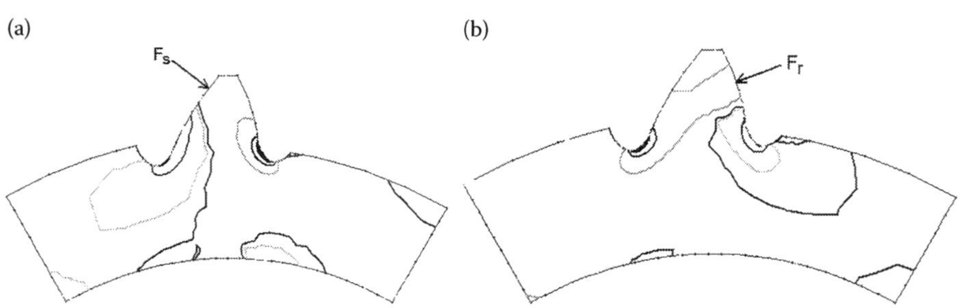

FIGURE 5.33
Planet gear stress isograms; (a) load applied from the sun gear, (b) load applied from the ring gear. (Kapelevich A.L. and Y.V. Shekhtman. 2018. Root Fillet Optimization of Thin Rim Planet Gears with Asymmetric Teeth. Gear Solutions, March, 52–55. With permission.)

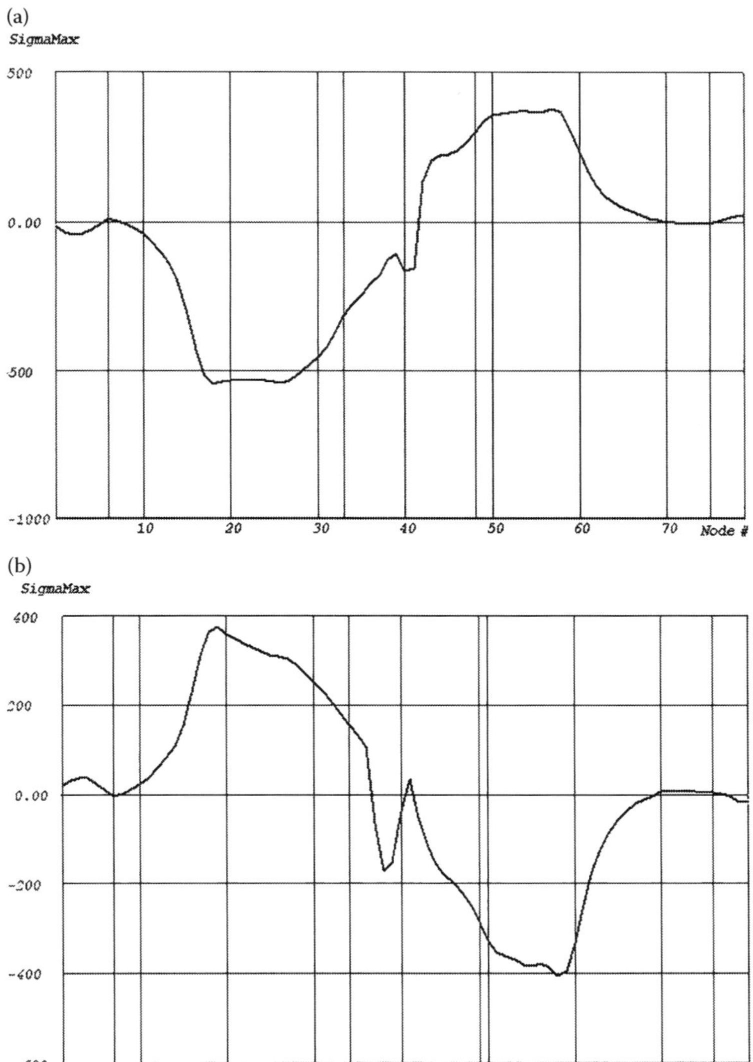

FIGURE 5.34
Stress distribution along the planet gear tooth profile; (a) load applied from the sun gear, (b) load applied from the ring gear. (Kapelevich A.L. and Y.V. Shekhtman. 2018. Root Fillet Optimization of Thin Rim Planet Gears with Asymmetric Teeth. Gear Solutions, March, 52–55. With permission.)

the maximum tensile stress in this case is just a bit greater compared to when a load is applied from the sun gear, but the maximum compressive stress is lower.

Table 5.6 presents the geometric data and torque for the epicyclic gear stage with asymmetric tooth gears.

TABLE 5.6

Epicyclic Gear Stage Geometric Data

Gear	Sun		Planet			Ring
Number of teeth	19		23			65
Module (m)	5.00		5.00			5.00
Drive pressure angle	36°		36°	19°		19°
Coast pressure angle	19°		–	–		36°
Pitch diameter (PD)	95.00		115.00			325.00
Tooth tip diameter	105.28		125.17			317.84
Root diameter	83.69		103.52*			336.04
Tooth thickness at PD	7.94		7.76			7.94
Face width	40.0		37.0			34.0
Center distance			105.00			
Drive contact ratio		1.25			1.60	
Torque per one mesh, Nm	1,000		1,210			3,421

The maximum planet gear root stress values for different bearing roller positions, load applications, and relative rim tooth thickness (rim thickness/module) values are shown in Table 5.7.

The charts of the maximum planet gear tooth root stresses vs. the relative rim thickness are provided in Figure 5.35. They indicate that when

TABLE 5.7

Tensile and Compressive Root Stresses of Thin Rim Planet Gears

Root Stress, MPa	Relative Rim Thickness	Tooth Is in Line with a Bearing Roller: Loaded by Sun Gear	Tooth Is in Line with a Bearing Roller: Loaded by Ring Gear	Tooth between Two Bearing Rollers: Loaded by Sun Gear	Tooth between Two Bearing Rollers: Loaded by Ring Gear
Tensile	4.0	273	295	261	281
	3.5	276	300	262	282
	3.0	282	308	263	284
	2.5	296	323	265	292
	2.0	330	359	270	312
	1.5	415	446	291	353
	1.0	623	662	367	468
Compressive	4.0	–398	–300	–414	–286
	3.5	–402	–306	–416	–290
	3.0	–407	–313	–423	–296
	2.5	–417	–331	–439	–308
	2.0	–456	–371	–468	–328
	1.5	–552	–456	–529	–360
	1.0	–763	–688	–730	–437

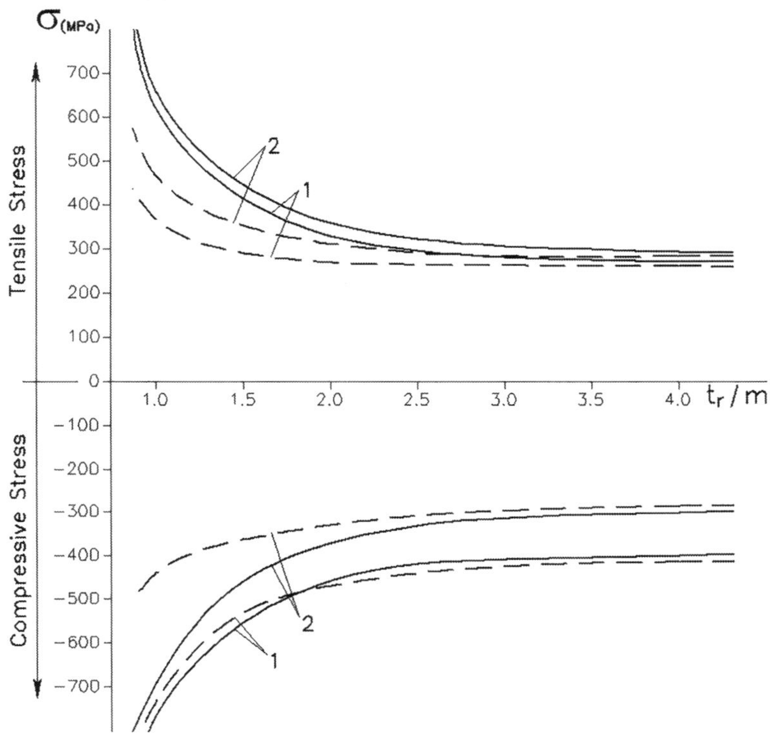

FIGURE 5.35
Maximum planet gear tooth root stress charts; 1: load applied from the sun gear, 2: load applied from the ring gear; solid lines: the loaded tooth is in line with a roller; dashed lines: the loaded tooth is between two rollers. (Kapelevich A.L. and Y.V. Shekhtman. 2018. Root Fillet Optimization of Thin Rim Planet Gears with Asymmetric Teeth. Gear Solutions, March, 52–55. With permission.)

a loaded tooth is in line with the bearing roller, the root stresses are generally greater than when the loaded tooth is between two rollers. If the rim thickness is greater than 3.5× module, the tooth root stresses do not depend on the rim thickness and practically are equal to the solid body gear stresses. If the rim thickness is less than 3.5× module, tooth root stresses grow exponentially as rim thickness is reducing. When the rim thickness that is less than 2.0× module, the maximum root stresses are getting very high. Such thin rim planet gear design is typically not recommended.

Table 5.8 shows how the planet gear root and inner diameters depend on rim thickness when the tooth root fillet is optimized.

Figure 5.36 demonstrates the optimized tooth root fillet profiles and inner diameters of the planet gear depending on the rim thickness.

TABLE 5.8

Root and Inner Diameters of Thin Rim Planet Gears

Rim Thickness	Root Diameter	Inner Diameter
20.00 (4.0 × *m*)	103.54	63.54
17.50 (3.5 × *m*)	103.59	68.59
15.00 (3.0 × *m*)	103.64	73.64
12.50 (2.5 × *m*)	103.70	78.70
10.00 (2.0 × *m*)	103.80	83.80
7.50 (4.0 × *m*)	103.99	88.99
5.00 (1.0 × *m*)	104.26	94.26

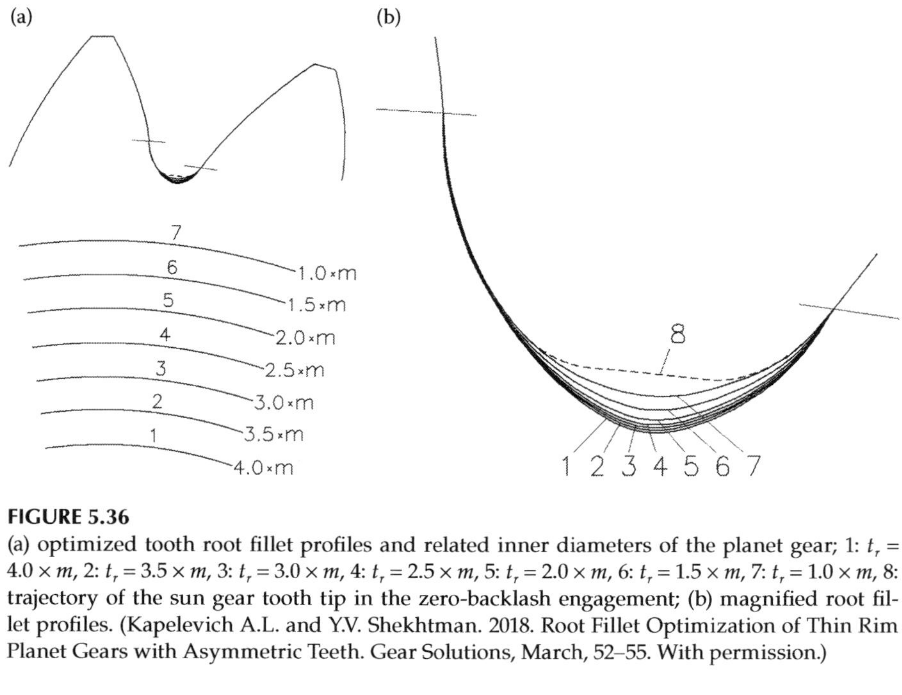

FIGURE 5.36
(a) optimized tooth root fillet profiles and related inner diameters of the planet gear; 1: t_r = 4.0 × *m*, 2: t_r = 3.5 × *m*, 3: t_r = 3.0 × *m*, 4: t_r = 2.5 × *m*, 5: t_r = 2.0 × *m*, 6: t_r = 1.5 × *m*, 7: t_r = 1.0 × *m*, 8: trajectory of the sun gear tooth tip in the zero-backlash engagement; (b) magnified root fillet profiles. (Kapelevich A.L. and Y.V. Shekhtman. 2018. Root Fillet Optimization of Thin Rim Planet Gears with Asymmetric Teeth. Gear Solutions, March, 52–55. With permission.)

5.2.3 Benefits of Root Fillet Optimization

A goal and main benefit of the root fillet optimization is bending stress concentration reduction. If load capacity of gears with conventional (trochoidal or circular) root fillet profiles is limited by maximum tooth bending stress, root fillet profile optimization increases gear load capacity proportionally to the bending stress reduction. However, quite often, the gear load capacity, and consequently the possible gear drive size and weight reduction, is limited by the tooth surface durability defined by pitting and scuffing resistance, which greatly depends on the contact stress, profile sliding, and tooth contact (flash) temperature. Then potential bending stress reduction provided by the fillet optimization can be used to improve other gear drive performance parameters. If the load capacity of gears with optimized root fillet profiles is not limited by maximum tooth bending stress and the root safety factors are excessive, numbers of gear teeth can be increased and a gear module reduced respectively to keep the constant center distance and approximately the same gear ratio. This will reduce the root safety factors to a sufficiently safe level, although simultaneously increasing the contact ratio and gear mesh efficiency, and reducing the Hertz contact stress, scuffing probability, and tooth contact (flash) temperature. Such gear performance parameter enhancement possibilities were described in [107] for standard symmetric tooth gears. Optimized asymmetric tooth gears additionally allow these benefits to be maximized for unidirectional gear drives, because they provide the root bending stress reduction not only by the root fillet optimization, but also by the significantly increased tooth thickness at the root area.

5.3 Bending Strength Balance

The mating gears typically have different tooth shapes and face widths, and they could be made out of dissimilar materials. In order to provide equally strong teeth of a pinion and gear, their maximum bending stress should be balanced or safety factors should be equalized [44]. This balance condition can be presented as follows:

$$\left|\sigma_{F\max1} - C_b\sigma_{F\max2}\right| \le \delta_F, \tag{5.30}$$

where $\sigma_{F\max1}$ and $\sigma_{F\max2}$ are the maximum bending tensile stresses in the root fillet of the pinion and the gear C_b is the bending stress balance coefficient reflecting the difference in the mating gear face widths, material properties, and in a number of tooth load cycles for the pinion and the gear that defined by their allowable bending stresses, and δ_F is the permissible stress balance tolerance (typically less than 2%–3%).

Direct Gear Design uses the stress balance approach that utilizes an iteration method with the FEA stress calculation to satisfy the bending stress balance condition (5.30) by adjusting the tooth thickness ratio (TTR), defined by Equation (5.27). This bending stress balance procedure (see Figure 5.37) works in combination with the tooth flanks and root fillet optimization.

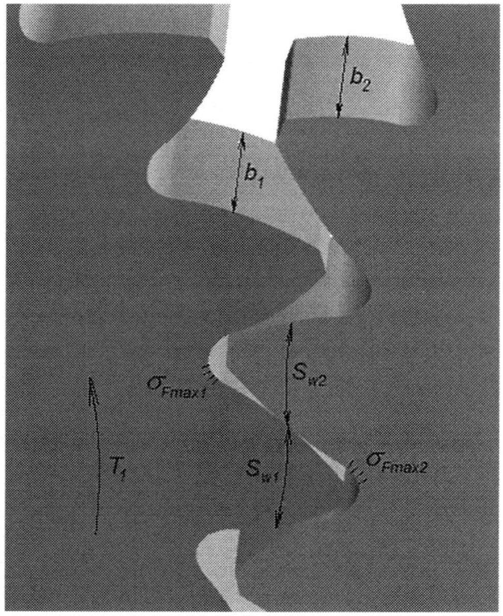

FIGURE 5.37
Bending stress balance; T_1: driving pinion torque, b_1 and b_2: pinion and gear face widths, S_{w1} and S_{w2}: pinion and gear tooth thicknesses at the operating pitch diameters, σ_{Fmax1} and σ_{Fmax2}: pinion and gear maximum tensile root stresses.

5.4 Transmission Error Minimization

This section describes an optimization of the microgeometry of the spur asymmetric gear tooth drive flank profile to minimize a transmission error. The goal of the tooth flank microgeometry optimization is to partially compensate influence of manufacturing tolerances, assembly misalignments, and operating conditions (including deflections of gears and other gearbox components under operating load, dynamic loads and inertia, temperature, etc.). All these factors distort a theoretically correct involute mesh by deviating the actual contact points from the ideal straight line of contact that increases a transmission error, and leads to increased noise and vibrations. Intend of the tooth microgeometry optimization is to alter the drive tooth flank from a theoretical involute profile to bring the actual contact points close to the ideal straight line of contact.

The tooth microgeometry optimization utilizes the same approach that was used in the effective contact ratio and transmission error calculation procedure (see Section 5.1.4), taking into account only the contact and bending gear tooth deflections. It is employed in combination with another iteration cycle that defines a tooth flank modification depth to achieve a minimal transmission error depending on a selected flank modification mode.

There are three most common tooth flank modification modes: a tip and root relief, an arc modification, and a parabolic crowning. Figure 5.38 shows the tooth flank modification modes as charts of the roll angle vs. the flank modification depth in microns. The roll angle can be presented as follows (see Figure 2.1):

$$\phi = \frac{180^o}{\pi}\tan\alpha, \tag{5.31}$$

where *a* is the involute profile angle.

The tip and root relief, and arc modification modes are applied to the low and medium contact ratio gears, because they alter only the double tooth pair contact zones where a load sharing between two pairs of teeth occur. The parabolic crowning modifies a complete involute flank and can be applied for the low and medium contact ratio gears as well as the HCR gears.

In a unidirectional asymmetric gear pair, a drive flank of the pinion is a subject for microgeometry optimization. A coast flank of the pinion and both flanks of the driven gear remain unmodified. However, when the optimization is done, a part of the modification depth can be transferred from a drive flank of the pinion to a drive flank of the mating gear. Most typical is a transfer of the pinion root relief depth to a tip relief depth of

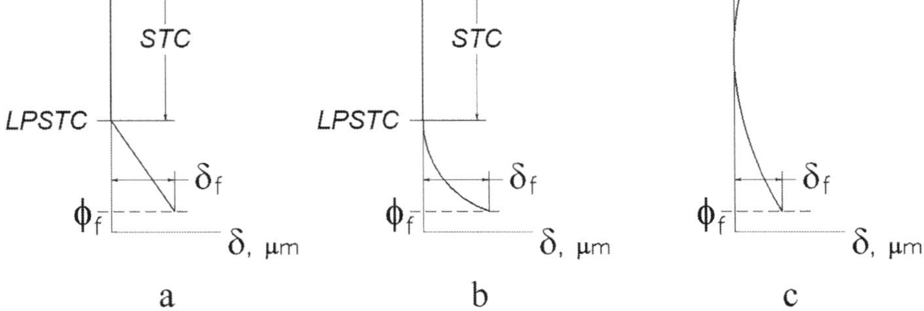

FIGURE 5.38
Tooth flank modification modes; (a) tip and root relief, (b) arc modification, (c) parabolic crowning; δ: flank modification depth in microns, δ_f: at the form diameter (near the tooth root), δ_e: at the effective tip diameter (at the highest involute profile point near the tooth tip); ϕ: roll angle in degrees, ϕ_f: at the form diameter, ϕ_e: at the effective tip diameter; LPSTC: lowest point of single tooth contact and HPSTC: highest point of single tooth contact for gears with a low medium contact ratio < 2.0.

the gear (see Figure 5.39). If the optimized pinion drive flank contact point X_1 is at the roll angle α_{xd} and its modification depth is δ_x, the mating gear drive flank contact point X_2 roll angle can be defined by altering Equations (2.44) and (2.48):

$$\tan\alpha_{xd1} \pm u\tan\alpha_{xd2} \mp (1 \pm u)\tan\alpha_{wd} = 0, \tag{5.32}$$

Here and in Equations (5.33) and (5.34), the top sign (+ or –) is for external gears and the bottom sign (+ or –) is for internal gears.

Then the drive involute profile angles α_{xd1}, α_{xd2}, and α_{wd} should be replaced with the related roll angles from Equation (5.31)

$$\phi_{xd1} \pm u\phi_{xd2} \mp (1 \pm u)\phi_{wd} = 0 \tag{5.33}$$

and finally, the mating gear drive flank contact point X_2 roll angle

$$\phi_{xd2} = \frac{u \pm 1}{u}\phi_{wd} \mp \frac{1}{u}\phi_{xd1}. \tag{5.34}$$

The modification depth transfer from the pinion drive flank dedendum to the drive addendum of the mating gear is illustrated in Figure 5.39. In this figure both gear pairs with drive flank modifications 1 and 2 should have an identical transmission error. This kind of a modification depth transfer could make sense for a technological reason; when the tooth dedendum modification is impractical, because, for example, it might affect a conjunction of the tooth flank with an optimized root fillet.

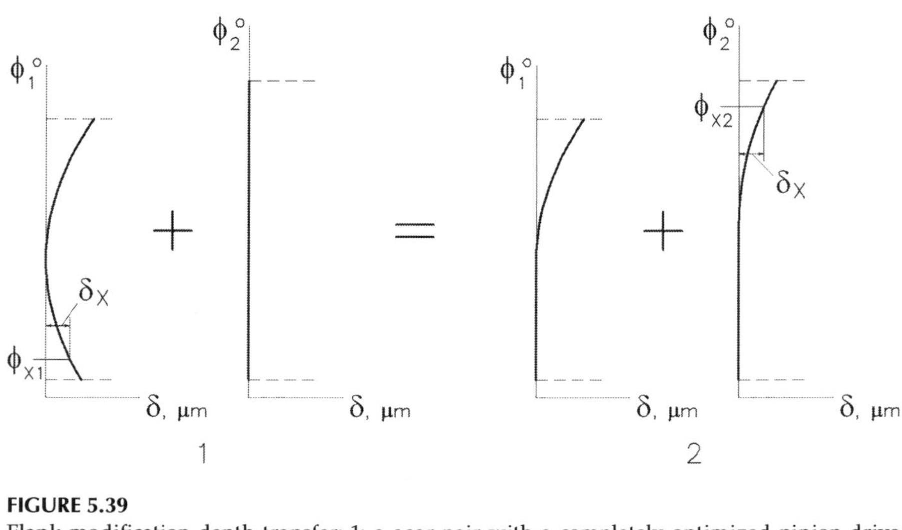

FIGURE 5.39
Flank modification depth transfer; 1: a gear pair with a completely optimized pinion drive flank, including the addendum and dedendum profile and an unmodified drive flank of the gear, 2: a gear pair with an optimized drive flank addendum of both the pinion and gear.

In some cases, like in an idler gear or a planet gear of an epicyclic gear stage, which have both drive flanks engaged with two different mating gears, these flanks' modifications should be optimized separately, considering differences in engagements and loading of opposite flanks. Besides, an asymmetric tooth of the idler gear or planet gear has different stiffness when loads are applied at the high and low pressure angle flanks.

5.4.1 Low-Medium Contact Ratio Gears

Table 5.9 presents the low and medium contact ratio gear pair data and tooth flank optimization results. Three tooth flank modification modes (see Figure 5.38) are considered.

Comparison of different modes of the drive flank modification optimization indicates that resulting contact stresses are practically identical, but a

TABLE 5.9

Low-Medium Contact Ratio Gear Pair Tooth Flank Optimization

	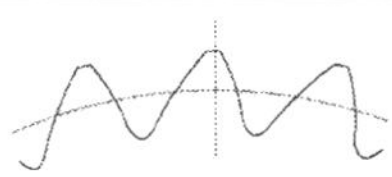		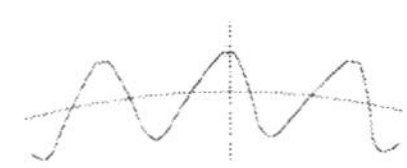
Gear	**Pinion**		**Gear**
Number of teeth	27		41
Module (m), mm		3.000	
Drive flank pressure angle		38°	
Coast flank pressure angle		19°	
Pitch diameter (PD), mm	81.000		123.000
Tooth tip diameter, mm	87.090		128.935
Root diameter, mm	74.393		116.230
Tooth thickness at PD, mm	4.807		4.618
Face width, mm	30.00		28.00
Torque, Nm	700		1,063
Root tensile stress, MPa	415		422
Drive flank mesh efficiency		98.8% (average friction coefficient = 0.1)	
Nominal drive contact ratio		1.24	
Effective drive contact Ratio		1.46 (no flank modification)	
Contact stress, MPa		1,394 (no flank modification)	
Transmission error, μm		7.3 (no flank modification)	
Type of flank modification	Tip and root relief	Arc modification	Parabolic crowning
Tip modification, μm	10	15	13
Root modification, μm	27	39	3
Effective drive contact ratio	1.28	1.28	1.40
Contact stress, MPa	1,398	1,400	1,400
Transmission error, μm	3.7	4.6	3.7
Transmission error reduction	49%	37%	49%

parabolic crowing produces high transmission error reduction and effective drive contact ratio, while keeping a minimal modification depth.

Figure 5.40 presents the transmission error charts for the gear pairs with the initial unmodified drive flanks and gears with the optimized drive pinion flanks utilizing three different modification modes.

Figure 5.41 shows the optimized drive flanks of the pinion for different modification modes.

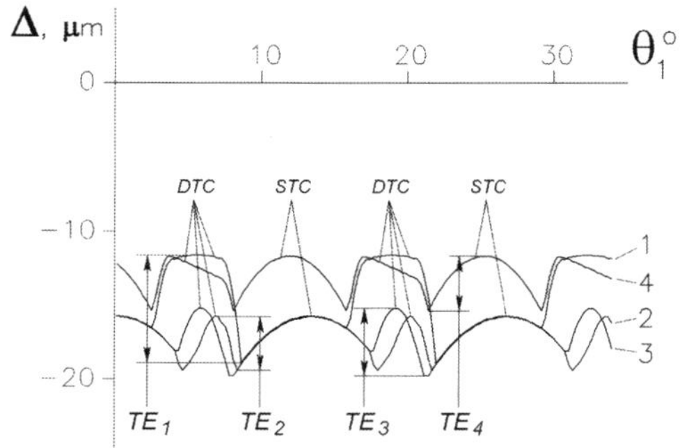

FIGURE 5.40
Transmission error charts; 1: initial unmodified drive flanks (transmission error TE_1), 2: drive flanks of the pinion with the tip and root relief (transmission error TE_2), 3: drive flanks of the pinion with the arc modification (transmission error TE_3), 4: drive flanks of the pinion with the parabolic crowning (transmission error TE_4); STC: single tooth pair contact, DTC: double tooth pair contact.

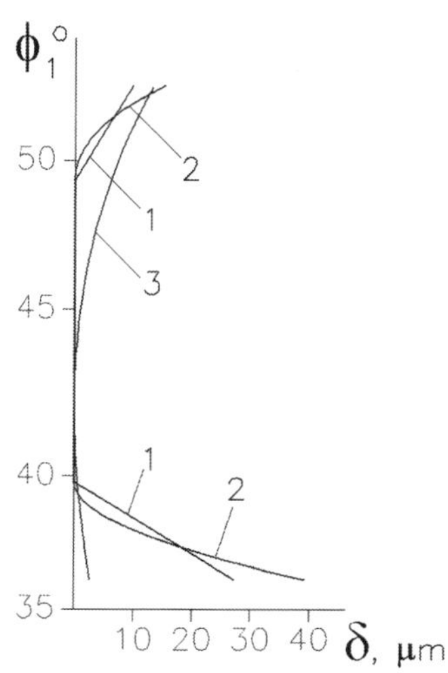

FIGURE 5.41
Optimized drive flanks of the pinion; 1: with the tip and root relief, 2: with the arc modification, 3: drive flanks of the pinion with the parabolic crowning.

5.4.2 High Contact Ratio Gears

Table 5.10 presents the HCR gear pair data and tooth flank optimization results for the parabolic crowning modification mode.

The HCR gears have several advantages over the low medium contact ratio gears due to a much greater effective contact ratio and load sharing between two and three gear tooth pairs. This results in significantly lower root bending stress and transmission error after the pinion drive flank microgeometry optimization. The contact stress is also slightly lower. However, the drive flank mesh efficiency is reduced, because of increased specific sliding velocities.

Figure 5.42 presents the transmission error charts for gear pairs with the initial unmodified drive flanks and gears with the optimized drive pinion flanks. The parabolic crowning of the pinion drive flank of the HCR gear pair is shown in Figure 5.43.

TABLE 5.10

High Contact Ratio Gear Pair Tooth Flank Optimization

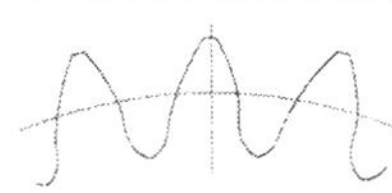
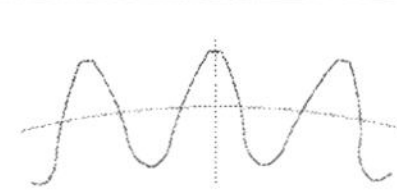

Gear	Pinion		Gear
Number of teeth	27		41
Module (m), mm		3.000	
Drive flank pressure angle		24°	
Coast flank pressure angle		14°	
Pitch diameter (PD), mm	81.000		123.000
Tooth tip diameter, mm	89.443		131.542
Root diameter, mm	71.636		113.877
Tooth thickness at PD, mm	4.807		4.618
Face width, mm	30.00		28.00
Torque, Nm	700		1,063
Root tensile stress, MPa	335		319
Drive flank mesh efficiency		98.0% (average friction coefficient = 0.1)	
Nominal drive contact ratio		2.04	
Effective drive contact ratio		2.24 (no flank modification)	
Contact stress, MPa		1,259 (no flank modification)	
Transmission error, μm		4.4 (no flank modification)	
Type of flank modification		Parabolic crowning	
Tip modification, μm		15	
Root modification, μm		12	
Effective drive contact ratio		2.2	
Contact stress, MPa		1,359	
Transmission error, μm		2.1	
Transmission error reduction		52%	

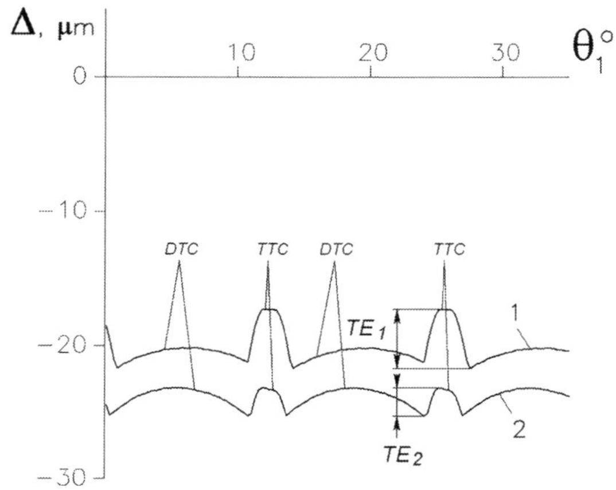

FIGURE 5.42
Transmission error charts; 1: initial unmodified drive flanks (transmission error TE_1), 2: drive flanks of the pinion with the parabolic crowning (transmission error TE_2); DTC: double tooth pair contact, TTC: triple tooth pair contact.

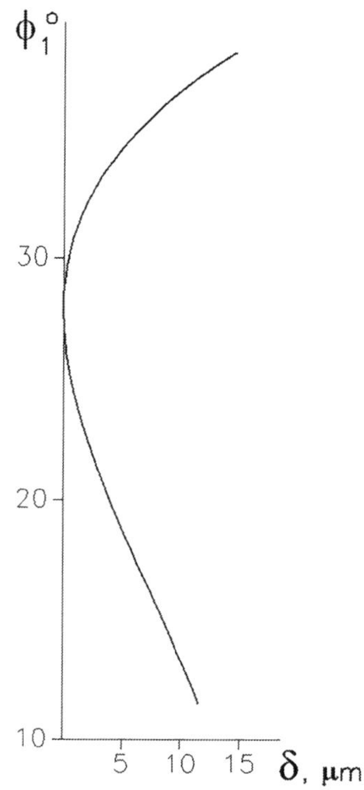

FIGURE 5.43
Parabolic crowning of the pinion drive flank of the HCR gear pair.

It is important to understand that the microgeometry optimization defines the shape and depth of the drive flank modification for a particular transmitted torque value, for which the resulting modified flank profile provides minimal transmission error. For any other torque value this flank profile is not optimal. For gear drives transmitting a constant torque, its value should be used for the driving flank microgeometry optimization. However, there are many gear drives that operate at variable load values. In such cases it is necessary to define which load condition is most damaging or critical for a specific gear drive application and use its value for the driving flank microgeometry optimization.

6

Stress Analysis and Rating of Asymmetric Gears

Over the last 100 years, a maximum gear bending stress, characterizing tooth breakage, has been defined based on the Lewis equation. Gear design standards specify the dimensionless Lewis form factor Y as a function of the number of teeth, generating rack parameters, and its X-shift or addendum modification. The form factor Y in combination with the stress correction factor K_f, accounting on stress concentration and load location, defines the bending strength geometry factor J.

Maximum gear contact stress, characterizing tooth surface pitting and wear breakage, is traditionally calculated by the Hertz equation. Gear design standards specify the pitting resistance geometry factor I accounting for the effects of radii of curvature, load sharing, and normal component of the transmitted load.

Asymmetric tooth gear stress definition is not covered by existing gear design standards. Nevertheless, there are publications [57,70,72,131] describing calculation of the maximum bending root stress based on the Lewis equation, modified for asymmetric tooth gears designed by the traditional rack generating method.

This chapter describes stress analysis and rating of asymmetric tooth gears defined by Direct Gear Design.

6.1 Stress Definition

Wide variability of parameters and shapes of directly designed asymmetric tooth gears makes it impossible to calculate their bending stress by the Lewis equation. Prior to the introduction of the finite element analysis (FEA), asymmetric tooth stresses were experimentally defined utilizing the optical method of photoelasticity for measuring and visualizing stresses of the loaded asymmetric tooth model made out of transparent material with birefringence properties (Figure 6.1a) [9,14]. Such asymmetric stress definition was very time-consuming and costly. When the FEA became available, it greatly accelerated asymmetric gear research and development. Figure 6.1b,c presents the FEA mesh and stress isograms.

(a) (b) (c)

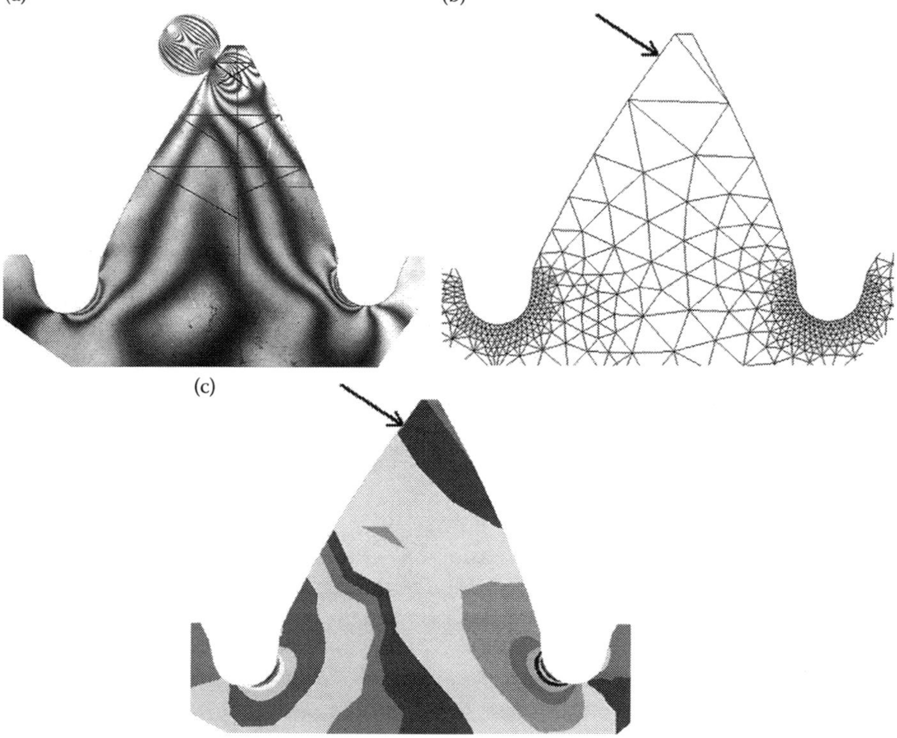

FIGURE 6.1
Asymmetric tooth stress definition; a: photoelastic modeling, b: FEA mesh, c: stress isograms.

Nowadays the FEA and boundary element method (BEM) are commonly used for asymmetric tooth gear stress analysis (see references in Table 1.1).

During the asymmetric tooth flank and root optimization (Chapter 5), the maximum root bending stress was preliminary defined by applying a 100% of the normal tooth load to the highest point of single tooth contact (HPSTC) for the low-medium contact ratio spur gears and about 2/3 of the normal tooth load at the highest point of double tooth contact for the high contact ratio spur gears [13]. A similar approach is used for helical gears, but, in such a case, a stress analysis is done for the virtual spur gear pair that represents a cross section of the helical gear pair that is normal to the tooth line.

After the tooth geometry of mating gears is finalized, the bending and contact stresses are defined considering the tooth deflection under the applied load. In this case, the tooth load value is defined at different contact points of the involute tooth flank. Similar to the effective contact ratio and transmission error calculation procedure (Section 5.1.4) and tooth microgeometry optimization (Chapter 5.4), each angular position of the driven gear relative to the driving gear is iteratively defined by equalizing the sum of the tooth contact load moments of each gear to its applied torque. The related tooth contact loads are also iteratively defined to conform to tooth bending and contact deflections, where the tooth bending deflection in each contact point is determined based on the FEA-calculated flexibility and the tooth contact deflection is calculated by the Hertz equation.

Unlike the low number of flank FEA nodes in preliminary stress calculation (see, for example, Figure 6.1b), a final stress analysis utilizes a much greater number of flank FEA nodes. Figures 6.2 and 6.3 show the FEA mesh and stress isograms for the low-medium, and high contact ratio gears.

The tooth load is consequently applied to each flank FE node from the lowest node near the root fillet to the highest one for the driving pinion and from the highest node to the lowest one near the root fillet for the driven gear. Each load value is defined considering load sharing with the neighboring

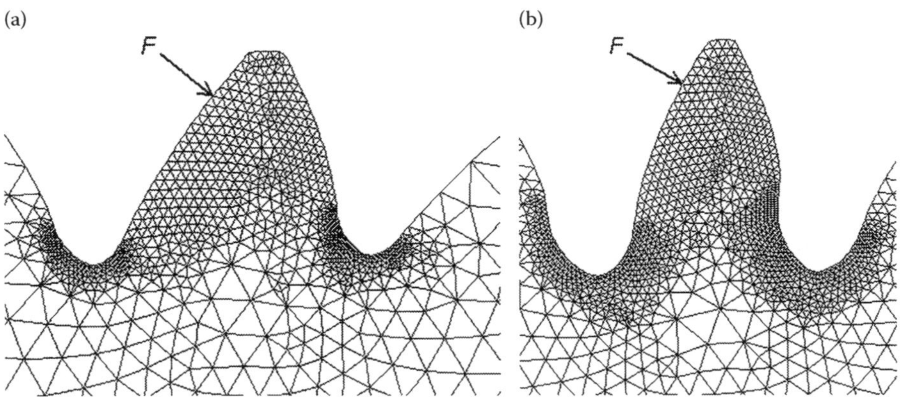

FIGURE 6.2
FEA mesh of the low-medium (a) and high (b) contact ratio gears.

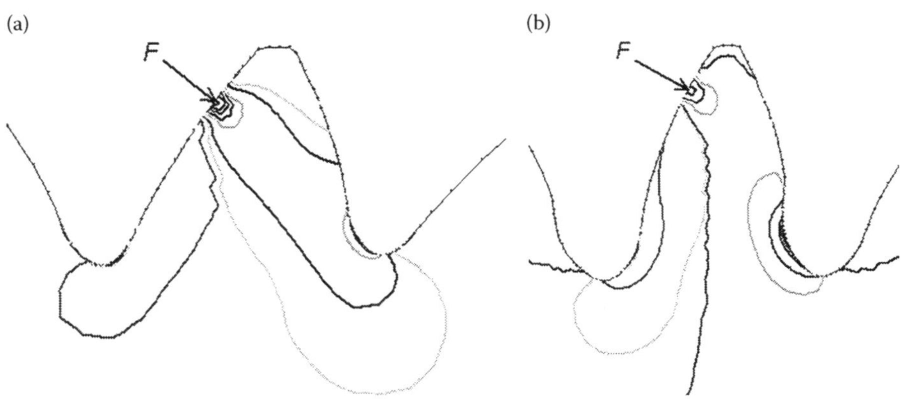

FIGURE 6.3
Stress isograms of the low-medium (a) and high (b) contact ratio gears.

teeth. The typical charts of the load distribution, and tensile bending and contact stresses along the tooth flank low-medium and high contact ratio gears are shown in Figures 6.4–6.6.

The charts in Figures 6.4–6.6 are defined considering the mating gear tooth bending and contact deflections. Manufacturing tolerances, assembly misalignments, other gear drive components' (housing, shafts, bearings, etc.) deflections, and operating conditions (temperature, humidity, etc.) are not considered in this book, although they also affect the gear tooth load and stress distribution.

These nominal static bending and contact stresses cannot be used for asymmetric gear safety factor definition and rating, because although asymmetric tooth gears are known for many years, they are not covered by any gear design and rating standards. This limits their broad implementation for various gear applications, despite potential performance advantages in comparison to symmetric tooth gears. In some industries, like aerospace, which are accustomed to using gears with nonstandard tooth shapes, rating of these gears is established by comprehensive testing [95]. However, such testing programs are not affordable for development of gear drives for other industries that could also benefit from application of asymmetric

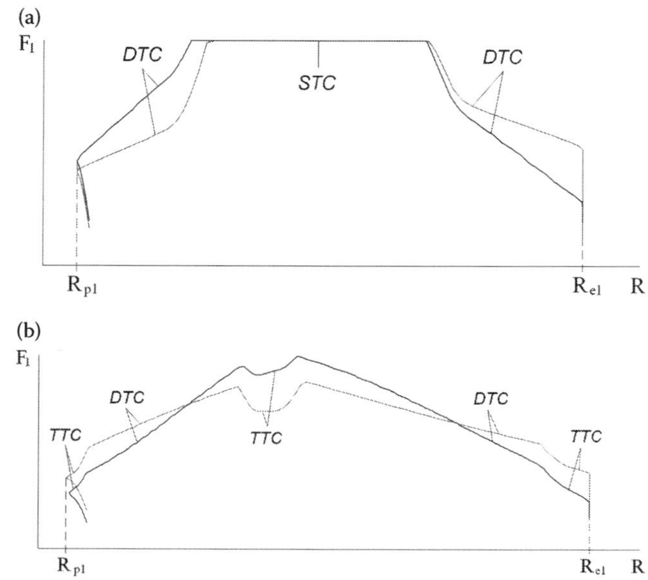

FIGURE 6.4
Normal load distribution along a drive flank of the pinion tooth; a: and low-medium contact ratio gears, b: and high contact ratio gears; think lines: without and thick lines: with optimized parabolic crowning; R_1: pinion radius, R_{p1}: pinion radius at the lowest point of contact near the root fillet, R_{e1}: pinion radius at the highest point of contact near the tooth tip, F_1: normal load applied to one tooth, STS: single tooth pair contact, DTC: double tooth pair contact, TTC: triple tooth pair contact.

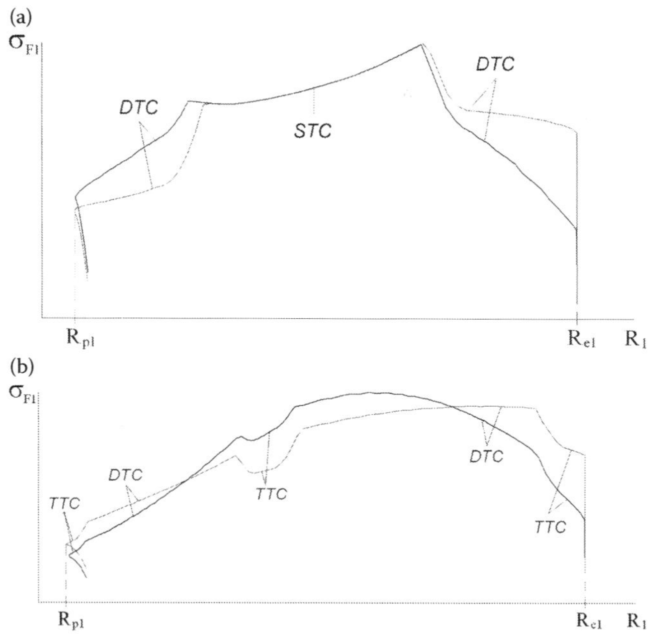

FIGURE 6.5
Tensile bending root stress distribution along a drive flank of the pinion tooth; a: for the low-medium contact ratio gears, b: for the high contact ratio gears; thin line charts: for a gear pair without the drive flank microgeometry optimization; thick line charts: for a gear pair with the drive pinion flank optimized by the parabolic crowning; R_1: pinion radius, R_{p1}: pinion radius at the lowest point of contact near the root fillet, R_{e1}: pinion radius at the highest point of contact near the tooth tip, σ_{F1}: tensile bending root stress, STS: single tooth pair contact, DTC: double tooth pair contact, TTC: triple tooth pair contact.

tooth gears. On the other hand, asymmetric teeth, although nonstandard, also have involute flank profiles like conventional symmetric tooth involute gears. The following paragraphs present an approach that makes it possible to apply the existing gear rating standards to asymmetric tooth gears.

6.2 Rating of Spur Gears

The suggested asymmetric gear rating approach assumes using the comparable symmetric gear pair, whose geometry is based on the predefined asymmetric gear parameters with the intention of evaluating this comparable symmetric gear pair by the existing gear rating standards [145]. Relations between the original asymmetric tooth gear pair and the comparable gear pair geometries should be established by definition of the stress conversion coefficients that are used for safety factors calculation of the asymmetric gear pair.

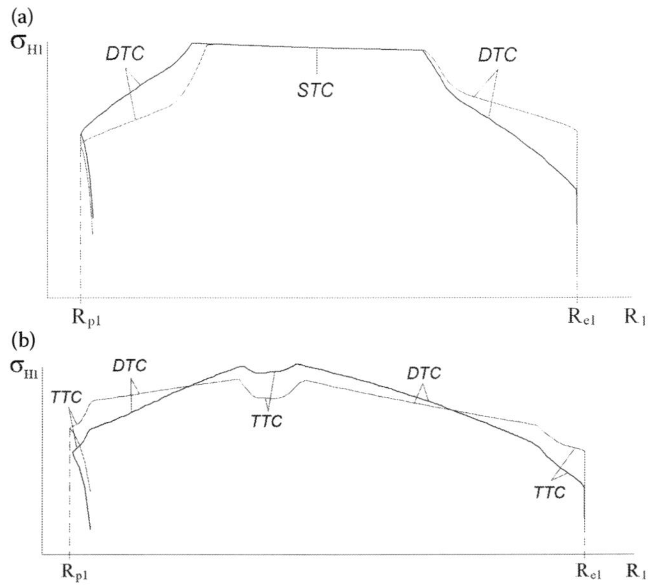

FIGURE 6.6
Contact stress distribution along a drive flank of the pinion tooth; a: and low-medium contact ratio gears, b: and high contact ratio gears; thin line charts: for a gear pair without the drive flank microgeometry optimization; thick line charts: for a gear pair with the drive pinion flank optimized by the parabolic crowning; R_1: pinion radius, R_{p1}: pinion radius at the lowest point of contact near the root fillet, R_{e1}: pinion radius at the highest point of contact near the tooth tip, σ_{C1}: contact stress, STS: single tooth pair contact, DTC: double tooth pair contact, TTC: triple tooth pair contact.

In order to make this comparable symmetric gear pair suitable for the standard rating, its geometry must be traditionally generated by the gear rack parameters and addendum modifications or X-shits (Figure 6.7), which are defined by the following equations:

Generating rack module (for the metric system) or diametral pitch (for the English system)

$$m = m_w = \frac{d_{w1}}{z_1} = \frac{d_{w2}}{z_2} \text{ or } DP = DP_w = \frac{z_1}{d_{w1}} = \frac{z_2}{d_{w2}}, \tag{6.1}$$

where m_w or DP_w is the operating module or operating diametral pitch of the asymmetric gear pair, $d_{w1,2}$ are the pitch diameters of the asymmetric gear pair, $z_{1,2}$ are numbers of teeth.

The rack profile or pressure angle is an average value of the drive α_{wd} and coast α_{wc} pressure angles of the asymmetric gear pair

$$\alpha = \frac{\alpha_{wd} + \alpha_{wc}}{2}. \tag{6.2}$$

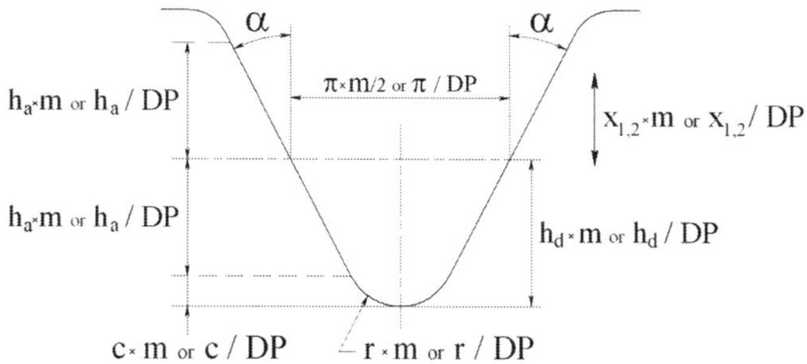

FIGURE 6.7
Gear rack parameters and addendum modification (X-shift) coefficients for the comparable symmetric gears.

The rack addendum coefficient is defined by the tooth tip diameters $d_{a1,2}$ and pitch diameters $d_{w1,2}$ of asymmetric gears:

$$h_a = \frac{(d_{a1} - d_{w1}) \pm (d_{a2} - d_{w2})}{4m} \quad \text{or} \quad h_a = \frac{(d_{a1} - d_{w1}) \pm (d_{a2} - d_{w2})}{4} \times DP, \quad (6.3)$$

where the "+" sign is for the external gears and the "−" sign is for the internal gears.

Gear rack should have a full tip radius to maximize root strength of the comparable symmetric gears. The rack full tip radius coefficient is

$$r = \frac{\pi/4 - h_a \tan\alpha}{\cos\alpha}. \quad (6.4)$$

Radial clearance coefficient is

$$c = r \times (1 - \sin\alpha). \quad (6.5)$$

Rack dedendum coefficient is

$$h_d = h_a + c. \quad (6.6)$$

Addendum modification (X-shift) coefficients are as follows:

$$x_1 = \frac{s_1 - s_2}{4m \times \tan\alpha} \text{ or } x_1 = \frac{s_1 - s_2}{4\tan\alpha} \times DP \text{ and } x_2 = -x_1. \quad (6.7)$$

All other input data for standard design of the comparable symmetric gear pair, such as tolerances, operating parameters (loads, RPM, and lifetime), materials and heat treatment, lubrication, etc. should be exactly the same as for the asymmetric tooth gear pair. Table 6.1 presents an example of asymmetric and comparable symmetric gear pair parameters.

TABLE 6.1

Asymmetric and Comparable Symmetric Gear Pair Parameters

Gear Pair	Asymmetric		Comparable Symmetric	
Design Method	**Direct Gear Design®**		**Rack Generation**	
Gear	**Pinion**	**Gear**	**Pinion**	**Gear**
Number of teeth	20	49	20	49
Module	5.000		5.000	
Pressure angle	35°/20°[a]		27.5°	
Asymmetry factor	1.147		1.0	
Pitch diameter (PD)	100.000	295.000	100.000	295.000
Base diameter	81.915/93.969[a]	200.692/230.225[a]	88.701	217.318
Tooth thickness at PD	8.168	7.540	8.168	7.540
Center distance	172.500		172.500	
Gear rack angle	–		27.5°	
Addendum coefficient	–		0.951	
Root radius coefficient	–		0.327	
Root clearance coefficient	–		0.176	
Dedendum coefficient	–		1.127	
X-shift coefficient	–	–	0.060	–0.060
Tip diameter	109.802	254.214	110.110	253.910
Root diameter	89.080	233.597	89.360	233.141
Root fillet profile	optimized	optimized	trochoidal	trochoidal
Face width	30.00	27.00	30.00	27.00
Contact ratio	1.20/1.55[a]		1.31	

[a] Drive/coast flanks.

The overlays of the asymmetric and comparable symmetric gears are shown in Figure 6.8.

The next step is calculation of the bending and contact stress conversion coefficients that reflect differences in the asymmetric and comparable symmetric gear geometry.

The bending stress conversion coefficients are defined as ratios of the maximum static tensile root stresses for the pinion and gear of the comparable symmetric and the analyzed asymmetric gear pairs:

$$C_{F1,2} = \frac{\sigma_{F\max(sym)1,2}}{\sigma_{F\max(asym)1,2}}, \tag{6.8}$$

where $\sigma_{F\max(sym)1,2}$ are the maximum static tensile root stresses for a pinion and gear of the comparable symmetric gear pair, and $\sigma_{F\max(asym)1,2}$ are the maximum static tensile root stresses for a pinion and gear of the asymmetric gear pair.

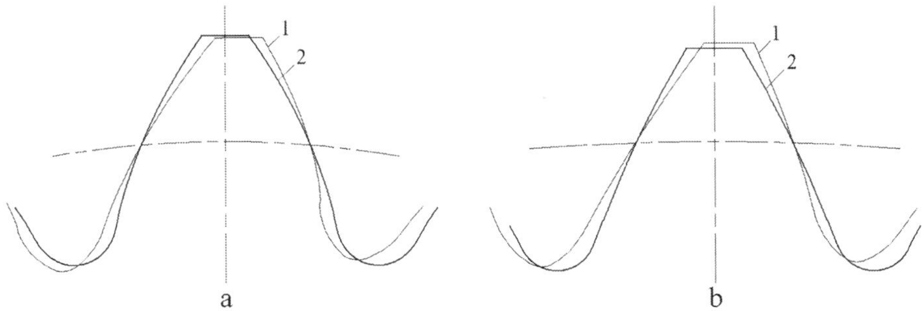

FIGURE 6.8
Overlays of the asymmetric and comparable symmetric gears; a: pinion, b: gear; 1: asymmetric gears, 2: comparable symmetric gears.

These maximum static tensile root stresses are calculated by FEA under the same load. If both gear pairs have a low-medium contact ratio (< 2.0), the load is applied to the HPSTC. If both gear pairs have a high contact ratio (≥ 2.0), the load is applied to the highest point of the double tooth contact. It is a rare probability when an asymmetric gear pair has a medium contact ratio near 2.0, but a comparable symmetric gear pair has a high contact ratio. It this case, a comparable symmetric gear pair needs a minor adjustment (for example, a little increase in a pressure angle) to reduce the contact ratio to below than 2.0.

The contact stress conversion coefficient is defined as the ratio of the maximum Hertz contact stresses of the comparable symmetric and the analyzed asymmetric gear pairs:

$$C_H = \frac{\sigma_{H\max(\text{sym})}}{\sigma_{H\max(\text{asym})}}, \tag{6.9}$$

where $\sigma_{H\max(\text{sym})}$ is the maximum Hertz contact stress for a pinion and gear of the comparable symmetric gear pair, $\sigma_{H\max(\text{asym})}$ is the maximum Hertz contact stress for a pinion and gear of the asymmetric gear pair.

Rating of involute gears with symmetric teeth is established by the rating standards, for example [175] or [185]. In order to apply these rating standards to the asymmetric tooth gears, the bending and contact safety factors that are defined for the comparable symmetric tooth gears by these standards should be multiplied by the contact and bending conversion coefficients accordingly. Then the bending safety factors of asymmetric tooth gears are as follows:

$$S_{F(\text{asym})1,2} = C_{F1,2} \times S_{F(\text{sym})1,2}, \tag{6.10}$$

where $S_{F(\text{sym})1,2}$ are the root bending safety factors of comparable symmetric tooth gears defined by a rating standard.

TABLE 6.2

Asymmetric and Comparable Symmetric Tooth Gear Stress Analysis

Gear Pair	Asymmetric		Comparable Symmetric	
Design Method	Direct Gear Design®		Rack Generation	
Gear	Pinion	Gear	Pinion	Gear
Number of teeth	20	49	20	49
Module	5.000		5.000	
Pressure angle	35°/20°[a]		27.5°	
Face width	30.00	27.00	30.00	27.00
Torque, Nm	900	2,205	900	2,205
RPM	1,000	408	1,000	408
Service life, h	2,000		2,000	
Material type	Carburized, case harden steel, AISI 8620			
Bending stress, MPa	–	–	448[b]	480[b]
Bending stress (FEA), MPa	276	277	309	334
Bending stress conversion Coefficients (FEA)	1.120	1.206	–	–
Contact stress, MPa	–	–	1,507[b]	1,407[b]
Contact stress (hertz), MPa	1,257		1,349	
Contact stress conversion coefficient (hertz)	1.073		–	
Bending safety factors	1.90	1.95	1.70[b]	1.62[b]
Contact safety factors	1.02	1.12	0.95[b]	1.04[b]

[a] Drive/coast flanks.
[b] Calculation method: per ISO 6336 standard.

The contact safety factor of asymmetric tooth gears is

$$S_{H(\text{asym})} = C_H \times S_{H(\text{sym})}, \tag{6.11}$$

where $S_{H(\text{sym})}$ is the flank contact (pitting) safety factor of comparable symmetric tooth gears defined by a rating standard.

An example of the asymmetric and comparable symmetric tooth gear stress analysis results is presented in Table 6.2.

6.3 Rating of Helical Gears

Rating of helical asymmetric gears [159] by the existing rating standards employs the same approach as for the spur asymmetric gears. The standard design of helical symmetric gears is based on parameters of the virtual spur gears that represent normal sections of the helical gears. Accordingly, all parameters of the generating rack of comparable helical symmetric gears can be defined by Equations (6.1)–(6.7) for the virtual spur gears in the normal section.

Table 6.3 presents an example of helical asymmetric and comparable symmetric gear pair parameters.

The teeth of the helical asymmetric gears are shown in Figure 6.9.

The teeth of the comparable symmetric gears are shown in Figure 6.10.

An example of the helical asymmetric and comparable symmetric tooth gear stress analysis results is presented in Table 6.4.

It is important to understand that the described comparable symmetric gear pairs represent virtual gears that are designed based on asymmetric

TABLE 6.3

Helical Asymmetric and Comparable Symmetric Gear Pair Parameters

Gear Pair	Asymmetric		Comparable Symmetric	
Design method	Direct gear design®		Rack generation	
Number of teeth	17	23	17	23
Normal module	4.000		4.000	
Normal pressure angle	35°/18°*		26.5°	
Asymmetry factor	1.179		1.0	
Helix angle	20°		20°	
Lead of helix	624.6074	845.057	624.6074	845.057
Helix hand	right	left	right	left
Pitch diameter (PD)	72.364	97.904	72.364	97.904
Base diameter	58.026/68.391[a]	78.506/92.529[a]	63.924	86.485
Normal tooth thickness at PD	6.390	6.176	6.390	6.176
Center distance	85.134		85.134	
Normal gear rack angle	–		26.5°	
Addendum coefficient	–		1.051	
Root radius coefficient	–		0.292	
Root clearance coefficient	–		0.162	
Dedendum coefficient	–		1.213	
X-shift coefficient	–	–	0.025	–0.025
Tip diameter	80.845	106.236	80.985	106.100
Root diameter	62.967	88.318	62.873	87.988
Root fillet profile	optimized	optimized	trochoidal	trochoidal
Face width	40.00	37.00	40.00	37.00
Transverse contact ratio	1.20/1.54[a]		1.32	
Axial contact ratio	1.01		1.01	
Total contact ratio	2.21/2.55		2.33	

[a] Drive/coast flanks.

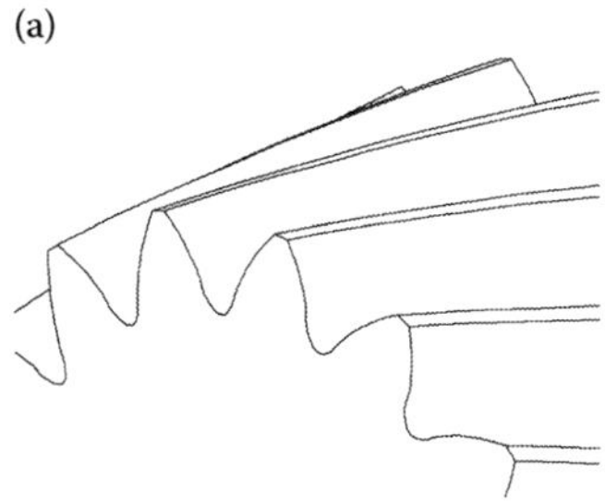

FIGURE 6.9
Helical asymmetric teeth; a: pinion, b: gear.

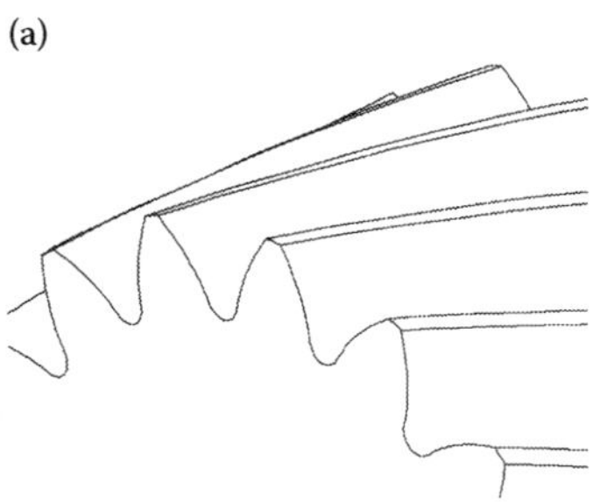

FIGURE 6.10
Comparable symmetric teeth; a: pinion, b: gear.

TABLE 6.4

Helical Asymmetric and Comparable Symmetric Tooth Gear Stress Analysis

Gear Pair	Asymmetric		Comparable Symmetric	
Design Method	Direct Gear Design®		Rack Generation	
Gear	Pinion	Gear	Pinion	Gear
Number of teeth	17	23	17	23
Module	4.000		4.000	
Normal pressure angle	35°/18°[a]		26.5°	
Helix angle	20°		20°	
Face width	40.00	37.00	40.00	37.00
Torque, Nm	700	947	700	947
RPM	1,000	739	1,000	739
Service life, h	2,000		2,000	
Material type	Carburized, case harden steel, AISI 8620			
Bending stress, MPa	–	–	369[b]	392[b]

(*Continued*)

TABLE 6.4 (*Continued*)

Helical Asymmetric and Comparable Symmetric Tooth Gear Stress Analysis

Gear Pair	Asymmetric		Comparable Symmetric	
Design Method	Direct Gear Design®		Rack Generation	
Gear	Pinion	Gear	Pinion	Gear
Bending stress (FEA), MPa	271	285	298	315
Bending stress conversion coefficients (FEA)	1.100	1.105	–	–
Contact stress, MPa	–	–	1,485[b]	1,485[b]
Contact stress (hertz), MPa	1,282		1,340	
Contact stress conversion coefficients (hertz)	1.045		–	
Bending safety factors	2.64	2.51	2.40[b]	2.27[b]
Contact safety factors	1.06	1.07	1.01[b]	1.02[b]

[a] Drive/coast flanks.

[b] Calculation method: per ISO 6336 standard.

gear parameters with a sole purpose to rate asymmetric gear pairs utilizing the existing rating standards for symmetric involute gears. Differences in safety factors of the asymmetric and comparable symmetric gear pairs must not be used to draw conclusions about potential load capacity and performance of asymmetric gear pairs. For this purpose, the asymmetric gears should be compared to the existing benchmark symmetric gears that are considered as the best gear solutions for a particular application.

7

Special Types of Asymmetric Gears

7.1 Self-Locking Asymmetric Gears

In most gear drives, when the driving torque applied to the input shaft is abruptly reduced to zero as a result of, for example, a motor shutdown or a power outage, the gears rotate either in the same direction driven by the system inertia or in the opposite direction driven by the resistant load, applied to the output shaft. The latter condition is known as backdriving. During the inertial motion or backdriving, the driven output shaft becomes the driving one and the driving input shaft becomes the driven one. There are many gear drive applications where such operating mode is not acceptable. In order to prevent it, different types of brake or clutch devices are used. However, there are solutions that prevent inertial motion or backdriving using self-locking gears without any additional devices. The most common one is a worm gear with a low lead angle. In a self-locking worm gear drive, the torque applied to the worm gear is blocked by friction force preventing its rotation. However, the worm gear drives have some limitations. They require a crossed axes shaft arrangement and relatively high gear ratio. They also have relatively low efficiency.

There are parallel axis self-locking gears that, unlike worm gears, can utilize practically any gear ratio. They have a driving mode and a self-locking mode when backdriving or inertial torque is applied to the output shaft. Driving mode efficiency of the parallel axes self-locking gears is substantially higher than that of the worm self-locking gears; it could reach 80%–85% [105]. Most studies [186–189] were done on the self-locking gears with symmetric teeth. Asymmetric self-locking gears were explored in [53,61].

Figure 7.1 shows a difference between conventional and self-locking asymmetric gears operating in driving and backdriving modes. Both gear pairs have the driving flank pressure angle α_{wd} significantly greater than a pressure angle of the opposite coast flank, which is not used in the gear drive operation at all. The high pressure angle drive flanks are engaged in both driving and backdriving modes.

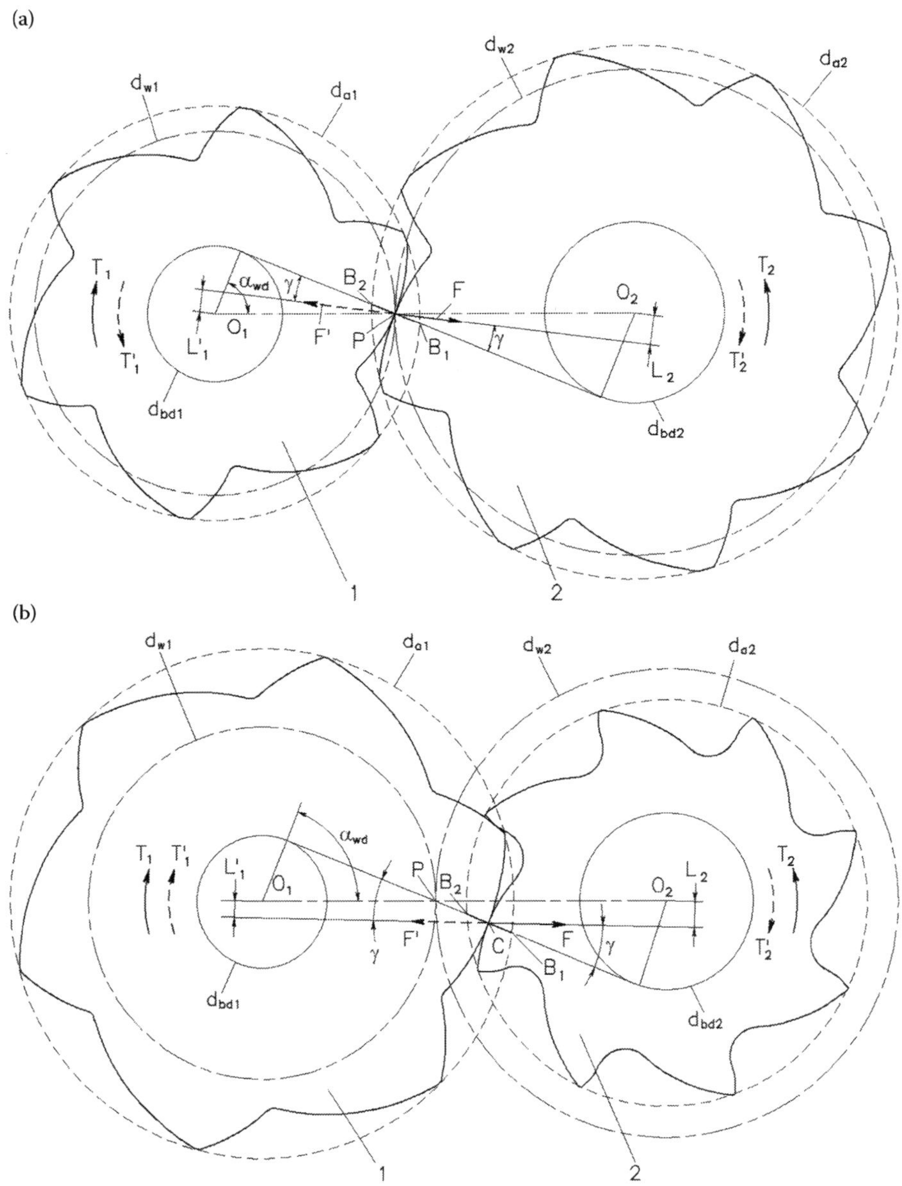

FIGURE 7.1
Conventional (a) and self-locking (b) asymmetric gears operating at the driving and backdriving modes; 1: pinion, 2: gear.

A conventional gear pair has a pitch point P located between the points B_1 and B_2 of the beginning and end of the line of contact. In driving mode, pinion 1 with a center O_1 transmits the clockwise torque T_1 to the driven gear 2. A tooth of pinion 1 touches a tooth of gear 2 in the contact point (in Figure 7.1a it is shown coinciding with the pitch point P) located on the line of contact B_1B_2. The driving contact force F is deflected by the friction angle γ from the line of contact B_1B_2. The lever L_2 is the distance from the gear center O_2 to the driving contact force F direction line. Then the counterclockwise gear torque $T_2 = F \times L_2$.

In backdriving mode, the gears rotate in an opposite direction to that of the driving mode. Gear 2 becomes driving and transmits the clockwise backdriving torque T'_2 to pinion 1 that becomes driven. The backdriving contact force F' is deflected by the friction angle γ from the line of contact B_1B_2. The lever L'_1 is the distance from the pinion center O_1 to the backdriving contact force F′ direction line that lies above the center O_1 in Figure 7.1a. Then the counterclockwise pinion torque $T'_1 = F' \times L_1$. As a result, the self-locking does not occur and gears are rotating freely by the torque T'_2 applied to gear 2.

The self-locking gear pair (Figure 7.1b) is designed in order to move the line of contact B_1B_2 completely beyond the pitch point P. This makes the driving pinion 1 larger in size with a root diameter greater than its pitch diameter d_{w1}, but the driven gear 2 becomes smaller with the outer diameter d_{a2} smaller than its pitch diameter d_{w2}. In driving mode, a self-locking gear pair operates exactly the same way as a conventional one. Although, because the line of contact B_1B_2 is completely shifted beyond the pitch point P, a self-locking gear pair has greater tooth flank sliding velocities, resulting in the lower gear mesh efficiency compared to the conventional gear pair.

In backdriving mode, gear 2 of the self-locking gear pair becomes driving and transmits the clockwise torque T'_2 to pinion 1. The backdriving contact force F' is deflected by the friction angle γ from the line of contact B_1B_2. However, unlike a conventional gear pair, a direction line of the backdriving contact force F' and the lever L'_1 lie below the center O_1 (see Figure 7.1b). Then the pinion torque $T'_1 = F' \times L_1$ is directed clockwise against the backdriving gear torque T'_2, locking the gear pair rotation.

Figure 7.2 compares the conventional and self-locking asymmetric gears operating in driving and inertial modes. Such gear drives operate differently than the ones that work in backdriving mode. In this case the drive tooth flank pressure angle α_{wd} is significantly lower than the coast flank pressure angle α_{wc} of the opposite tooth flanks, which are engaged during the inertial mode. The base diameters $d_{bd1,2}$ of the drive flanks and $d_{bc1,2}$ of the cost flanks of pinion 1 and gear 2 are shown in Figure 7.2.

In driving mode, the conventional gear pair (Figure 7.2a) operates exactly the same way as the conventional gear pair in Figure 7.1a, except the driving flanks have the low pressure angle α_{wd}. It has a pitch point P located between the points B_1 and B_2 of the beginning and end of the line of contact. The driving pinion 1 with a center O_1 transmits the clockwise torque T_1 to the driven gear 2. A tooth of pinion 1 touches a tooth of gear 2 in the contact point (in Figure 7.2a it is shown coinciding with the pitch point P) located on the line of contact B_1B_2. The driving contact force F is deflected by the friction angle γ

from the line of contact B_1B_2. The lever L_2 is the distance from the gear center O_2 to the driving contact force F direction line. Then the counterclockwise gear torque $T_2 = F \times L_2$.

In inertial mode, gear 2 becomes driving and transmits the counterclockwise inertial torque T'_2 to pinion 1 that becomes driven. The gears remain rotating in the same direction as in driving mode. In this inertial mode the opposite high pressure angle coast flanks along the line of contact $B'_1B'_2$. The inertial contact force F' is deflected by the friction angle γ from the line of contact $B'_1B'_2$. The lever L'_1 is the distance from the pinion center O_1 to the inertial contact force F' that lies below the center O_1 in Figure 7.2a. Then the clockwise pinion torque $T'_1 = F' \times L_1$. As a result, the self-locking does not occur and gears rotate freely by the torque T'_2 applied to gear 2.

Similar to the self-locking gear pair preventing the backdriving (Figure 7.1b), the self-locking gear pair preventing the inertial driving (Figure 7.2b) is designed to have the lines of contact B_1B_2 and $B'_1B'_2$ completely beyond the pitch point P. This makes the driving pinion 1 much larger in size with a root diameter greater than its pitch diameter d_{w1}, but the driven gear 2 becomes much smaller with the outer diameter d_{a2} smaller than its pitch diameter d_{w2}.

In driving mode, the self-locking gear pair operates in exactly the same way as a conventional one. Although, because the line of contact B_1B_2 is completely shifted beyond the pitch point P, a self-locking gear pair has greater tooth flank sliding velocities, resulting in lower gear mesh efficiency compared to a conventional gear pair.

In inertial mode, gear 2 of the self-locking gear pair becomes driving and transmits the clockwise torque T'_2 to pinion 1. The backdriving contact force F' is deflected by the friction angle γ from the line of contact $B'_1B'_2$. However, unlike the conventional gear pair, a direction line of the backdriving contact force F' and the lever L'_1 lie above the center O_1 (see Figure 7.2b). Then the pinion torque $T'_1 = F' \times L_1$ is directed counterclockwise against the inertial gear torque T'_2, locking gear pair rotation.

Application of asymmetric tooth shape for the "anti-inertial" self-locking gears with a low drive flank pressure angle allows an increase in gear mesh efficiency in driving mode and significantly reduces the bearing radial load compared to the symmetric "anti-inertial" self-locking gears with a very high drive flank pressure angle.

It should be understood that despite the benefits of asymmetric gear application, the "anti-backdriving" asymmetric self-locking gears will not stop a gear drive rotation, if the inertial driving condition occurs, as well as the "anti-inertial" asymmetric self-locking gears will not stop a gear drive rotation, if the backdriving driving condition occurs. It is possible that some self-locking gear applications may require a self-locking operation mode against both backdriving (for example, at low speed) and inertia (for example, at high speed). In such cases, application of the symmetric self-locking gear may need to be considered or the asymmetric self-locking gears should be optimized based on actual loads and number of cycles in each operating condition.

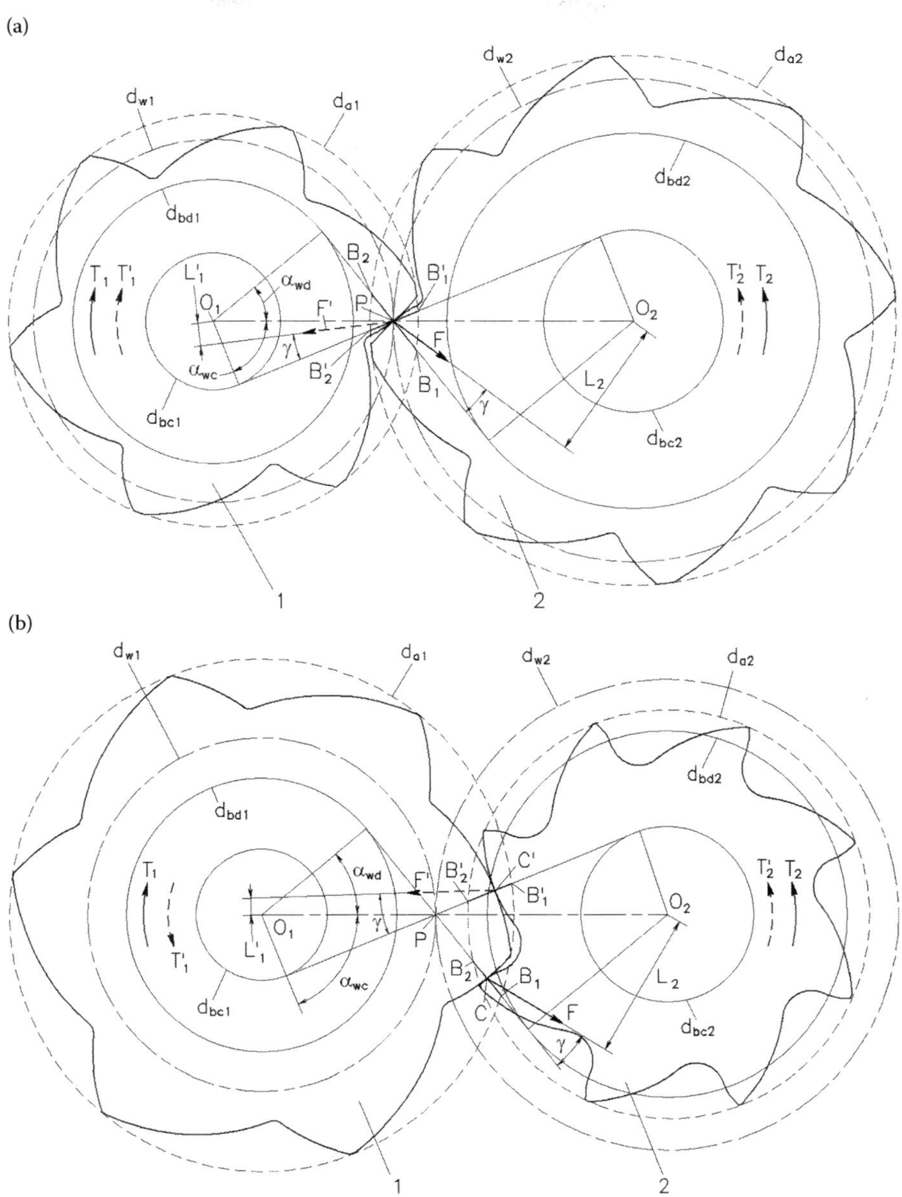

FIGURE 7.2
Conventional (a) and self-locking (b) asymmetric gears operating at the driving and inertial modes; 1: pinion, 2: gear.

There are two necessary requirements for a reliable self-locking:

- locating the lines of contact B_1B_2 and $B'_1B'_2$ completely beyond the pitch point P, enlarging the pinion and reducing the gear size;
- sufficiently large friction angle γ to deflect the force F' beyond the center of the pinion O_1 and create the resisting self-locking moment (torque) $T'_1 = F' \times L'_1$.

The second condition can be presented as $L'_{1\min} > 0$ or

$$\gamma > \arctan\left[\frac{1}{(1+u)\times\tan\alpha_{wsl} - u\times\tan\alpha_{e2}}\right] \tag{7.1}$$

or

$$f > \frac{1}{(1+u)\times\tan\alpha_{wsl} - u\times\tan\alpha_{e2}}, \tag{7.2}$$

where f is the average friction coefficient; $u = z_2/z_1$ is the gear ratio, z_1 and z_2 are the pinion and gear number of teeth; α_{wsl} is the pressure angle at the locking tooth flank. It is equal to α_{wd} for the self-locking gears preventing a backdriving and it is equal to α_{wc} for the self-locking gears preventing an inertial driving; α_{e2} is the effective involute angle at the tooth tip (see Equations 2.19, 2.20) It is equal to α_{ed2} for the self-locking gears preventing a backdriving and α_{ec2} for the self-locking gears preventing an inertial driving.

Self-locking gears are custom and suitable for using Direct Gear Design® to define their gear tooth geometry [107]. Self-locking conditions (7.1) and (7.2) require a high pressure angle and high sliding friction in the tooth contact. If the average sliding friction coefficient is $f = 0.1$–0.3, it requires the transverse operating pressure angle to be $\alpha_w = 75°$–$85°$. For the asymmetric self-locking gears this results in a low transverse contact ratio $\varepsilon_{ad} < 1.0$. Lack of a transverse contact ratio should be compensated by an overlap or face contact ratio ε_β to guarantee the total contact ratio $\varepsilon_{\gamma d} = \varepsilon_{ad} + \varepsilon_\beta \geq 1.0$ (see Section 2.3.5). This can be achieved by using helical gears (Figure 7.3a) with a high helix angle (65°–75° or more). Such a high helix angle produces substantial axial thrust load that must be supported.

Double helical (or herringbone) gears (Figure 7.3b) allow this thrust load to be neutralized.

Besides the external asymmetric self-locking gear arrangements shown in Figures 7.1 and 7.2, the internal (Figure 7.4) and planetary (Figure 7.5) asymmetric self-locking gears are also possible.

Another axial force free self-locking gear design option is application of laminated spur gears (Figure 7.6). They have several layers assembled with an angular shift. These layers are engaged in gear mesh with the mating

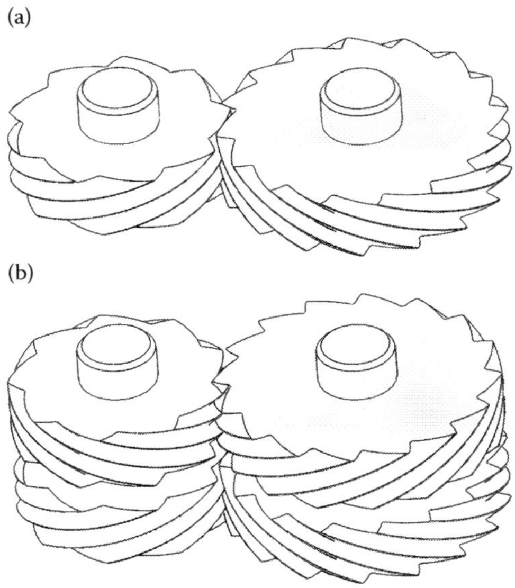

FIGURE 7.3
Helical (a) and double helical (b) self-locking gears.

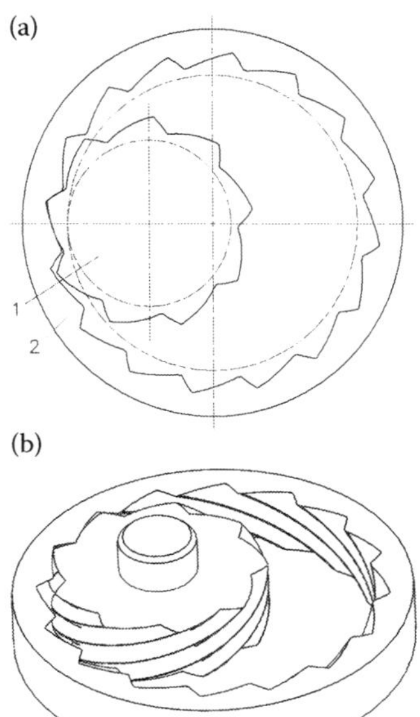

FIGURE 7.4
Internal self-locking gear stage; (a) arrangement (1: pinion, 2: gear), (b) assembly model.

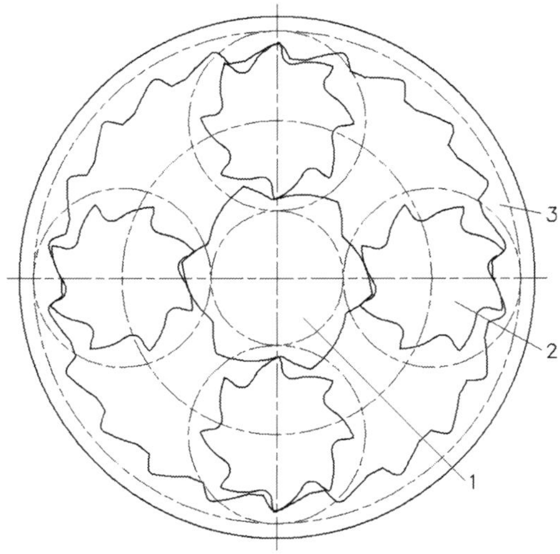

FIGURE 7.5
Planetary self-locking gear arrangement; 1: sun gear, 2: planet gears, 3: ring gear.

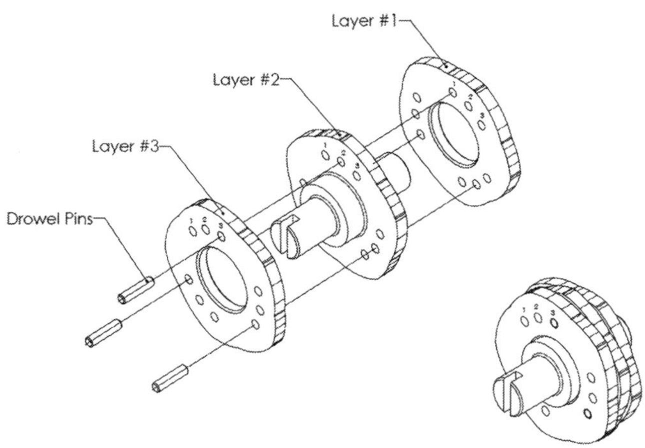

FIGURE 7.6
Laminated self-locking gear.

gear layers consecutively, providing a necessary overlap or face contact ratio, like in helical gears, but without axial force. A minimum number of layers depends on the type of gear operation. In an "anti-backdriving" self-locking gear stage, the driving high pressure angle flanks have a relatively high contact ratio $0.5 \ll \varepsilon_{\alpha d} < 1.0$ and two layers are sufficient to have a total contact ratio $\varepsilon_{\gamma d} > 1.0$. In an "anti-inertial" self-locking gear stage, the driving low pressure angle flanks have a low contact ratio $\varepsilon_{\alpha d} < 0.5$ and three or more layers are required. Since these gears are engaged in mesh, one layer at a time, every layer face width must provide sufficient tooth surface endurance in both driving and self-locking modes. This is especially important considering very high transverse pressure angles resulting in increased bearing loads that could be up to ten times greater than for conventional 20° pressure angle gears with the same driving torque. Bearing selection, and shafts and gearbox housing design should be done accordingly to support this increased load.

An example of the asymmetric "anti-backdriving" self-locking gear data is presented in Table 7.1.

Initially, self-locking gears were used in the textile industry [187]. However, this type of gears has many potential applications like, for example, in lifting mechanisms, and other gear drives where the backdriving or inertial driving is not permissible. One such potential application of the self-locking gears is for a continuously variable valve lift system in an automotive engine [190].

TABLE 7.1

Asymmetric "Anti-Backdriving" Self-Locking Gear Data

Gear	Pinion		Gear
Number of teeth	7		11
Normal module, mm	1.000		1.000
Normal pressure angle		70°/20°[a]	
Helix angle		75°	
Hand of helix	right		left
Helix lead, mm	22.767		35.777
Transverse pressure angle		84.62°/54.58°[a]	
Pitch diameter (PD), mm	27.046		42.501
Base diameter, mm	2.537/15.674		3.986/24.63
Tooth tip diameter, mm	29.514		41.968
Root diameter, mm	27.28		39.73
Normal tooth thickness at PD, mm	3.770		–0.620
Center distance, mm		34.773	
Face width, mm	6.5		6.5
Drive flank transverse contact ratio		0.85	
Axial contact ratio		2.0	
Drive flank total contact ratio		2.85	
Drive flank efficiency		80%	
Torque, Nm	0.60		0.75
Drive flank tangent load, N		44.4	
Drive flank bearing radial load, N		473	
Drive flank axial thrust load, N		166	

[a] Drive/coast flank.

7.2 Spiral Face Gears

Face gears typically have a straight or skewed tooth line and a varying normal section tooth profile along the tooth line. These face gears are engaged with spur or helical involute pinions at intersecting or crossed axes. Such face gears with asymmetric teeth are described in [39].

This section presents spiral face gears with an involute tooth line and an identical normal section tooth profile along the involute tooth line [136]. These spiral face gears can be engaged with the involute helical pinion or involute worm at the intersecting or crossed axes. Such gear engagement is used in the Helicon® type gears. Advantages of the spiral face gears include increased power transmission density and possibility of utilizing a highly productive hobbing method to fabricate them.

The geometry of an involute spiral face gear with asymmetric teeth is explained in Figure 7.7.

Spiral angle ϕ at the reference diameter $d > d_b$ is

$$\phi = \arccos\left(\frac{d_b}{d}\right). \tag{7.3}$$

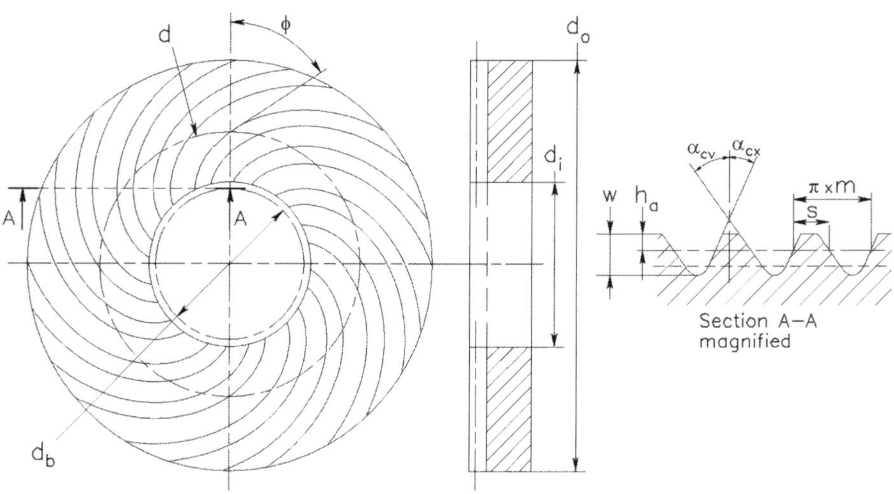

FIGURE 7.7
Face involute spiral gear; d_b, d_i, d, and d_o are base, minor, reference and major diameters, ϕ is spiral angle at reference diameter, m: module, α_{cc} and α_{cv}: concave and convex tooth flank profile (pressure) angles, s: tooth thickness at reference or pitch line, h_a: tooth addendum, w: tooth whole depth. (Kapelevich A.L. and S.D. Korosec. Involute Spiral Face Coupling and Gears: Design Approach and Manufacturing Technique, Gear Solutions, February 2015, 33–37. With permission.)

Unlike conventional spur or helical gears that have involute tooth flank profiles and a straight or helical tooth line, these spiral face gears have the straight tooth flank profiles in the normal section and the involute tooth line. Any normal cross-sectional plane to the involute tooth line of the spiral face gear is tangent to the base cylinder diameter d_b and presents a straight flank gear rack (Figure 7.7, section A-A). This gear rack can be generated by hobbing. Schematics of the spiral face gear hobbing is shown in Figure 7.8.

The centerline offset distance of the generating worm is

$$a = \frac{d_b}{2\cos\phi} \times \cos(\phi \pm \gamma), \tag{7.4}$$

where γ is the generating worm lead angle; the "+" sign is used if the spiral face gear and generating worm have the opposite left-right (Figure 7.8a) or right-left hands, and the "–" sign is used if the spiral face gear and generating worm have the same right-right (Figure 7.8b) or left-left hands.

A normal section of the spiral gear tooth profile is an impression of the hob tooth profile in the normal section (Figure 7.8, section A-A).

Figure 7.9 presents the engagement of the generating worm and spiral face gear.

With the purpose of avoiding interference and undercut of the concave tooth flank the minimal gear flank curvature radius r_g must be greater than the maximum radius r_c of the generating worm helical surface measured parallel to its axis (see Figure 7.8a, section B-B) from base diameter d_b to the generating worm centerline offset by distance "a" or

$$r_{g\,\min} > r_{c\,\max}. \tag{7.5}$$

The minimal gear flank curvature radius $r_{g\,\min}$ at the minor diameter d_i is

$$r_{g\,\min} = \frac{d_i}{2} \times \tan\left(\arccos\left(\frac{d_b}{d_i}\right)\right). \tag{7.6}$$

The profile of the generating worm helical surface section that is parallel to its axis presents a complex curve and providing an exact definition of its maximum radius $r_{c\,\max}$ at a contact point with the concave flank of the spiral face gear is a difficult task. However, it could be defined with sufficient for a practical solution accuracy by the equation

$$r_{c\,\max} = \frac{d_c}{2\tan\alpha_{cc}}. \tag{7.7}$$

A combination of a high concave tooth flank profile (pressure) angle α_{cc} and a small generating worm major diameter d_c allows the small spiral face gear minor diameter to be achieved without a concave flank undercut.

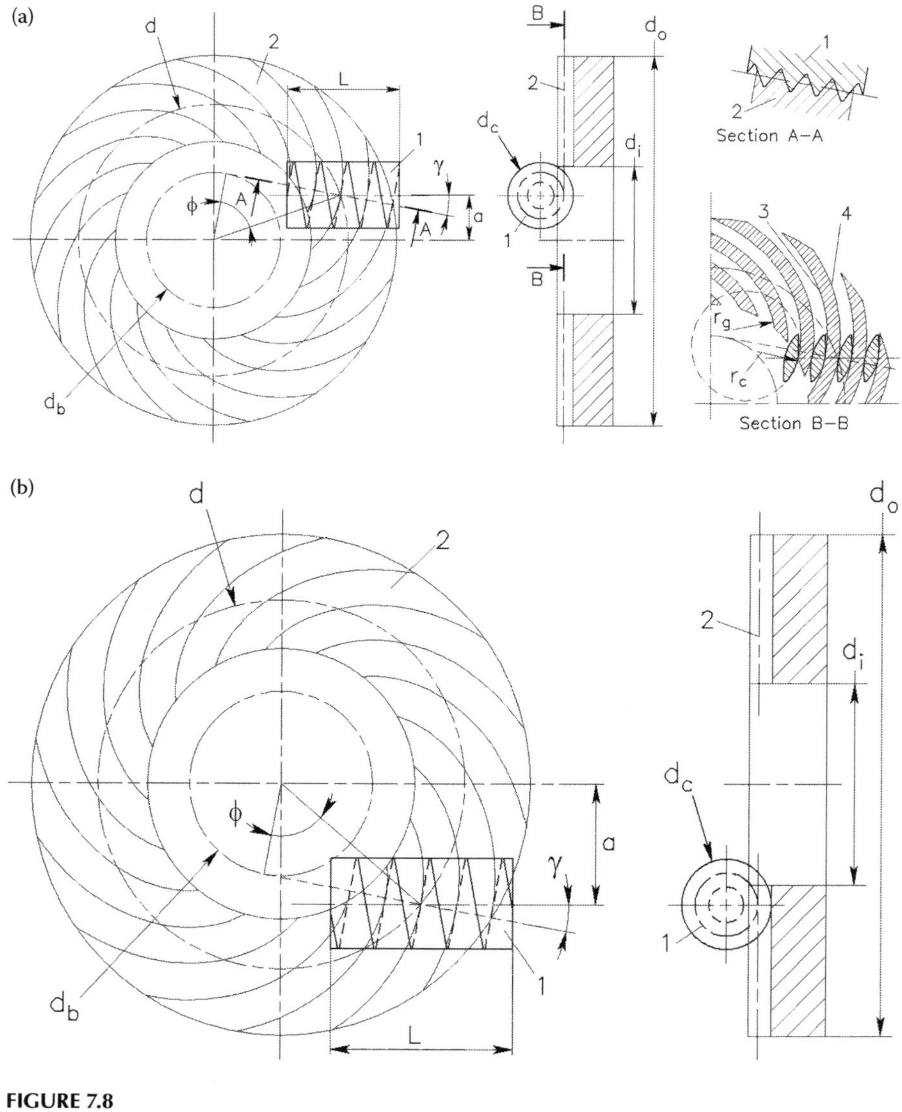

FIGURE 7.8
Schematics of the left (a) and right (b) hand spiral face gear generating by the right-hand worm (thick lines); 1: generating worm, 2: spiral face gear, γ: generating worm lead angle, L: generating worm length, d_c: generating worm major diameter, d_i: minor diameter of the spiral face, a: generating worm centerline offset distance; section A-A is tangent to the base circle d_b and normal to the gear tooth line; section B-B is parallel to the generating worm axis, 3: concave gear tooth flank, 4: convex gear tooth flank. (Kapelevich A.L. and S.D. Korosec. Involute Spiral Face Coupling and Gears: Design Approach and Manufacturing Technique, Gear Solutions, February 2015, 33–37. With permission.)

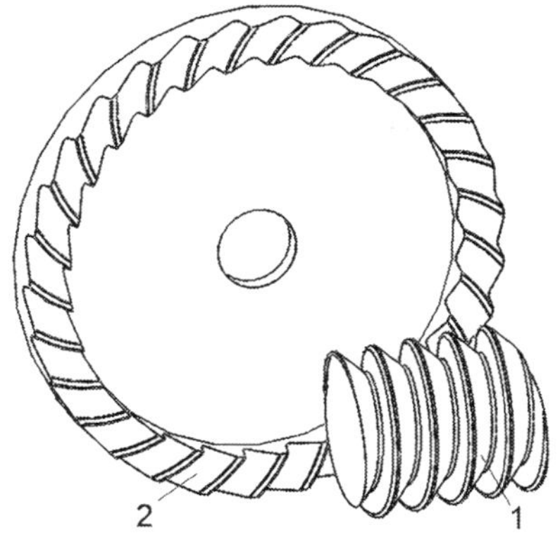

FIGURE 7.9
Engagement of the generating worm 1 and spiral face gear 2.

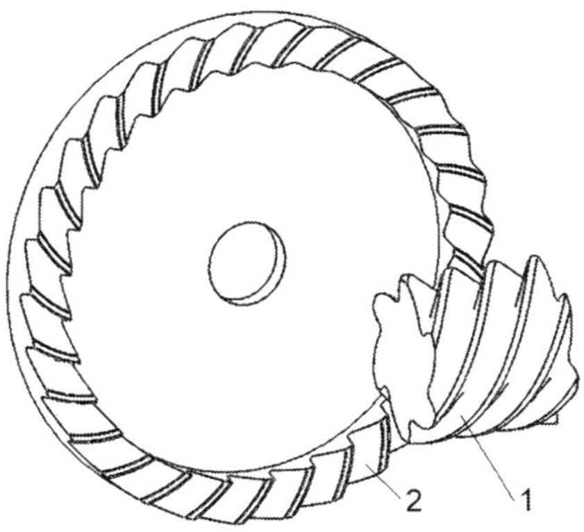

FIGURE 7.10
Engagement of the generating involute helical pinion 1 and spiral face gear 2.

Figure 7.10 presents the engagement of the involute helical pinion and spiral face gear.

In order to avoid an interference of the helical pinion tooth (or the worm thread) tip with the spiral face gear tooth tip at its concave flank, the pinion and spiral gear geometry must satisfy condition (7.5). This condition requires a high pressure angle in mesh with the spiral face gear tooth concave flank and the mating helical pinion flank. At the same time the effective tooth height in the gear mesh must provide a minimal contact ratio greater than 1.0. In most cases both these conditions cannot be satisfied with a symmetric tooth profile in the normal section of the spiral face gear. Application of asymmetric gear teeth with the high pressure angle concave flanks and the low pressure angle

FIGURE 7.11
Spiral face gear pair. (Kapelevich A.L. and S.D. Korosec. Involute Spiral Face Coupling and Gears: Design Approach and Manufacturing Technique, Gear Solutions, February 2015, 33–37. With permission.)

convex flanks of the spiral face gear allows a minimal contact ratio greater than 1.0 to be achieved in both tooth flank meshes. The load capacity of the concave flanks of the spiral face gear is much greater than the convex ones for two reasons, because of a greater pressure angle and a larger relative contact radius in the convex-concave engagement of the helical pinion and spiral face gear. Accordingly, the convex-concave engagement with a high pressure angle should be used for the drive flanks of the unidirectional spiral face gears.

A sample of the spiral face gear pair and its data are shown in Figure 7.11 and Table 7.2.

TABLE 7.2
Spiral Face Gear Data

Gear	Helical Pinion	Spiral Face Gear
Number of teeth	5	26
Normal module, mm	0.85	0.85
Normal drive pressure angle	45°	45° (concave flank)
Normal coast pressure angle	10°	10° (convex flank)
Helix/spiral angle	57.9°	32.0°
Helix/spiral hand	LEFT	RIGHT
Pitch diameter (PD), mm	9.600	29.068
Base diameter, mm	4.504/9.111[a]	24.650
Major diameter, mm	11.00	36.00
Minor diameter, mm	7.74	28.00
Addendum, mm	0.70	0.70
Whole depth, mm	2.00	2.00
Normal tooth thickness at PD, mm	1.270	1.270
Face width, mm	10.00	4.00
Center distance, mm	6.350 ± 0.025	
Contact ratio	1.05/2.60[a]	

[a] Drive/coast flank.

FIGURE 7.12
Asymmetric tooth hobs; a: for helical pinion, b: for spiral face gear. (Kapelevich A.L. and S.D. Korosec. Involute Spiral Face Coupling and Gears: Design Approach and Manufacturing Technique, Gear Solutions, February 2015, 33–37. With permission.)

If a generating worm shown in Figure 7.8 presents a special gear hob with asymmetric tooth profiles, a spiral face gear can be hobbed in the same way and with the same accuracy as spur or helical gears utilizing conventional gear hobbing machines. Figure 7.12 presents the asymmetric tooth hobs for the helical pinion and spiral face gear.

Spiral face gears that utilize a convex-concave contact at the high pressure angle drive tooth flanks have potentially greater load capacity in comparison to conventional face gears. This makes them suitable for different power drives and actuators, as well as for positioning systems for different industries and applications.

7.3 Plastic Gear Design Specifics

Nowadays, the progress in polymer materials and injection molding processing has allowed drastic expansion of plastic gear applications. They are used not only for lightly loaded motion transmissions, but also in moderate load power drives in automotive, agriculture, medical, robotics, and many other industries.

Comprehensive books about plastic gears are written by C. Adams [191] and V.E. Starzhinsky with coauthors [192]. The AGMA standards [193,194] present tooth proportions for plastic gears and the AGMA standard [195] gives a polymer material selection guideline. The VDI 2736 standard [196] describes material selection, design and tolerancing, and production methods of thermoplastic gears. The paper [197] describes polymer gear wear behavior and its performance prediction based on the extensive investigations on the gear thermal mechanical contact both numerically and experimentally. The papers [72,80] studied the plastic asymmetric gears.

Benefits of polymer gears in comparison to metal ones include the following:

- low cost injection molding processing;
- low vibration and noise;
- low weight and inertia;
- no corrosion;
- no electric current conductivity;
- ability to work without external (oil or grease) lubrication in some applications.

These advantages made possible the usage of polymers for a wide variety of gear drives. However, there are limitations that must be taken into account considering plastic gears:

- low strength and wear resistance;
- low thermal conductivity and operating temperature;
- wide deviation of material property parameters;
- sensitivity to operating conditions (temperature and humidity);
- low modulus of elasticity and increased tooth deflection;
- limited injection molding process accuracy;
- creep.

Main polymer gear materials are acetals (POM) and nylons, polyesters, and polycarbonates. They can be used with operating temperatures up to 150°C. For elevated temperatures (<170°C) suitable gear polymers are polyphthalamide (PPA), nylon 46, and similar. High temperature (<200°C) plastic materials include polyetherimide (PEI), polyetheretherketone (PEEK), and liquid crystal polymers (LCPs).

Some drawbacks of gear plastics properties can be reduced by using different additives to polymer composition. Additives for higher flexural strength include glass, carbon, and aramid (Kevlar) fibers. Tooth flank wear resistance of nonlubricated plastic gears can be increased by antiwear and antifriction additives: silicone, polytetrafluoroethylene (PTFE), graphite powders, molybdenum disulfide (MoS_2), etc.

Although there are machined plastic gears, the most common and cost-effective plastic gear fabrication technology is injection molding. This process does not use the generating rack type tooling to form gear teeth. Besides, molding tool cavity shape is unique for every single gear profile. This makes plastic molded gears very suitable for Direct Gear Design that maximizes performance of plastic gearing [55] by compensating some polymer property limitations. Design guidelines for polymer gears include the following:

- increased safety factors to guarantee sufficient tooth strength considering a wide range of material properties deviation and dependence on molding process parameters, operating temperature, humidity, etc.;
- increased tooth size (larger module or coarser diametral pitch) to reduce bending stress;
- lower number of teeth to fit the increased tooth size into required center distance with given gear ratio to reduce tolerance sensitivity;
- higher operating pressure angle and contact ratio to reduce bending and contact stress, and increase tooth flank wear resistance;
- bending stress balance that equalizes safety factors of mating gears;
- specific sliding balance for higher efficiency and tooth flank wear resistance;
- root fillet profile optimization for bending stress reduction;
- asymmetric tooth profile for unidirectional gear drives.

Besides original plastic gear designs, applications of plastic gears are often considered to replace the already existing relatively lightly loaded metal gears usually for cost and noise reduction. In this case, exact replication of a metal gear design typically does not work, because of material property difference, mostly because of low strength of polymers in comparison to metals. Design of plastic gears should compensate for the lack of polymer strength following the described above guidelines.

TABLE 7.3

Example of Metal-to-Plastic Gear Conversion

		Metal Gear Pair		**Plastic Gear Pair**	
Design Method		**Traditional (Standard 20° Pressure Angle Tool)**		**Direct Gear Design**	
Gear		**Driving**	**Driven**	**Driving**	**Driven**
Number of teeth		32	52	16	26
Normal module, mm		0.500		0.900	
Generating rack coefficients	Addendum	1.0		–	
	Dedendum	1.25		–	
	Tip radius	0.3		–	
	Radial clearance	0.2		–	

(Continued)

TABLE 7.3 (*Continued*)
Example of Metal-to-Plastic Gear Conversion

Design Method	Metal Gear Pair: Traditional (Standard 20° Pressure Angle Tool)		Plastic Gear Pair: Direct Gear Design	
Gear	**Driving**	**Driven**	**Driving**	**Driven**
X-shift coefficient	0.0	0.0	–	–
Normal pressure angles	20°		35°/18°[a]	
Helix angle	0°		25.8°	
Hand of helix	–	–	right	left
Helix lead, mm	–	–	103.942	168.906
Pitch diameter (PD), mm	16.000	26.000	16.000	26.000
Tooth tip diameter, mm	16.956	26.956	18.000	27.945
Root diameter, mm	13.636	23.636	13.828	23.746
Root fillet	Trochoidal	Trochoidal	Optimized	Optimized
Normal tooth thickness at PD, mm	0.769	0.769	23.746	1.365
Center distance, mm	21.0		21.0	
Face width, mm	7.0	6.5	7.0	6.5
Transverse contact ratio	1.62		1.20/1.52[a]	
Axial contact ratio	–		1.00	
Total contact ratio	1.62		2.20/2.52[a]	
Mesh efficiency, %	98.7		98.5	
Maximum temperature, °C	110		110	
Maximum driving torque, Nm	1.0		1.0	
Tensile bending stress, MPa	104	107	50	52
Gear material	Steel 4140 normalized		Amodel AS-1133 HS	
Yield tensile strength, MPa	635		193	
Min. bending safety factor	6.1	5.9	3.9	3.7
Image of driving gear			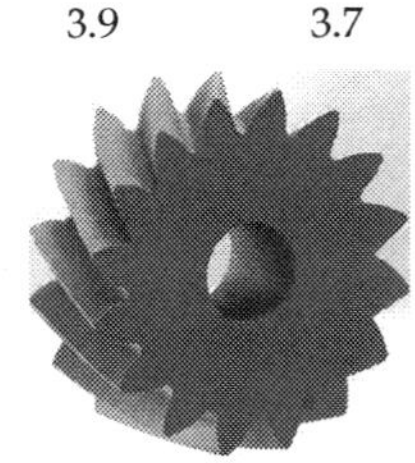	

[a] Drive/coast tooth flank.

Table 7.3 presents an example of metal-to-plastic gear conversion. It shows how Direct Gear Design makes it possible to replace the spur metal machined gears with the plastic helical asymmetric gears providing sufficient safety factors and increasing the total contact ratio. This replacement allows not only a match in load carrying capacity, but also reduces gear noise and vibration.

8

Tolerancing and Tolerance Analysis

Tolerances that are seemingly negligibly small in comparison to nominal gear dimensions greatly affect gear drive performance and product cost. Incorrect tolerancing can turn a potentially successful project into a total failure.

This chapter considers a tolerance selection approach and shows how tolerancing influences some gear pair performance parameters.

8.1 Gear Specification

Comprehensive gear drawing specification is critically important. With an incompletely specified drawing a designer practically delegates his responsibilities to a gear supplier, letting him guess about the designer's actual intentions. The main problem here is that the supplier is not in a position to do this, even if he has gear design experience, because he usually does not know enough about gear drive application specifics. As a result of such guessing, gears are typically made the most cost-efficient way that may compromise required gear drive performance. The 3D CAD gear model also cannot be a replacement for the properly specified gear drawing.

A gear drawing specification must completely define gear geometry, accuracy level (standard accuracy grade or accuracy parameters), reference dimensions, all critical dimensions with tolerances, gear inspection dimensions (for example, measurement over/between pins or balls, span measurement, etc.), surface finish, material data including its grade, heat treatment (surface and core hardness, harden case depth), and post-machining surface treatment (for example, shot peening, super finishing, coating, etc.). It should also describe gear tooth flank microgeometry shape and parameters, and contain information about tooling, processing, material specimen condition, and some additional information. Gear geometry parameters and inspection dimensions must be included in the gear specification table. Examples of proper gear drawing specification tables for most types of involute gears are described in [198,199].

Asymmetric gear dimensions and parameters defined by the Direct Gear Design® method are specified by a very similar way as for traditionally designed gears to make it understandable to a gear supplier and exclude any guesswork or confusion, although such specification typically has some additional data. For example, it should have geometric parameters of both drive and coast tooth flanks as well as a description of the optimized root fillet profile (as a coordinate point table or as a part of the CAD tooth profile that accompanies a gear drawing) and its tolerance. Drive and coast tooth flanks of an asymmetric gear should be identified in the gear drawing to avoid potential assembly problems (Figure 8.1). A gear specification table for gears with asymmetric teeth contains geometry parameters of both drive and coast tooth flanks (see for example Table 8.1).

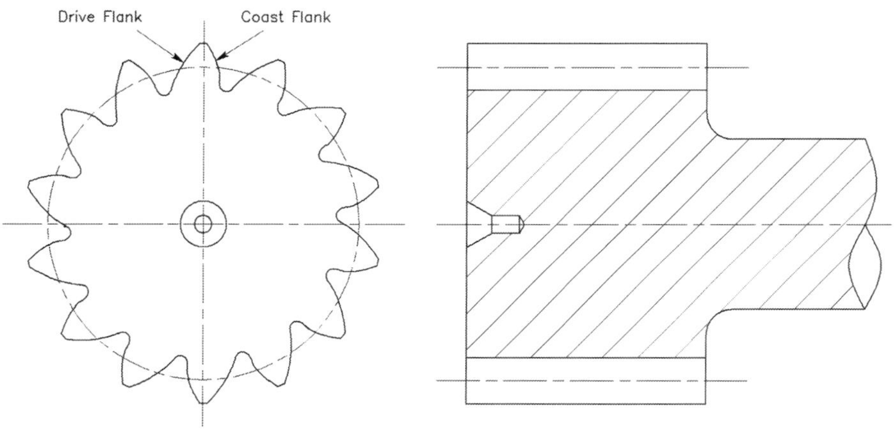

FIGURE 8.1
Asymmetric tooth flank identification in the gear drawing.

TABLE 8.1
Gear Specification Table Example

Number of teeth		[a]
Module		[b]
Reference diameter		[b]
Pressure angle at reference pitch diameter	Drive flank	[b]
	Coast flank	[b]
Base diameter	Drive flank	[b]
	Coast flank	[b]
Form diameter	Drive flank	min/max[a]
	Coast flank	min/max[a]
Tooth tip diameter		min/max[a]
Root diameter		min/max[a]
Tooth thickness at reference pitch diameter		min/max[a]

(*Continued*)

TABLE 8.1 ***(Continued)***

Gear Specification Table Example

Tooth tip radius		min/max[a]
Face width		min/max[a]
Pin diameter		[b]
Measurement over two pins		min/max[a]
Accuracy grade per AGMA 2015-A01		[b]
Runout tolerance, F_r		[a]
Total cumulative pitch tolerance, F_p		[a]
Single pitch tolerance, f_{pt}		[a]
Profile tolerance	Total, F_α	[a]
	From, $f_{f\alpha}$	[a]
	From, $f_{H\alpha}$	[a]
Helix tolerance	Total, F_β	[a]
	From, $f_{f\beta}$	[a]
	From, $f_{H\beta}$	[a]
Root fillet profile tolerance		[a, c]
Mating gear part number		[b]
Mating gear number of teeth		[b]
Center distance		min/max[b]

[a] Critical parameter or dimension.

[b] Reference parameter or dimension.

[c] Root fillet profile tolerance is not defined by gear accuracy standards and should be assigned by designer.

8.2 Accuracy Selection

Gear accuracy selection depends on gear drive application, operating conditions, and technical and market performance requirements. This selection is also affected by chosen gear fabrication technology, materials, heat treatment, etc. Gear accuracy is defined by a set of tolerances. Types of gear tolerances and their effects on fabrication, cost, and function of a gear drive are described in [165]. For high performance gear drives, such as, for example, aerospace and racing transmissions, the functional requirements are primary factors for tolerance selection. For less demanding gear drive applications, other factors, such as cost, manufacturability, availability of fabrication equipment, and tooling, can prevail. In many cases gear accuracy standards, like, for example, those in [200–203], are used for tolerance selection. There are also gear dimensions, for which tolerances are not defined by accuracy standards. Such dimensions are the tooth tip diameter, root diameter, gear face width, tooth tip radius or chamfer, etc. The standards also do not cover the tooth microgeometry tolerances defining acceptable limits of the tooth profile and lead modification. This tolerance selection is based on gear drive application, operating conditions, experience with the previously designed similar drives, and the prototype performance testing results.

Asymmetric tooth gears calculated by the Direct Gear Design® method utilize the same tolerance selection criteria and the same standards as traditionally designed symmetric tooth gears. At the same time, they have some specific tolerance requirements. One such requirement is the root fillet profile tolerancing. Although the gear tooth root fillet, as an area of maximum bending stress concentration, is critically important, existing gear accuracy standards do not define its tolerances, and in traditional gear design its profile and accuracy are marginally defined on the gear drawing by typically very generous root diameter tolerance and, in some cases, by the minimum fillet radius. Direct Gear Design optimizes the gear tooth root fillet profile for minimum bending stress concentration. This requires that the root fillet must be comprehensibly specified and then inspected. If the whole gear profile, including the root fillet, is shaped by the same fabrication process and tooling, the fillet profile tolerance can be the same as the involute flank tolerance that is specified by the standards. However, some gear designs require different fabrication methods, and separate machining processes and tooling for final shaping of the tooth involute flank and the root fillet. One such gear design case assumes using the gear cutters (hobs or shaper cutters) with a protuberance for final machining of the tooth root fillet profile and preliminary machining of the tooth flanks, leaving a stock for grinding or shaving after heat treatment. As a result, the involute tooth flank and root fillet accuracy and surface finish can be very different. In this case, the fillet profile tolerance is selected according to the accuracy grade achievable by the fillet profile machining. The root diameter tolerance of directly designed asymmetric tooth gears is defined considering the root fillet profile tolerance.

8.3 Tolerance Analysis

The goal of this tolerance analysis is to verify the mating gear pair design data in order to guarantee the adequate normal backlash, sufficient root clearance, and minimal acceptable contact ratio at any possible tolerance combinations and operating conditions.

Operating conditions include operating temperature and humidity ranges. A wide operating temperature range noticeably changes gear drive dimensions (particularly for large gear transmissions) and may greatly affect tolerance analysis results in case of dissimilar materials for gears, gear housing, and other gear drive assembly parts. This is especially critical for gear drives made with polymer components. Besides, some gear polymers like nylons absorb moisture, resulting in increased gear dimensions. Tolerance analysis defines the mating gear normal backlash, contact ratio, and root clearances at two extreme (minimum and maximum) value combinations of tolerances and parameters presented in Table 8.2.

TABLE 8.2

Tolerance Analysis Input Data

Dimensions and Tolerances	Symbol	Units
Number of teeth of mating gears	z_1, z_2	–
Normal module or diametral pitch at reference pitch diameter	m_n or DP_n	mm or 1/in
Normal pressure angle at reference pitch diameter (for asymmetric gears: drive and coast pressure angles)	A α_d and α_c	°
Helix angle at reference pitch diameter	β	°
Tooth tip diameters (minimum/maximum values)	d_{amin1}, d_{amax1}, d_{amin2}, d_{amax2}	mm or in
Root diameters (minimum/maximum values)	d_{rmin1}, d_{rmax1}, d_{rmin2}, d_{rmax2}	mm or in
Normal tooth thickness at reference pitch diameter (minimum/maximum values)	S_{nmin1}, S_{nmax1}, S_{nmin2}, S_{nmax2}	mm or in
Tooth tip radius (minimum/maximum values)	R_{amin1}, R_{amax1}, R_{amin2}, R_{amax2}	mm or in
Gear face width (minimum/maximum values)	b_{min1}, b_{max1}, b_{min2}, b_{max2}	mm or in
Runout tolerances[a]	F_{r1}, F_{r2}	mm or in
Single pitch tolerance	f_{pt1}, f_{pt2}	mm or in
Total profile tolerance	$F_{\alpha 1}$, $F_{\alpha 2}$	mm or in
Total helix tolerance	$F_{\beta 1}$, $F_{\beta 2}$	mm or in
Housing center distance	a_{min}, a_{max}	mm or in
Bearing radial gap	δ_{min1}, δ_{max1}, δ_{min2}, δ_{max2}	mm or in
Operating conditions		
Temperature (minimum/ambient[b]/maximum values)	T_{min}, T_{amb}, T_{max}	°C or °F
Humidity (minimum/ambient[b]/maximum values)	RH_{min}, RH_{amb}, RH_{max}	%
Material properties		
Linear coefficient of thermal expansion (CTE)	δ_t	mm/mm-°C or in/in-°F
Moisture expansion coefficient (CME)[c]	δ_m	mm/mm-% or in/in-%

[a] In some cases the total radial composite deviation (F_i'') that is also known as the total composite error (TTE) is used instead the runout tolerance.

[b] Ambient temperature or humidity in this content is inspection lab temperature or humidity. Typical ambient temperature is 20 C or 68 F and ambient humidity is about 50% RH.

[c] For moisture absorbing materials.

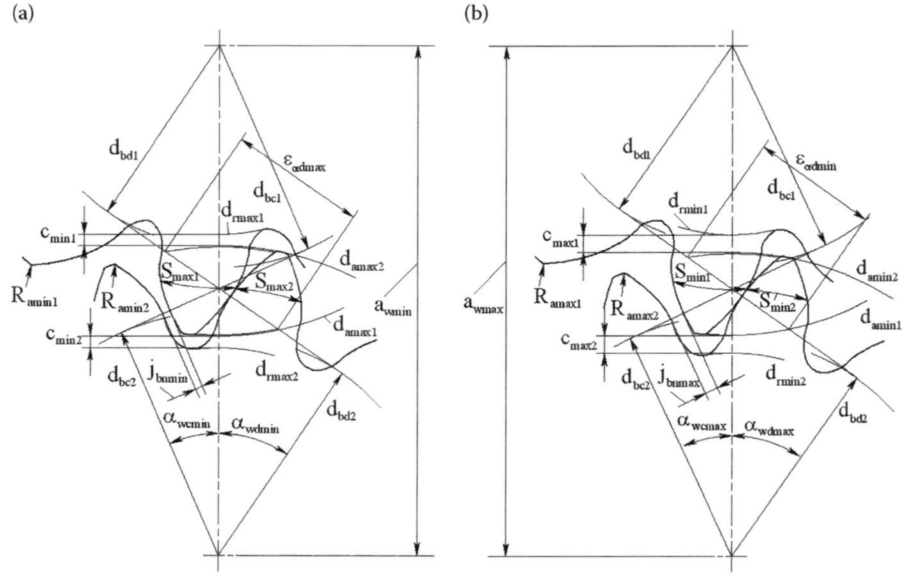

FIGURE 8.2
Cases 1 (a) and 2 (b) of extreme tolerance combinations for external gear pair with asymmetric teeth.

Tolerance analysis for an external gear pair considers two cases:

Case 1 (Figure 8.2a): Resulting in minimum normal backlash and radial clearances, and maximum contact ratio. This gear tolerance buildup includes the minimum gear drive housing center distance, bearing radial plays, and tooth tip radii, and the maximum tooth tip diameters, root diameters, and normal tooth thicknesses at reference pitch diameters of mating gears. In this case, runouts of both mating gears should reduce the effective center distance, and the tooth profile, pitch, and helix tolerances increase the effective tooth thickness. Minimum or maximum operating temperature value is selected depending on a gear and housing material combination to reduce the effective center distance. If, for example, the material combination is the steel gears and aluminum housing, the minimum effective center distance is achieved at the minimum operating temperature, because the coefficient of *thermal expansion* (CTE) of steel is lower than that of aluminum. If the material combination is the plastic gears and aluminum housing, the minimum effective center distance is achieved at the maximum operating temperature, because the CTE of plastics is usually greater than that of aluminum. If gears or gear drive housing are made out of the moisture absorbing materials (nylon, for example), the minimum or maximum value of operating humidity is selected depending on the gear and housing material combination to reduce the effective center distance.

Case 2 (Figure 8.2b): Resulting in the maximum normal backlash and radial clearances, and minimum contact ratio. This gear tolerance buildup includes the maximum housing center distance, bearing radial plays, and tooth tip radii, and the minimum tooth tip diameters, root diameters, and normal tooth thicknesses at reference pitch diameters of mating gears. In this case, runouts of both mating gears should increase the effective center distance, and the tooth profile, pitch, and helix tolerances do not change the effective tooth thickness. Minimum or maximum operating temperature and humidity values are selected depending on gear and housing material combination to increase the effective center distance.

For an internal gear pair, a tolerance analysis approach is the same, but selection of parameters limits is different.

Case 1 (Figure 8.3a): Resulting in the minimum normal backlash and radial clearances, and maximum contact ratio. This gear tolerance buildup includes the maximum housing center distance, minimum bearing radial plays and tooth tip radii, maximum pinion tooth tip and root diameters of the pinion, minimum ring gear tooth tip

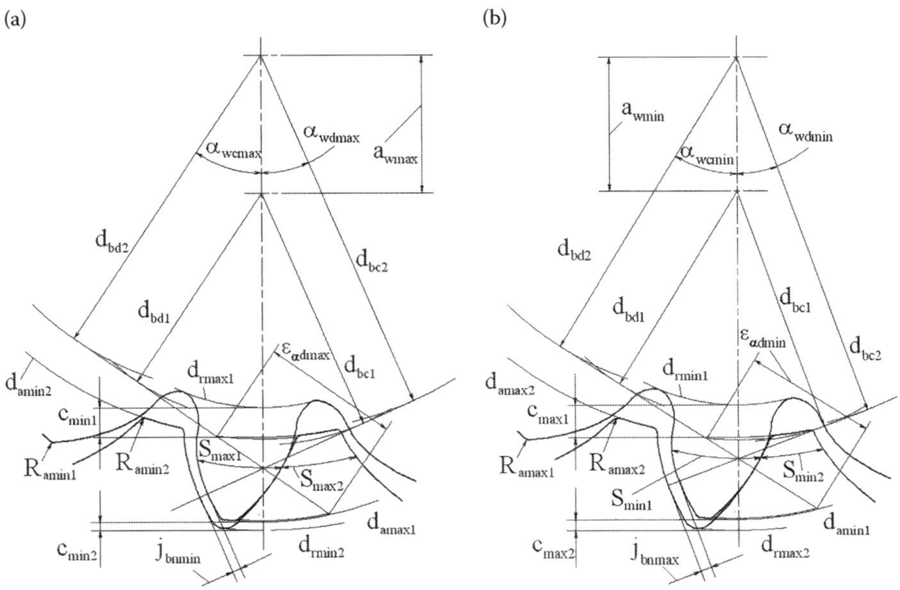

FIGURE 8.3
Cases 1 (a) and 2 (b) of extreme tolerance combinations for internal gear pair with asymmetric teeth.

and root diameters, and maximum normal tooth thickness at reference pitch diameters of mating gears. In this case, the gear runouts should increase the effective center distance, and the tooth profile, pitch, and helix tolerances increase the effective tooth thickness. Minimum or maximum operating temperature and humidity values are selected depending on the gear and housing material combination to increase the effective center distance.

Case 2 (Figure 8.3b): Resulting in the maximum normal backlash and radial clearances, and minimum contact ratio. This gear tolerance buildup includes the minimum housing center distance, maximum bearing radial plays and tooth tip radii, minimum pinion tooth tip and root diameters of the pinion, maximum ring gear tooth tip and root diameters, and minimum normal tooth thickness at reference pitch diameters of mating gears. In this case, the gear runouts should reduce the effective center distance, and the tooth profile, pitch, and helix tolerances combination does not affect the effective tooth thickness. Minimum or maximum operating temperature and humidity values are selected depending on the gear and housing material combination to reduce the effective center distance.

The tolerance analysis set of equations that allows the definition of the minimum and maximum values of the normal backlash, radial clearances, and minimum contact ratio is presented below. It is described for helical gears with asymmetric teeth. For spur gears these equations are used assuming the helix angle equal to zero. Here is an assumption that the mating gears and gear drive housing are made out of different materials, and those materials could be moisture absorbing. The operating condition coefficients are as follows:

For gears:

$$\lambda_{o1,2} = 1 + \delta_{t1,2}(T_o - T_{amb}) + \delta_{m1,2}(RH_o - RH_{amb}). \tag{8.1}$$

For gear drive housing:

$$\lambda_{oh} = 1 + \delta_{th}(T_o - T_{amb}) + \delta_{mh}(RH_o - RH_{amb}), \tag{8.2}$$

where T_{amb} and RH_{amb} are ambient temperature and humidity, T_o and RH_o are operating temperature and humidity, which can have minimum, ambient, or maximum values, and 1,2, and h are for the pinion, gear, and housing, accordingly.

For ambient conditions coefficients $\lambda_{o1,2}$ and λ_{oh} are equal to 1.0. For actual operating conditions major gear and housing dimensions should be adjusted by these coefficients:

Center distance:

$$a_{\min} = \lambda_{oh} \times a_{\min} \quad \text{and} \quad a_{\max} = \lambda_{oh} \times a_{\max}. \tag{8.3}$$

Tooth tip diameters:

$$d_{a\min 1,2} = \lambda_{o1,2} \times d_{a\min 1,2} \quad \text{and} \quad d_{a\max 1,2} = \lambda_{o1,2} \times d_{a\max 1,2}. \tag{8.4}$$

Root diameters:

$$d_{r\min 1,2} = \lambda_{o1,2} \times d_{r\min 1,2} \quad \text{and} \quad d_{r\max 1,2} = \lambda_{o1,2} \times d_{r\max 1,2}. \tag{8.5}$$

Normal tooth thicknesses:

$$s_{n\min 1,2} = \lambda_{o1,2} \times s_{n\min 1,2}$$

and

$$s_{n\max 1,2} = \lambda_{o1,2} \times s_{n\max 1,2} + f_{pt1,2} + F_{\alpha 1,2} + F_{\beta 1,2}. \tag{8.6}$$

Gear face widths:

$$b_{\min 1,2} = \lambda_{o1,2} \times b_{\min 1,2} \text{ and } b_{\max 1,2} = \lambda_{o1,2} \times b_{\max 1,2}. \tag{8.7}$$

It is not necessary to apply operating condition coefficients $\lambda_{o1,2}$ and λ_{oh} to small dimensions like the tooth tip radii or gear and gear housing accuracy tolerances, because typically it does not make any noticeable difference.

Then the operating center distance values $a_{w\min}$ and $a_{w\max}$ are as follows:

For external gearing:

$$a_{w\min} = a_{\min} - \frac{F_{r1}}{2} - \frac{F_{r2}}{2} + \frac{\delta_{\min 1}}{2} + \frac{\delta_{\min 2}}{2}$$

and

$$a_{w\max} = a_{\max} + \frac{F_{r1}}{2} + \frac{F_{r2}}{2} + \frac{\delta_{\max 1}}{2} + \frac{\delta_{\max 2}}{2}. \tag{8.8}$$

For internal gearing:

$$a_{w\min} = a_{\min} - \frac{F_{r1}}{2} - \frac{F_{r2}}{2} - \frac{\delta_{\max 1}}{2} - \frac{\delta_{\max 2}}{2}$$

and

$$a_{w\max} = a_{\min} + \frac{F_{r1}}{2} + \frac{F_{r2}}{2} - \frac{\delta_{\min 1}}{2} - \frac{\delta_{\min 2}}{2}. \tag{8.9}$$

The reference pitch diameters $d_{p1,2}$ are as follows:
In the metric system:

$$d_{p1,2} = \frac{m_n z_{1,2} \lambda_{o1,2}}{\cos\beta}. \tag{8.10}$$

In the English system:

$$d_{p1,2} = \frac{z_{1,2} \lambda_{o1,2}}{DP_n \cos\beta}. \tag{8.11}$$

The transverse pressure angles α_{td} and α_{tc} at the reference pitch diameters are as follows:
For drive flanks:

$$\alpha_{td} = \arctan\left(\frac{\tan\alpha_d}{\cos\beta}\right). \tag{8.12}$$

For coast flanks:

$$\alpha_{tc} = \arctan\left(\frac{\tan\alpha_c}{\cos\beta}\right). \tag{8.13}$$

The base diameters $d_{b1,2}$ are as follows:
For drive flanks:

$$d_{bd1,2} = d_{p1,2}\cos\alpha_{td}. \tag{8.14}$$

For coast flanks:

$$d_{bc1,2} = d_{p1,2}\cos\alpha_{tc}. \tag{8.15}$$

The helix angles β_{bd} and β_{bc} at the base diameters are as follows:
For drive flanks:

$$\beta_{bd} = \arctan\left(\tan\beta\cos\alpha_{td}\right). \tag{8.16}$$

For coast flanks:

$$\beta_{bc} = \arctan\left(\tan\beta\cos\alpha_{tc}\right). \tag{8.17}$$

The tooth tip involute angles $\alpha_{admin1,2}$ and $\alpha_{admax1,2}$ are as follows:

$$\alpha_{admin1,2} = \arccos\left(\frac{d_{bd1,2}}{d_{amin1,2}}\right)$$

and

$$\alpha_{admax1,2} = \arccos\left(\frac{d_{bd1,2}}{d_{amax1,2}}\right). \tag{8.18}$$

The helix angles at the tooth tip diameters $\beta_{amin1,2}$ and $\beta_{amax1,2}$ are as follows:

$$\beta_{amin1,2} = \arctan\left(\tan\beta \frac{\cos\alpha_{td}}{\cos\alpha_{admin1,2}}\right)$$

and

$$\beta_{amax1,2} = \arctan\left(\tan\beta \frac{\cos\alpha_{td}}{\cos\alpha_{admax1,2}}\right). \tag{8.19}$$

The effective tooth tip involute angles for gears with external teeth from (2.19) and (2.20) are as follows:

For drive flanks $\alpha_{edmin1,2}$ and $\alpha_{edmax1,2}$:

$$\alpha_{edmin1,2} = \arctan\left(\tan\left(\arccos\left(\frac{d_{bd1,2}}{d_{amin1,2} - 2R_{amax1,2}}\right)\right) + \frac{2R_{amax1,2}}{d_{bd1,2}}\right)$$

and

$$\alpha_{edmax1,2} = \arctan\left(\tan\left(\arccos\left(\frac{d_{bd1,2}}{d_{amax1,2} - 2R_{amin1,2}}\right)\right) + \frac{2R_{amin1,2}}{d_{bd1,2}}\right). \tag{8.20}$$

For coast flanks $\alpha_{ecmin1,2}$ and $\alpha_{ecmax1,2}$:

$$\alpha_{ecmin1,2} = \arctan\left(\tan\left(\arccos\left(\frac{d_{bc1,2}}{d_{amin1,2} - 2R_{amax1,2}}\right)\right) + \frac{2R_{amax1,2}}{d_{bc1,2}}\right)$$

and

$$\alpha_{ecmax1,2} = \arctan\left(\tan\left(\arccos\left(\frac{d_{bc1,2}}{d_{amax1,2} - 2R_{amin1,2}}\right)\right) + \frac{2R_{amin1,2}}{d_{bc1,2}}\right). \tag{8.21}$$

The tooth tip effective involute angles for a gear with internal teeth from (2.19) and (2.20) are as follows:

For drive flanks $\alpha_{edmin1,2}$ and $\alpha_{edmax1,2}$:

$$\alpha_{edmin2} = \arctan\left(\tan\left(\arccos\left(\frac{d_{bd2}}{d_{amin2} + 2R_{amax2}}\right)\right) - \frac{2R_{amax2}}{d_{bd2}}\right)$$

and

$$\alpha_{aedmax2} = \arctan\left(\tan\left(\arccos\left(\frac{d_{bd2}}{d_{amax2} + 2R_{amin2}}\right)\right) - \frac{2R_{amin2}}{d_{bd2}}\right). \tag{8.22}$$

For coast flanks $\alpha_{ecmin1,2}$ and $\alpha_{ecmax1,2}$:

$$\alpha_{ecmin2} = \arctan\left(\tan\left(\arccos\left(\frac{d_{bc2}}{d_{amin2} + 2R_{amax2}}\right)\right) - \frac{2R_{amax2}}{d_{bc2}}\right)$$

and

$$\alpha_{ecmax2} = \arctan\left(\tan\left(\arccos\left(\frac{d_{bc2}}{d_{amax2} + 2R_{amin2}}\right)\right) - \frac{2R_{amin2}}{d_{bc2}}\right). \tag{8.23}$$

From (2.3) the asymmetry factor K is

$$K = d_{bc1,2}/d_{bd1,2} = \cos\alpha_{tc}/\cos\alpha_{td}. \tag{8.24}$$

The involute intersection profile angles $\nu_{dmin1,2}$ and $\nu_{dmax1,2}$ for gears with external teeth are defined from a system of equations:

$$inv(\nu_{dmin1,2}) + inv(\nu_{cmin1,2}) = inv(\alpha_{td}) + inv(\alpha_{tc}) + \frac{2s_{nmin1,2}\tan\alpha_{td}}{d_{bd1,2}\cos\beta} \tag{8.25}$$

and

$$\cos\nu_{cmin1,2} = K\cos\nu_{dmin1,2}, \tag{8.26}$$

$$inv(\nu_{dmax1,2}) + inv(\nu_{cmax1,2}) = inv(\alpha_{td}) + inv(\alpha_{tc}) + \frac{2s_{nmax1,2}\tan\alpha_{td}}{d_{bd1,2}\cos\beta} \tag{8.27}$$

and

$$\cos\nu_{cmax1,2} = K\cos\nu_{dmax1,2}. \tag{8.28}$$

The involute intersection profile angles ν_{dmin2} and ν_{dmax2} for gears with internal teeth are defined from a system of equations:

$$inv(\nu_{dmin2}) + inv(\nu_{cmin2}) = inv(\alpha_{td}) + inv(\alpha_{tc}) + \frac{2(P_n - s_{nmax1,2})\tan\alpha_{td}}{d_{bd2}\cos\beta} \tag{8.29}$$

and (8.24), and

$$inv(\nu_{dmax2}) + inv(\nu_{cmax2}) = inv(\alpha_{td}) + inv(\alpha_{tc}) + \frac{2(P_n - s_{nmin2})\tan\alpha_{td}}{d_{bd2}\cos\beta} \tag{8.30}$$

and (8.26), where P_n is a normal circular pitch at the reference pitch diameter that is equal to $P_n = \pi m_n$ or $P_n = \pi/DP_n$ in the metric or English system, accordingly.

Then operating pressure angles are as follows:

For drive flanks α_{wdmin} and α_{wdmax}:

$$\alpha_{wdmin} = \arccos\left(\frac{d_{bd1}(u \pm 1)}{2a_{wmin}}\right)$$

and

$$\alpha_{wd\max} = \arccos\left(\frac{d_{bd1}(u \pm 1)}{2a_{w\max}}\right), \quad (8.31)$$

where $u = z_2/z_1$ is gear ratio, and the + sign is for the external gearing and the – sign is for the internal gearing.

For coast flanks $\alpha_{wc\min}$ and $\alpha_{wc\max}$:

$$\alpha_{wc\min} = \arccos(K\cos\alpha_{wd\min})$$

and

$$\alpha_{wc\max} = \arccos(K\cos\alpha_{wd\max}). \quad (8.32)$$

The operating transverse tooth thicknesses $S_{w\min 1,2}$ and $S_{w\max 1,2}$ are as follows:

For gears with external teeth:

$$s_{w\min 1,2} = \frac{d_{bd1,2}}{2\cos\alpha_{wd\max}}\left(inv(\nu_{d\min 1,2}) + inv(\nu_{c\min 1,2}) - inv(\alpha_{wd\max}) - inv(\alpha_{wc\max})\right)$$

and

$$s_{w\max 1,2} = \frac{d_{bd1,2}}{2\cos\alpha_{wd\min}}\left(inv(\nu_{d\max 1,2}) + inv(\nu_{c\max 1,2}) - inv(\alpha_{wd\min}) - inv(\alpha_{wc\min})\right). \quad (8.33)$$

For gears with internal teeth:

$$s_{w\min 2} = \frac{d_{bd2}}{2\cos\alpha_{wd\min}}\left(\frac{2\pi}{z_2} - inv(\nu_{d\max 2}) - inv(\nu_{c\max 2}) + inv(\alpha_{wd\min}) + inv(\alpha_{wc\min})\right)$$

and

$$s_{w\max 2} = \frac{d_{bd2}}{2\cos\alpha_{wd\max}}\left(\frac{2\pi}{z_2} - inv(\nu_{d\min 2}) - inv(\nu_{c\min 2}) + inv(\alpha_{wd\max}) + inv(\alpha_{wc\max})\right). \quad (8.34)$$

For gears with asymmetric teeth the normal backlash is defined between the coast tooth flanks, when the drive tooth flanks are in contact. The normal backlash values $j_{bn\min 1,2}$ and $j_{bn\max 1,2}$ are as follows:

$$j_{bn\min} = \left(\frac{\pi d_{bd1,2}}{z_1\cos\alpha_{wd\min}} - s_{w\max 1} - s_{w\max 2}\right)\cos\alpha_{wc\min}\cos\beta_{bc}$$

and

$$j_{bnmax} = \left(\frac{\pi d_{bd1,2}}{z_1 \cos\alpha_{wdmax}} - s_{wmin1} - s_{wmin2} \right) \cos\alpha_{wcmax} \cos\beta_{bc}. \tag{8.35}$$

The normal backlash must be greater than the maximum possible gear tooth tip deflection under the load. This allows the avoidance of simultaneous contact of the opposite tooth flanks with mating gear tooth flanks.

The root clearances $c_{min1,2}$ and $c_{max1,2}$ are as follows:

For external gearing:

$$c_{min1} = a_{min} - \frac{d_{amin2}}{2} - \frac{d_{rmax1}}{2} \quad \text{and} \quad c_{max1} = a_{max} - \frac{d_{amax2}}{2} - \frac{d_{rmin1}}{2}, \tag{8.36}$$

$$c_{min2} = a_{min} - \frac{d_{amin1}}{2} - \frac{d_{rmax2}}{2} \quad \text{and} \quad c_{max2} = a_{max} - \frac{d_{amax1}}{2} - \frac{d_{rmin2}}{2}. \tag{8.37}$$

For internal gearing:

$$c_{min1} = \frac{d_{a\,min2}}{2} - \frac{d_{r\,max1}}{2} - a_{max} \quad \text{and} \quad c_{max1} = \frac{d_{a\,max2}}{2} - \frac{d_{r\,min1}}{2} - a_{min}, \tag{8.38}$$

$$c_{min2} = \frac{d_{r\,min2}}{2} - \frac{d_{a\,max1}}{2} - a_{max} \quad \text{and} \quad c_{max2} = \frac{d_{r\,max2}}{2} - \frac{d_{a\,min1}}{2} - a_{min}. \tag{8.39}$$

The minimal root clearances must be greater than zero to avoid the tooth tip/root interference. Actually, this condition is already taken into consideration during the tooth root fillet construction and optimization (see Section 5.2). However, operating conditions, including temperature and humidity, may reduce root clearances. Besides, low radial clearances may result in trapping lubricant in the tooth root area, increased hydraulic losses, and reduced gear efficiency, especially for relatively wide spur gears. This may require designing the tooth fillet with increased root clearances even with some compromise of bending stress reduction.

The transverse operating contact ratios are as follows:

For drive flanks of external gear mesh $\varepsilon_{\alpha dmin}$ and $\varepsilon_{\alpha dmax}$:

$$\varepsilon_{\alpha dmin} = \frac{z_1}{2\pi} \left(\tan\alpha_{aedmin1} + u \tan\alpha_{aedmin2} - (u+1) \tan\alpha_{wdmax} \right)$$

and

$$\varepsilon_{\alpha dmax} = \frac{z_1}{2\pi} \left(\tan\alpha_{aedmax1} + u \tan\alpha_{aedmax2} - (u+1) \tan\alpha_{wdmin} \right). \tag{8.40}$$

For coast flanks of external gear mesh $\varepsilon_{\alpha cmin}$ and $\varepsilon_{\alpha cmax}$:

$$\varepsilon_{\alpha cmin} = \frac{z_1}{2\pi} \left(\tan\alpha_{aecmin1} + u \tan\alpha_{aecmin2} - (u+1) \tan\alpha_{wcmax} \right)$$

and

$$\varepsilon_{\alpha c\max} = \frac{z_1}{2\pi}\left(\tan\alpha_{aec\max 1} + u\tan\alpha_{aec\max 2} - (u+1)\tan\alpha_{wc\min}\right). \tag{8.41}$$

For drive flanks of internal gear mesh $\varepsilon_{\alpha d\min}$ and $\varepsilon_{\alpha d\max}$:

$$\varepsilon_{\alpha d\min} = \frac{z_1}{2\pi}\left(\tan\alpha_{aed\max 1} - u\tan\alpha_{aed\min 2} + (u-1)\tan\alpha_{wd\min}\right)$$

and

$$\varepsilon_{\alpha d\max} = \frac{z_1}{2\pi}\left(\tan\alpha_{aed\min 1} - u\tan\alpha_{aed\max 2} + (u-1)\tan\alpha_{wd\max}\right). \tag{8.42}$$

For coast flanks of internal gear mesh $\varepsilon_{\alpha c\min}$ and $\varepsilon_{\alpha c\max}$:

$$\varepsilon_{\alpha c\min} = \frac{z_1}{2\pi}\left(\tan\alpha_{aec\max 1} - u\tan\alpha_{aec\min 2} + (u-1)\tan\alpha_{wc\min}\right)$$

and

$$\varepsilon_{\alpha c\max} = \frac{z_1}{2\pi}\left(\tan\alpha_{aec\min 1} - u\tan\alpha_{aec\max 2} + (u-1)\tan\alpha_{wc\max}\right). \tag{8.43}$$

The axial operating contact ratio values $\varepsilon_{\beta\min}$ and $\varepsilon_{\beta\max}$ from (2.99) are as follows:

$$\varepsilon_{\beta\min} = \frac{b_{w\min}\tan\beta_{bd}}{p_{bd}} = \frac{b_{w\min}\tan\beta_{bc}}{p_{bc}}$$

and

$$\varepsilon_{\beta\max} = \frac{b_{w\max}\tan\beta_{bd}}{p_{bd}} = \frac{b_{w\min}\tan\beta_{bc}}{p_{bc}}, \tag{8.44}$$

where p_{bd} and p_{bc} are transverse base circular pitches of the drives and coast tooth flanks, accordingly, and $b_{w\min}$ and $b_{w\max}$ are the minimum and maximum values of the operating face width of gear engagement (see Figure 8.4).

In asymmetric gearing the operating axial contact ratio is the same for the drive and coast tooth flanks.

The total operating contact ratios are defined from Equations (2.92) and (2.93):

For drive flanks $\varepsilon_{\gamma d\min}$ and $\varepsilon_{\gamma d\max}$:

$$\varepsilon_{\gamma d\min} = \varepsilon_{\alpha d\min} + \varepsilon_{\beta\min} \quad \text{and} \quad \varepsilon_{\gamma d\max} = \varepsilon_{\alpha d\max} + \varepsilon_{\beta\max}. \tag{8.45}$$

For coast flanks $\varepsilon_{\gamma c\min}$ and $\varepsilon_{\gamma c\max}$:

$$\varepsilon_{\gamma c\min} = \varepsilon_{\alpha c\min} + \varepsilon_{\beta\min} \quad \text{and} \quad \varepsilon_{\gamma c\max} = \varepsilon_{\alpha c\max} + \varepsilon_{\beta\max}. \tag{8.46}$$

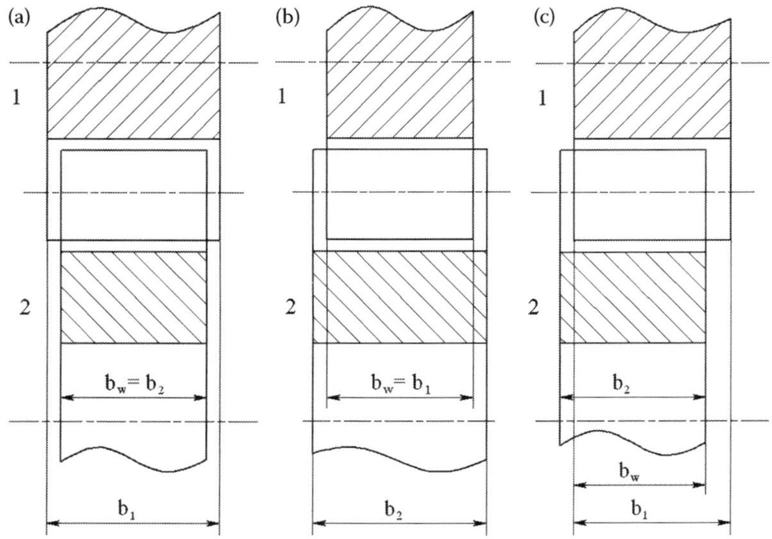

FIGURE 8.4
Operating face width b_w definition; (a) $b_w = b_2$, (b) $b_w = b_1$, (c) gears are assembled with axial shift.

The minimum total operating contact ratio must be greater than 1.0 for smooth tooth pair engagement.

The equations above describe the absolute tolerance analysis approach with output parameters (normal backlash, radial clearances, and contact ratio) defined at two extreme tolerance combinations, described above as cases 1 and 2. As a result, differences between these parameters' minimum and maximum values can be considerable. At the same time, a substantial number of dimensions involved in a tolerance analysis create a very low probability of simultaneous coincidence of extreme tolerance combinations of their minimum or maximum values. Application of the statistical tolerance analysis allows the reduction of tolerance ranges of output parameters or achievement of required ranges with the larger input tolerances, which reduces production cost. According to paper [204], "statistical tolerance analysis can be used by designers and manufacturing personnel to take advantage of statistical averaging over assemblies of parts, allowing the use of less restrictive tolerances in exchange for admitting the small probability of non-assembly." Reduction of this probability can be provided by the statistical process control (SPC) that establishes the technological tolerances for critical dimensions that are lower than the drawing tolerances. The statistical tolerance analysis approach is also described in [205,206].

9

Asymmetric Gear Manufacturing

9.1 Fabrication Technologies and Tooling

Direct Gear Design® of asymmetric tooth gears provides improved performance for unidirectional gear drives. This is achieved by gear tooth macro- and microgeometry enhancement that also requires certain customization of manufacturing technology and tooling. This chapter considers some essential issues of asymmetric gear fabrication.

There are many gear fabrication processes. They can be divided into two groups:

- machining processes that shape gear and tooth profiles by material removal;
- forming processes that shape gear and tooth profiles by distortion of material or changing material state.

Other manufacturing processes that are used for gear as well as many other mechanical component fabrications, like heat treatment, surface engineering, coating, etc., are not considered in this book.

9.1.1 Gear Machining

Table 9.1 presents the main machining processes used for spur and helical gears.

9.1.1.1 Form Machining

A form gear cutting or grinding tool profile (Figure 9.1) is the same as a space profile between gear teeth. This process is applicable for spur and helical [207] gears. Accuracy of form cutting tool positioning relative to the gear blank is very important.

Gear form machining process is typically removing material from one space between teeth after another. When machining of one space between

TABLE 9.1

Spur and Helical Gear Machining Processes

Type of Process		Type of Tooling
Form machining	Cutting	Form disk cutter, broach, etc.
	Grinding	Grinding wheel
Generating machining	Cutting	Hob, shaper cutter, rack cutter, shaver cutter
	Grinding	Grinding wheel
Contour machining	CNC milling	Cylinder or ball mill cutter
	Wire cut EDM[a]	Wire
	Laser cutting	Laser beam
	Water Jet cutting	Water or water and abrasive media mixture stream

[a] EDM: electric discharge machining.

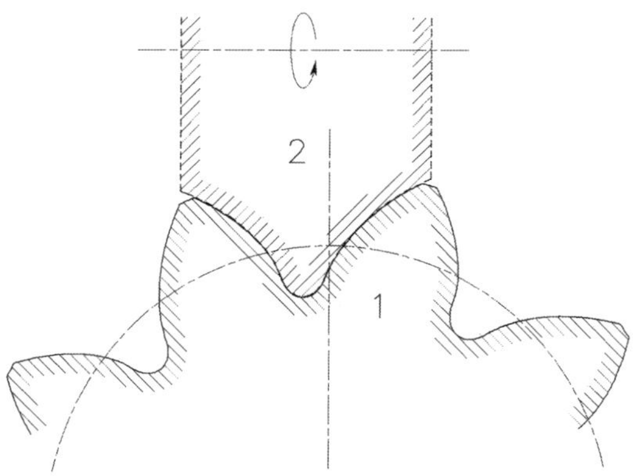

FIGURE 9.1
Gear form machining; 1: gear profile (solid contour), 2: tool profile (dashed contour).

teeth is completed the indexing device (usually rotary table) positions a gear blank for the next space cutting.

Figure 9.2 presents the asymmetric gear form grinding with the cubic boron nitride (CBN) form grinding wheels.

But there is one type of the form machining processes—gear fly cutting [208] that has a tool (fly cutter) in a mesh with the gear blank (Figures 9.3 and 9.4). Fly cutting uses conventional gear hobbing machines that operate when a cutter and a gear blank are in constant synchronized rotation. However, unlike the gear rack generating process, a space between teeth is shaped by a whole cutter tooth profile exactly as in conventional form machining with the disk cutter (Figure 9.3).

The gear fly cutter also looks similar to the conventional gear disk cutter, but all cutter edges are turned at the start angle φ, which is defined as follows:

FIGURE 9.2
Asymmetric gear form grinding setup with CBN grinding wheel. (Courtesy of Aero Gear, South Windsor, Connecticut). From Brown, F.W., et al. Analysis and Testing of Gears with Asymmetric Involute Tooth Form and Optimized Fillet Form for Potential Application in Helicopter Main Drives, Gear Technology, June/July 2011, 46–55. With permission.)

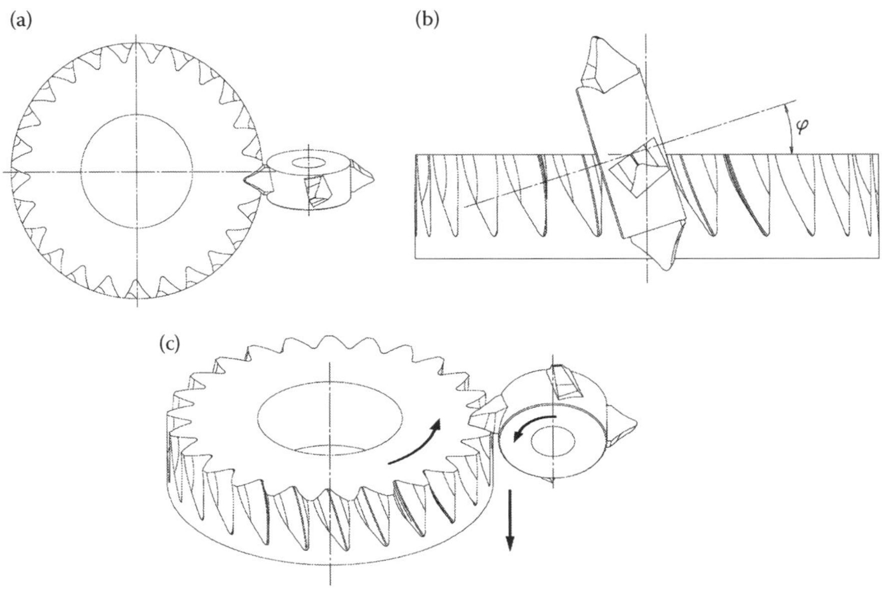

FIGURE 9.3
Gear fly cutting schematics: (a) top view, (b) right view, (c) isometric view; φ: cutter setup angle.

$$\phi = \arcsin \frac{m_n n_t}{d_t}, \tag{9.1}$$

where m_n is the normal module of the gear, n_t is the number of the fly cutter teeth, and d_t is the fly cutter pitch diameter.

A gear fly cutter can also be considered a multi-start gear hob that has just one tooth in each start. The cutter setup angle ϕ relative to the gear plane is defined in the same way as for conventional gear hobbing;

$$\varphi = \beta \pm \phi, \tag{9.2}$$

where β is the gear helix angle, the + sign indicates that the gear helix and cutter start angle have the same directions (right-right or left-left), and the – sign indicates that the gear helix and cutter start angle have opposite directions (right-left or right-left).

The adjustable gear fly cutter is shown in Figure 9.4. It makes it possible to machine symmetric and asymmetric gears with different numbers of teeth and modules by using replaceable cutting and angle inserts. Such gear fly cutter is very suitable for prototyping practically any custom tooth shape gears.

9.1.1.2 Generating Machining

The schematics of the rack generating asymmetric gear machining process and asymmetric gear hob cutter are shown in Figure 9.5. In this process a gear blank and tool are engaged in a mesh, and all gear teeth are machined simultaneously. In traditional gear design, the tooling (a hob or rack cutter) profile is known at the beginning of gear design. In combination with its position relative to the gear center, established by the addendum modification (or X-shift), the tooling rack profile defines the gear tooth and whole gear profiles.

Direct Gear Design applies a different approach. The optimized asymmetric tooth and whole gear profiles are described without using any preselected

(a) (b)

FIGURE 9.4
Gear fly cutting: (a) hobbing machine setup, (b) adjustable fly cutter with the cutting inserts.

(a)

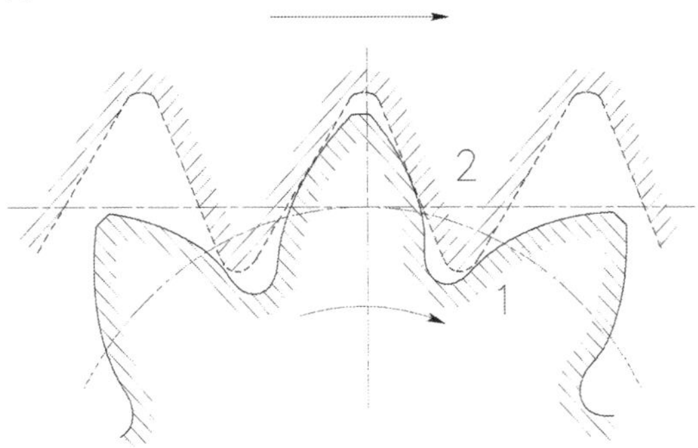

(b)

FIGURE 9.5
(a) Rack generating. 1: gear profile (solid contour); 2: hob or rack cutter profile (dashed contour); (b) Asymmetric gear hob cutter.

tooling rack parameters or its position relative to a gear center. Instead the reverse generating technique is applied to find the tooling rack profile using the known gear profile. This technique assumes that in the rack/gear mesh every point of the gear tooth profile has its mating point on the rack tooth profile.

Figure 9.6 demonstrates how the tooling rack profile point A_t position is defined from the gear tooth profile point A_g position. In order to find the generating rack profile point A_t corresponding to the gear tooth profile point A_g, the line A_gB_g perpendicular to the tooth profile in point A_g is constructed. Point B_g lies in an intersection of the line A_gB_g with the gear pitch circle.

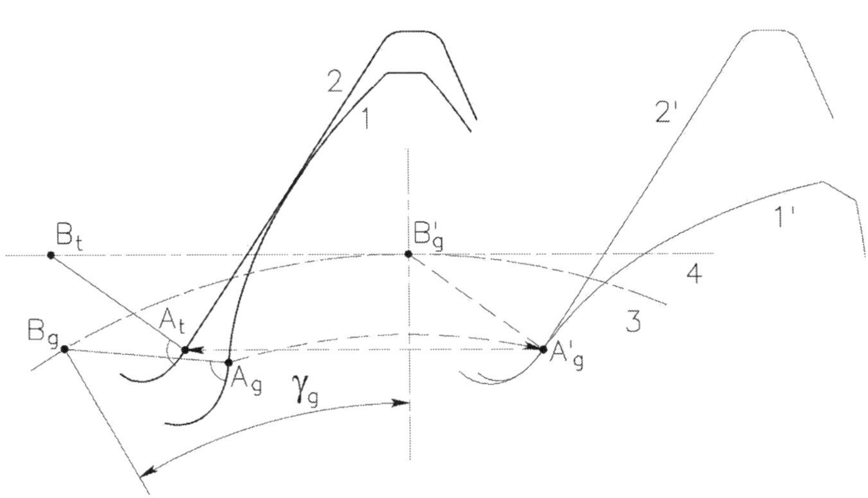

FIGURE 9.6
Generating rack profile definition; 1 and 1′: gear profile positions; 2 and 2′: rack cutter profile positions; 3: gear pitch circle in mesh with rack; 4: rack pitch line in mesh with gear.

The gear tooth profile and line A_gB_g are rotated on the angle γ_g relative to the gear center until point B_g reaches its pitch point position B_g', where the gear pitch circle 3 is tangent to the rack pitch line 4. Then point A_g is in position A_g', where gear tooth profile 1′ is tangent to rack profile 2′. The line $A_g'B_g'$ is moved parallel to the rack pitch line 4 on distance $B_g'B_t$, which is equal to the length of the arc B_gB_g'. This movement puts point A_g' in position A_t at the rack profile that corresponds to point A_g at the gear tooth profile. This approach allows the definition of any generating rack profile point that is related to a certain gear tooth profile point.

Figure 9.7 shows the asymmetric pump gears and its hob, which rack profile is defined by the reverse generation.

In many applications, including automotive transmissions, gears are machined by the protuberance hobs that leave a required tooth flank grinding stock and finally cut the tooth root fillet. After the hobbing operation, a gear is heat treated (typically carburized) and harden, and in some cases a shot peening process is applied as well. This results in a compressive residual stress that strengthens the tooth root. Then the tooth flank grinding operation delivers the specified gear accuracy and flank surface finish.

Figure 9.8 presents the reverse generation of the protuberance hob cutter profile. The hob module m_n and pressure angles α_{nd} and α_{nc} are not necessarily equal to the operating gear module and pressure angles, but they are defined to match the drive and coast base diameters of the gear. Values of the protuberance depths *prd* and *prc* should provide smooth transition from the ground tooth flanks to the root fillets. They should be slightly greater than Δ_d and Δ_c—tooth flank grinding stocks. The protuberance addendums h_{prd} and h_{prc}, and angles α_{prd} and α_{prc} are defined to guarantee the full grinding stocks Δ_d and Δ_c at the lowest tooth contact points near to the root fillet. This requirement results in a smaller root diameter and

(a) (b)

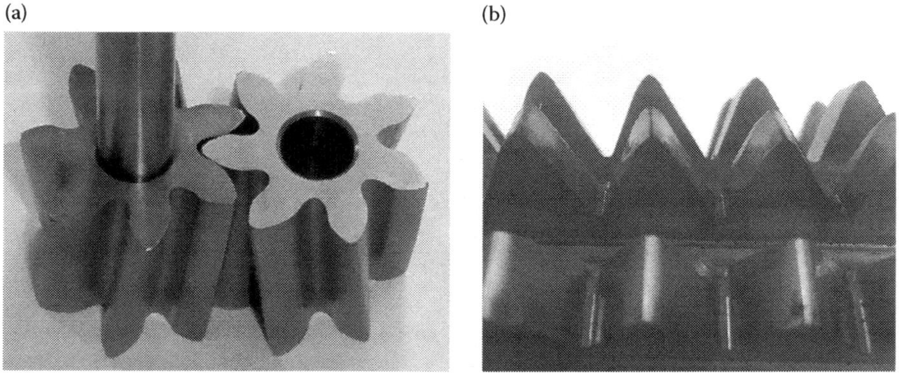

FIGURE 9.7
A sample of the gear rack reverse generation; (a) asymmetric pump gears, (b) hob teeth. (Courtesy of Melling Engine Parts, Jackson, Michigan. With permission.)

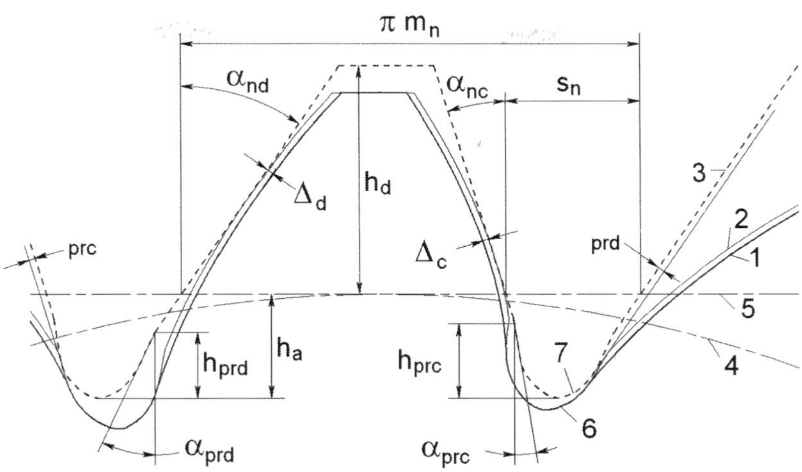

FIGURE 9.8
Protuberance gear hob cutter profile reverse generation; 1: finished gear profile with ground flanks, 2: prefinished hobbed gear profile, 3: protuberance hob profile, 4: hobbing pitch diameter, 5: pitch line of the hob, 6: optimized root fillet profile, 7: a tip of the hob tooth that produces the optimized root fillet; Δ_d and Δ_c: tooth flank grinding stocks, m_n: hob module, s_n: hob pitch line tooth thickness, h_a: hob addendum; h_d: hob dedendum, h_{prd} and h_{prc}: protuberance addendums, α_{nd} and α_{nc}: hob pressure angles, α_{prd} and α_{prc}: protuberance angles, *prd* and *prc*: protuberance depths, indexes *d* and *c* are for drive and coast gear and hob tooth flanks.

higher root stress level in comparison to the completely ground (including the root fillet) tooth gears that are used, for example, in modern aviation gearboxes. Figure 9.9 shows the asymmetric protuberance hub cutter, tooth flank grinding wheel, and the asymmetric gear.

If a gear is to be made by shaping cutting, traditional gear design suggests that the shaper cutter parameters and profile should be known prior to gear design. In combination with its position relative to the gear center (addendum modification or X-shift), the shaper cutter profile defines the gear tooth and whole gear profiles. The schematics of the shaper generating gear cutting process are shown in Figure 9.10.

Direct Gear Design defines a shaper cutter profile also after the gear profile is already known. The gear profile is used for the shaper cutter profile reverse generation using a technique similar to that for the rack type tooling profile definition. It assumes that in the shaper cutter/gear mesh every point of the gear tooth profile has its mating point on the shaper cutter tooth profile.

Figure 9.11 demonstrates how the shaper cutter profile point A_t position is defined from the gear tooth profile point A_g position. In order to find the shaper cutter profile point A_t corresponding to the gear tooth profile point A_g, the line A_gB_g perpendicular to the tooth profile is constructed. Point B_g lies in an intersection of the line A_gB_g with the gear pitch circle. The gear tooth profile and line A_gB_g are rotated on the angle γ_g relative to the gear center

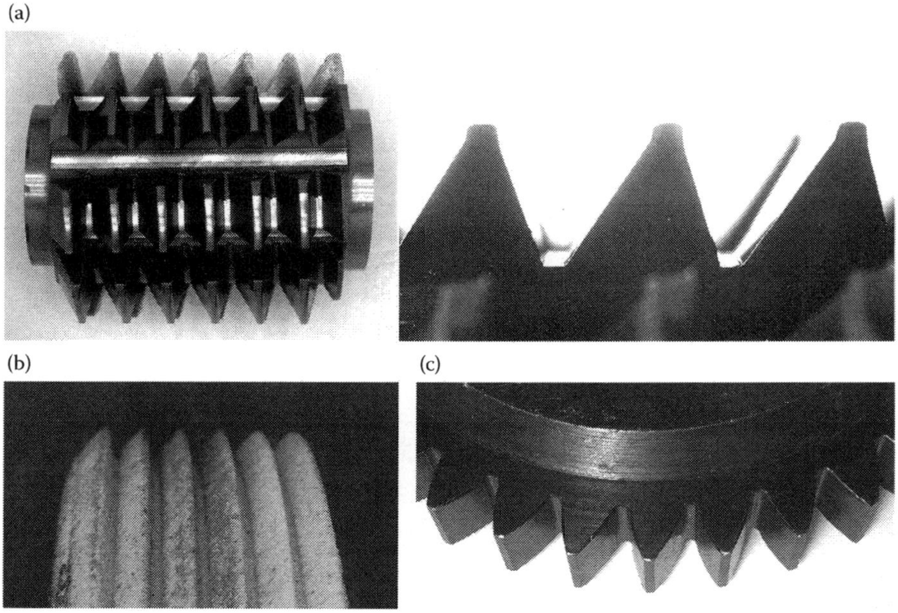

FIGURE 9.9
(a) asymmetric protuberance hub cutter and its teeth (magnified), (b) generating helical grinding wheel, (c) asymmetric gear.

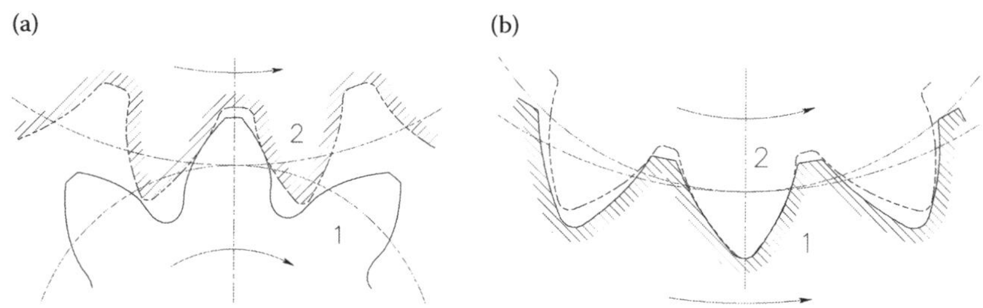

FIGURE 9.10
Shaper cutter profile generating for external (a) and internal (b) gears; 1: gear profile (solid contour); 2: shaper cutter profile (dashed contour).

until point B_g reaches its pitch point position B_g', where gear pitch circle 3 is tangent to shaper cutter pitch circle 4. The angle γ_g is

$$\gamma_g = \frac{2B_gB_g'}{d_{pg}}, \tag{9.3}$$

where B_gB_g' is the length of arc B_gB_g', and d_{pg} is the gear pitch diameter in the mesh with the shaper cutter.

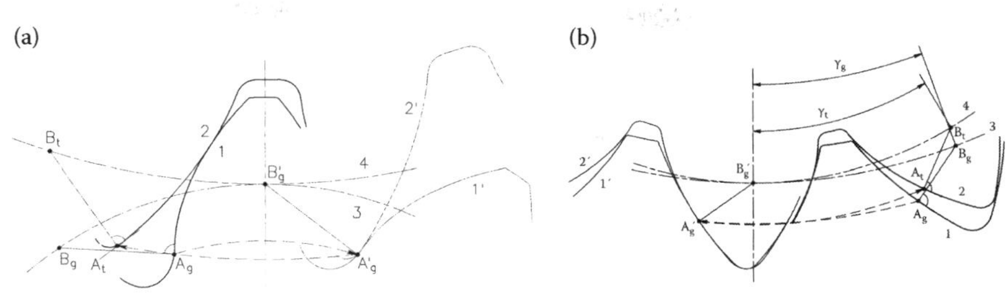

FIGURE 9.11
Generating gear shaper cutter profile definition for external (a) and internal (b) gears. 1 and 1′: gear profile positions; 2 and 2′: shaper cutter profile positions; 3: gear pitch circle in mesh with shaper cutter; 4: shaper cutter pitch circle in mesh with gear.

Point A_g is in position A_g' where gear tooth profile 1′ is tangent to shaper cutter profile 2′. Line $A_g'B_g'$ is rotated back relative to the shaper cutter center on angle γ_t, that is,

$$\gamma_t = \gamma_g \frac{d_{pg}}{d_{pt}}, \tag{9.4}$$

where d_{pg} is the gear pitch diameter in the mesh with the shaper cutter.

This movement puts point A_g' in position A_t at the shaper cutter profile that corresponds to point A_g at the gear tooth profile. This approach allows any shaper cutter profile point corresponding to the certain gear tooth profile point to be defined.

The gear pitch diameter in a mesh with a generating tooling (a hob or shaper cutter) is not necessarily equal to the operating pitch diameter with the mating gear. The pressure angles in an asymmetric mesh with generating tooling can be different than the gear operating pressure angles. The pressure angles in mesh with generating tooling selection are important. Too large pressure angles result in a tooling profile that will not be able to generate the required gear profile, because, for example, of the pointed tooling tooth tip or its too small radius. A too small (close to zero) coast pressure angle negatively affects the cutting condition, gear tooth profile surface finish, and tooling life.

One of the benefits of the gear generating machining process is the possibility of using one tool (hob, rack, or shaper cutter) for machining gears with different numbers of teeth. This allows the reduction of tooling inventory and cost of low and medium volume gear production where a tooling share per one gear is relatively high. In general, traditionally designed mating gears are machined with the same generating gear cutter. However, in mass gear production gear cutting machines are typically set up to machine one gear, and they use a set of tools, including the gear cutters, dedicated to that particular gear.

Directly designed asymmetric tooth gears require a custom dedicated generating tool for every gear with different numbers of teeth. This increases tooling inventory and related gear cost when production volume is low. Although, even in this case, application of directly designed gears is beneficial, if their improved performance justifies some production cost increase, like, for example, in aerospace and racing transmissions. In mass gear production a share of custom dedicated machining tool per gear is low, and as a result, the cost of the directly designed asymmetric gears becomes practically equal to that of the standard ones. This makes them applicable for automotive, agriculture, and many other industries.

Rack generation asymmetric gear grinding by a helical grinding wheel is described in [137, 140, 153].

9.1.1.3 Contour Machining

Contour machining (Figure 9.12) is not a highly productive gear fabrication method, but unlike form or generating machining, it does not require special tooling. This makes it very useful for asymmetric gear prototyping and quick fabrication of relatively small quantity of gears.

Contour gear machining processes include the CNC milling, wire-cut EDM, laser cutting, water jet cutting, etc. In many cases, in order to achieve higher accuracy and surface finish, the contour machining process may require several passes. The final pass typically removes a tiny amount of material.

Most of these processes are suitable only for spur gears, although the CNC ball cutter contour milling can be used for helical gears as well. The contour cutting path (Figure 9.13) is defined with offset S from the nominal (average material condition) gear profile. This offset contains a half of the cut width W_c and some additional offset that depends on the machining process. This additional offset may include a stock for final machining, overcut or overburn (for wire-cut EDM and laser machining), and a defective layer that should be removed by the following tooth surface treatment operation, for example, polishing.

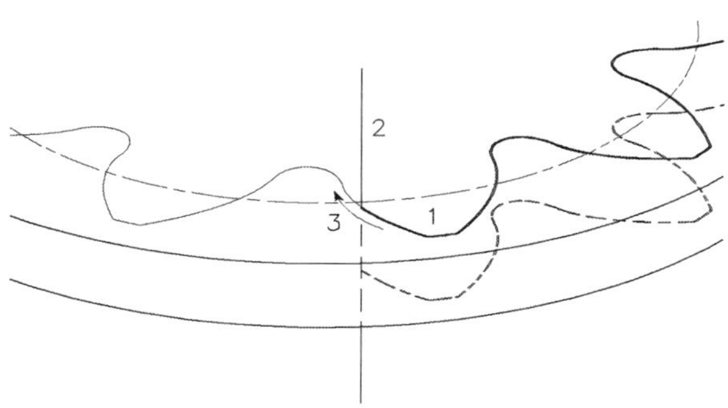

FIGURE 9.12
Schematic of contour gear machining. 1: gear profile; 2: cutting tool; 3: cutting direction.

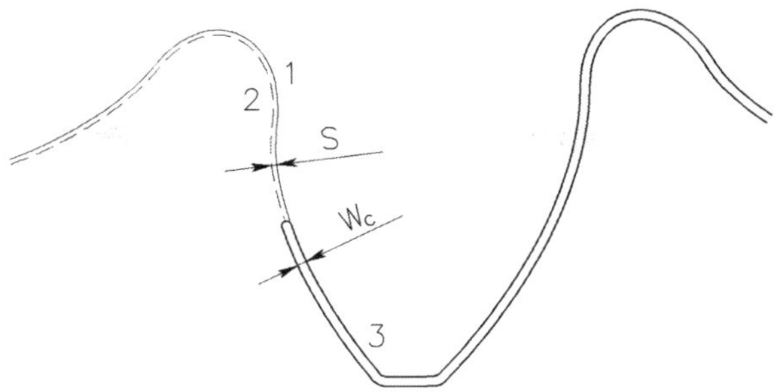

FIGURE 9.13
Contour cutting path: 1: nominal (average material condition) gear profile; 2: tool path; machined gear profile; W_c: cut width; S: tool path offset.

9.1.2 Gear Forming

Forming gear fabrication technologies like plastic and metal injection molding, powder metal (PM) processing, net forging, die casting, extrusion, gear and worm rolling, etc. have gained popularity in the last several decades, providing a high benefit-cost ratio for mass-produced gear drives. Progress of these technologies allowed a significant increase in their accuracy. As a result, usage of formed gears has also increased considerably. Most of these very different forming gear fabrication processes have a similar tooling component that actually defines a gear shape and its accuracy—the tooling (mold or die) cavity. Any gear forming process cavity is dedicated to a particular gear profile. This makes the Direct Gear Design method very acceptable for gear forming technology, because production cost of the custom optimized gears, in this case, is practically the same as that of the similar size standard gears [209].

The forming tool cavity has a profile similar to that of the gear, but adjusted for shrinkage and warpage, which greatly affect gear size and shape accuracy. This makes proper prediction of shrinkage and warpage critical for all gear forming technologies, particularly for the plastic injection molding process. Plastic gears often have an intricate body shape, including ribs or spokes to maintain limited maximum material thickness to exclude voids, and for weight and cost reduction. They are also often incorporated as one piece with other mechanism components, like, for example, shafts, cams, etc. These design specifics, in combination with a huge variety of available gear polymers and enhancing additives (for increased strength, thermoresistance, lubricity, etc.), make prediction of the gear shrinkage and warpage a task of utmost importance. There are several methods to develop a gear molding tool cavity compensating for shrinkage and warpage [210,211]. The most promising approach is an application of a mapping technique for achieving most accurate plastic gears [50,212].

Figure 9.14 shows the gear pump with the asymmetric polymer gears.

PM processing is a cost-effective gear forming technology that is used in many high production volume applications where plastic gears cannot deliver required load capacity due their natural strength limitation.

FIGURE 9.14
Gear pump with the asymmetric tooth polymer gears. (From Kapelevich A.L. and R.E. Kleiss. Gear Technology, September/October 2002. 29–35. With permission.)

Nowadays, this technology is expanding to highly loaded gear drives, for example, for automotive transmissions. For such applications, a conventional PM processing that includes compaction to net shape, sintering, and hardening does not provide required gear accuracy. A final gear tooth grinding is necessary. Figure 9.15 presents an experimental asymmetric gear stage of

FIGURE 9.15
Asymmetric tooth gear pair; 1: machined and ground pinion, 2: powder metal gear with ground teeth (Courtesy of Höganäs AB, Höganäs, Sweden. With permission)

Saab 9-5 that was used as a demonstrator [156]. The driving shaft pinion was made conventionally out of solid steel and the driven gear out of the PM alloy produced by Höganäs AB.

An additional reserve of increasing load capacity of PM gears is the tooth profile densification that increases gear tooth strength and wear resistance. This additional operation allows PM gear performance to be close to the performance of solid gear steels [151]. A study on the roll-burnishing densification process of asymmetric PM gears is described in [139].

9.2 Gear Measurement

Inspection is a critical stage of the gear production process. It is also absolutely essential for development of new gear transmissions. Comprehensive inspection prior to assembly and prototype testing allows the potential design issues to be isolated from manufacturing errors and makes it possible to draw correct conclusions based on the prototype testing results.

Manufacturing of directly designed asymmetric tooth gears not only requires custom tooling, but also affects gear measurement. This chapter presents definitions of main inspection dimensions and parameters for directly designed spur and helical, external, and internal gears with asymmetric teeth.

Gear measurement of asymmetric gears is basically the same as for symmetric gears. However, gear span measurement is not applicable for gears with asymmetric teeth, because it is impossible to have a common tangent line to two concentric base circles of asymmetric tooth flanks.

9.2.1 Measurement over (between) Balls or Pins

Measurement over (for gears with external teeth) or between (for gears with internal teeth) balls or pins is an indirect way to inspect the tooth thickness at the given reference diameter.

9.2.1.1 Measurement of Spur Gears

The position of the measuring ball or pin center is shown in Figure 9.16. The involute angles α_{gd} and α_{gc} at the ball or pin center location circle diameter d_g are defined by [15,22,94]:

- For external gear:

$$inv(\alpha_{gd}) + inv(\alpha_{gc}) = inv(\nu_d) + inv(\nu_c) + \frac{D}{d_{bd}} + \frac{D}{d_{bc}} - \frac{2\pi}{z}. \tag{9.5}$$

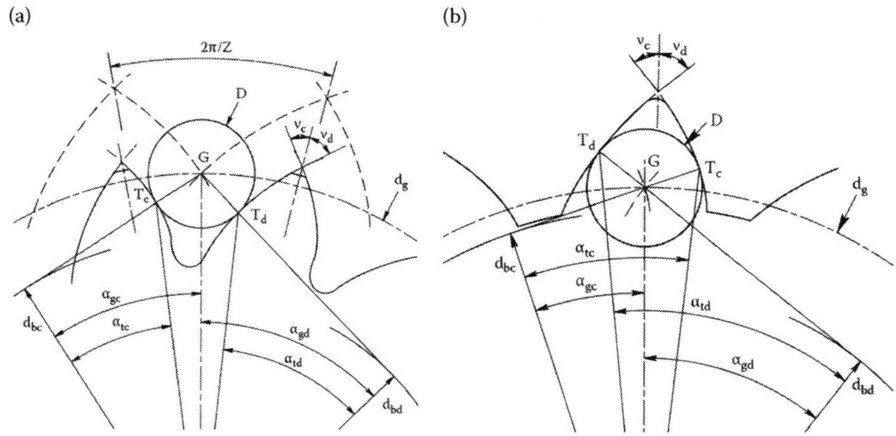

FIGURE 9.16
Ball or pin position: (a) external gear, (b) internal gear. D: ball or pin diameter; G - center of ball or pin. (From Kapelevich A.L., Gear Technology, January/February 2011, 60–65. With permission.)

- For internal gear:

$$inv(\alpha_{gd}) + inv(\alpha_{gc}) = inv(\nu_d) + inv(\nu_c) - \frac{D}{d_{bd}} - \frac{D}{d_{bc}}. \tag{9.6}$$

The ball or pin center location circle diameter d_g is

$$d_g = \frac{d_{bd}}{\cos\alpha_{gd}} = \frac{d_{bc}}{\cos\alpha_{gc}}. \tag{9.7}$$

The involute angles α_{td} and α_{tc} in the ball or pin contact points T_d and T_c are as follows:

$$\alpha_{td} = \arctan\left(\tan\alpha_{gd} \mp \frac{D}{d_{bd}}\right) \tag{9.8}$$

and

$$\alpha_{tc} = \arctan\left(\tan\alpha_{gc} \mp \frac{D}{d_{bc}}\right), \tag{9.9}$$

where the sign "–" is for an external gear and the sign "+" is for an internal gear.

Then a measurement over two balls or pins for an external gear is:

- For even number of teeth (Figure 9.17a):

$$M = d_g + D. \tag{9.10}$$

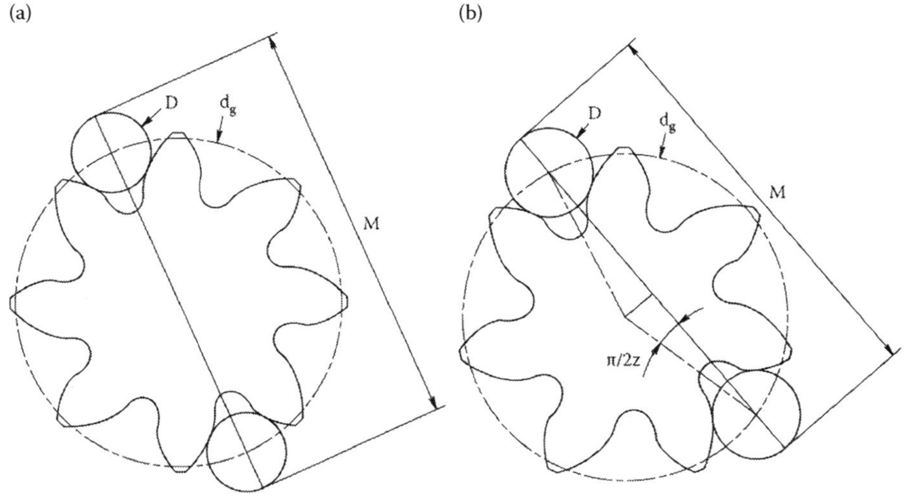

FIGURE 9.17
Measurement over balls or pins for external spur gears: (a) even number of teeth, (b) odd number of teeth. (From Kapelevich A.L., Gear Technology, January/February 2011, 60–65. With permission.)

- For odd number of teeth (Figure 9.17b):

$$M = d_g \cdot \cos\frac{\pi}{2z} + D. \tag{9.11}$$

A measurement between two balls or pins for an internal gear is:

- For even number of teeth (Figure 9.18a):

$$M = d_g - D. \tag{9.12}$$

- For odd number of teeth (Figure 9.18b):

$$M = d_g \cdot \cos\frac{\pi}{2z} - D. \tag{9.13}$$

The ball or pin contact points T_d and T_c should always be located on the involute flanks. They must not contact the tooth profile at the tooth tip radius or chamfer and at the root fillet profile (Figure 9.19). These conditions are described by the contact point involute angle limits:

- For external gears:

$$\arccos\frac{d_{bd}}{d_{fd}} < \alpha_{td} < \arccos\frac{d_{bd}}{d_a} \text{ and } \arccos\frac{d_{bc}}{d_{fc}} < \alpha_{tc} < \arccos\frac{d_{bc}}{d_a}. \tag{9.14}$$

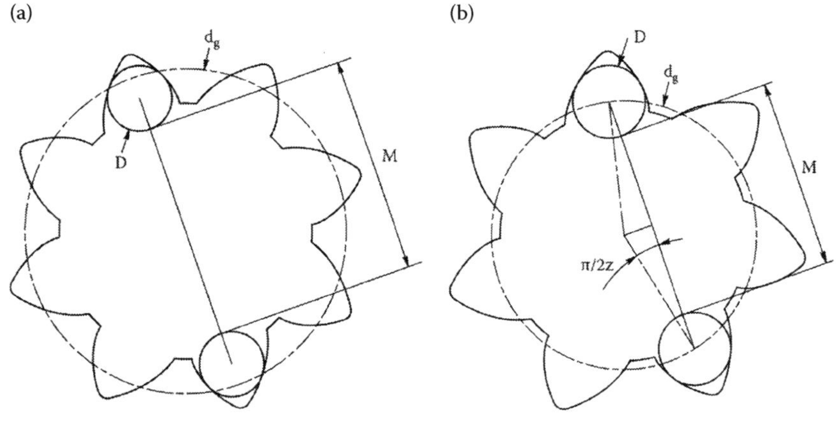

FIGURE 9.18
Measurement over balls or pins for internal spur gears: (a) even number of teeth, (b) odd number of teeth. (From Kapelevich A.L., Gear Technology, January/February 2011, 60–65. With permission.)

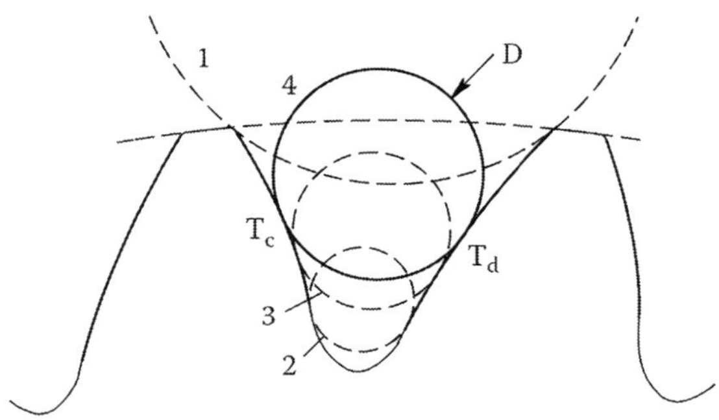

FIGURE 9.19
Ball or pin diameter selection. 1: diameter is too large resulting in a contact point at the tooth tip corner; 2: diameter is too small resulting in a contact point at the tooth root fillet; 3: diameter is acceptable, but the small tip caliper (or micrometer) is required to fit between gear teeth; 4: suitable ball or pin diameter.

- For internal gears:

$$\arccos\frac{d_{bd}}{d_a} < \alpha_{td} < \arccos\frac{d_{bd}}{d_{fd}} \text{ and } \arccos\frac{d_{bc}}{d_a} < \alpha_{tc} < \arccos\frac{d_{bc}}{d_{fc}}, \tag{9.15}$$

  where d_a is the tooth tip diameter, and d_{fd} and d_{fc} are the drive and coast tooth flank form diameters.

For measurement convenience the ball or pin surface should be above the gear tooth tips; otherwise the caliper (or micrometer) should have small tips to fit between gear teeth. This condition can be presented as follows:

$$D > \left| d_a - d_g \right|. \tag{9.16}$$

For external gears this inequality should be solved with Equations (9.5) and (9.7), and for internal gear with Equations (9.6) and (9.7), to define a proper ball or pin diameter.

9.2.1.2 Measurement of Helical Gears

The involute angles α_{gd} and α_{gc} at the ball or pin location circle diameter d_g are defined by [22]:

- For an external gear:

$$inv(\alpha_{gd}) + inv(\alpha_{gc}) = inv(\nu_d) + inv(\nu_c) + \frac{D}{d_{bd} \times \cos\beta_{bd}} + \frac{D}{d_{bc} \times \cos\beta_{bc}} - \frac{2\pi}{z}. \tag{9.17}$$

- For an internal gear:

$$inv(\alpha_{gd}) + inv(\alpha_{gc}) = inv(\nu_d) + inv(\nu_c) - \frac{D}{d_{bd} \times \cos\beta_{bd}} - \frac{D}{d_{bc} \times \cos\beta_{bc}}, \tag{9.18}$$

where β_{bd} and β_{bc} are the helix angles at the drive and coast base diameters that are equal to

$$\beta_{bd} = \arctan(\tan\beta \times \cos\alpha_d), \tag{9.19}$$

$$\beta_{bc} = \arctan(\tan\beta \times \cos\alpha_c). \tag{9.20}$$

Then the ball or pin center location circle diameter d_g is defined by (9.7). This equation is used for the pin center location circle definition only for external helical gears. Cylindrical pins cannot be used to measure the internal helical gears, because the pin surface cannot be tangent to the internal helical gear tooth flanks. They should be inspected by a measurement between balls. The ball or pin diameters should also satisfy conditions (9.14)–(9.16). When diameter d_g is known, measurements over two balls for external helical gears (Figure 9.20) and between two balls for internal helical gears (Figure 9.21) are defined by Equations (9.10) and (9.11), and (9.12) and (9.13), accordingly.

Measurement over two pins for external helical gears with an even number of teeth is also defined by Equation (9.10). Measurement over two pins for external helical gears with an odd number of teeth is considered inappropriate by some gear experts, because the shortest distance between the pin centers does not intersect the gear axis. However, this type of measurement is commonly used in gear production, and it is necessary to provide a correct definition of measurement over two pins for external helical gears with an odd number of teeth.

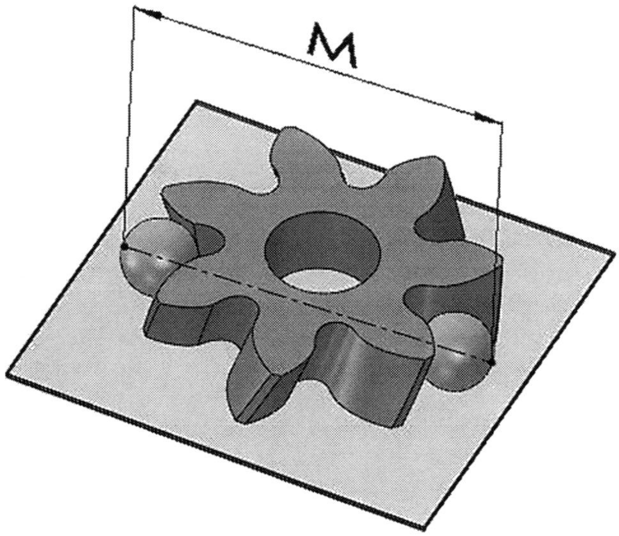

FIGURE 9.20
Measurement over balls of external helical gear. (From Kapelevich A.L., Gear Technology, January/February 2011, 60–65. With permission.)

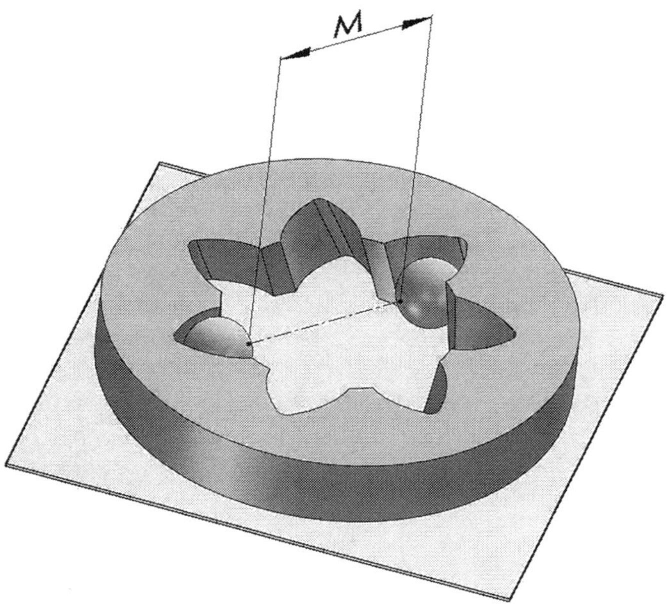

FIGURE 9.21
Measurement between balls of internal helical gear. (From Kapelevich A.L., Gear Technology, January/February 2011, 60–65. With permission.)

For external helical gears with an odd number of teeth, the shortest distance L between the pin centers does not lie in the transverse section of the circle diameter d_p (Figure 9.22). This distance definition is described in [213] as follows:

$$L = \frac{d_g}{2 \times \tan\beta_g}\sqrt{\lambda^2 + 4 \times \left(\tan\beta_g \times \cos\left(\frac{\pi}{2z} + \frac{\lambda}{2}\right)\right)^2}, \tag{9.21}$$

where the helix angle at the pin center diameter β_g is

$$\beta_g = \arctan\left(\frac{\tan\beta_{bd}}{\cos\alpha_{gd}}\right) = \arctan\left(\frac{\tan\beta_{bc}}{\cos\alpha_{gc}}\right) \tag{9.22}$$

and the angle λ is a solution of the equation

$$\frac{\lambda}{\tan\beta_g} - \sin\left(\frac{\pi}{z} + \lambda\right) = 0. \tag{9.23}$$

Then the measurement over two pins for external helical gears with an odd number of teeth (Figure 9.23) is

$$M = L + D. \tag{9.24}$$

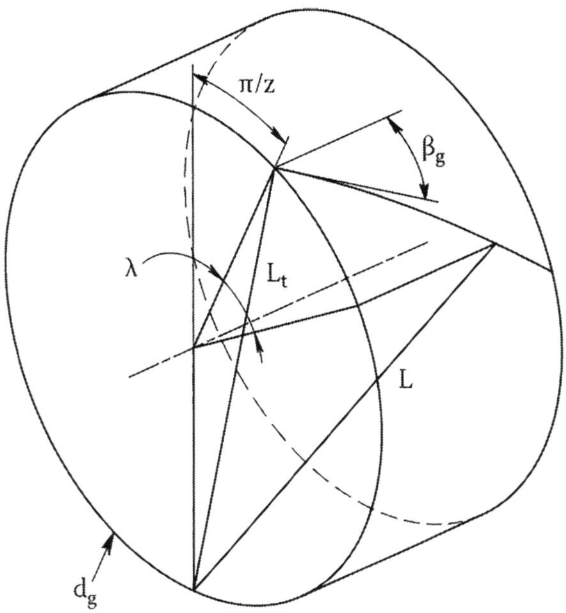

FIGURE 9.22
Definition of the distance between the pin centers for the helical gears with an odd number of teeth. (From Kapelevich A.L., Gear Technology, January/February 2011, 60–65. With permission.)

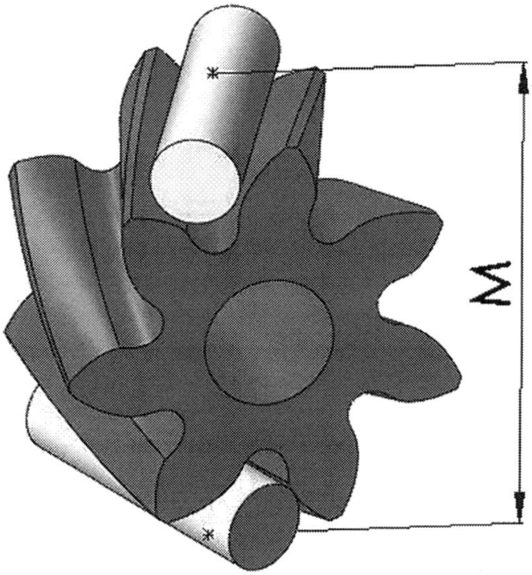

FIGURE 9.23
Measurement over pins of the external helical gear with an odd number of teeth. (From Kapelevich A.L., Gear Technology, January/February 2011, 60–65. With permission.)

9.2.2 Composite Gear Inspection

There are two types of composite gear inspection: single- and double-flank composite testing [214]. Single-flank composite testing is used for the mating gears at a fixed center distance for transmission error component measurement that includes adjacent pitch variation, total accumulated pitch variation, tooth-to-tooth transmission variation, and total transmission variation. This type of testing applied to custom directly designed gears is practically the same as for conventional gears, except for the gears with asymmetric teeth that use both flanks for torque or motion transmission. In this case, opposite flanks require separate testing. This method provides a good indication of gear pair functionality because it checks two mating gears.

Double-flank composite testing has the inspected gear mounted on a rolling fixture (roll tester) with a tight spring-loaded mesh with a master gear. Deviations of the center distance during gear rotation indicate the tooth-to-tooth composite error (TTE) and total composite error (TCE). Modern roll testers with a computerized data acquisition system also allow the evaluation of functional tooth thickness and radial runout. Double-flank roll testing is a quick and inexpensive way to separate acceptable and defective gears. However, it does not indicate which gear dimension or accuracy parameter is responsible for the excessive TTE and TCE. It also does not recognize which gear tooth flank is a major contributor to the composite errors. Double-flank composite testing does not provide sufficient data about actual gear pair functionality because it checks only one gear in mesh with the master gear.

FIGURE 9.24
CMM measurement of asymmetric gear. (From Kapelevich A.L., Gear Technology, January/February 2011, 60–65. With permission.)

This type of testing is also applied to asymmetric gears. However, it requires the special asymmetric master gears.

9.2.3 Elemental Gear Inspection

Coordinate measuring machines (CMM) are used for elemental gear inspection (Figure 9.24). It allows the mapping a surface of all teeth, including the fillet profiles. It provides measurement results for involute flank elemental accuracy parameters: radial runout tolerance, pitch variation, profile

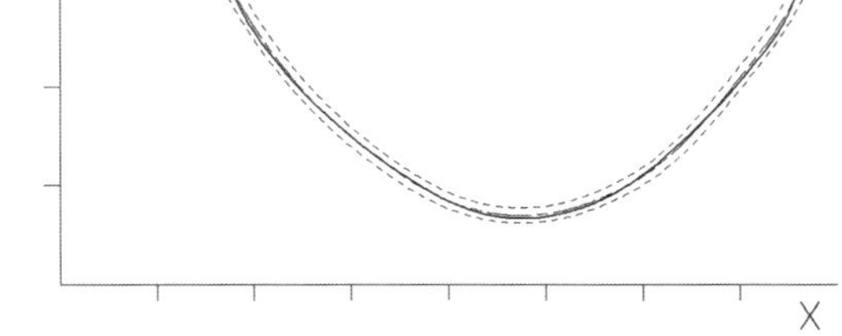

FIGURE 9.25
CMM inspection chart of the asymmetric gear root fillet profile. 1: nominal (average material condition) profile (dash-dotted contour); 2: tolerance corridor (dashed contour); 3: actual inspected profile (solid fat contour).

tolerance, and lead or tooth alignment tolerance. Directly designed asymmetric gears have an optimized tooth profile, including the root fillet. Therefore, actual accuracy of the root fillet profile should also be inspected by a CMM. The gear tooth (including the root fillet) CAD profile for the average material condition is used for the CMM inspection. The data set also includes the involute flank and fillet profile tolerances that are established, depending on required gear accuracy and also manufacturing technology. The CMM is programmed to indicate if the inspected tooth profile points lie within the corridor defined by the CAD tooth profile ± a profile tolerance. An example of the CMM inspection chart of the optimized root fillet profile of the asymmetric tooth gear is shown in Figure 9.25.

10

Symmetric vs. Asymmetric Gears

10.1 Analytical Comparison

In some publications, performance results of new advanced gear geometry have been compared to those of standard 20° pressure angle gears, indicating remarkable advantages of a new gear geometry in, for example, load carrying capacity. While such a comparison makes some sense, because baseline performance of the standard gears is well known, it cannot be considered a fair one. Although the standard gears are universally applicable, but they are not suitable for applications that require high performance. New advanced performance gear geometry should be compared with the best existing solutions to evaluate its true benefits. The comparable, for load carrying capacity, existing baseline gear geometry can be found, for example, in the aerospace gear transmissions. Although such a comparison approach does not usually result in very impressive numbers, indicating new gear geometry advantages, it puts this geometry side by side with the best existing one to assess its true benefits. Table 10.1 presents the directly designed optimized gears with asymmetric teeth in comparison to the traditionally designed aerospace type gears that have a 25° pressure angle and a full circular root fillet. High transverse contact ratio (HCR) gears also have potentially greater load capacity in comparison to the low and medium transverse contact ratio gears (Section 5.1.1). Table 10.2 presents a comparison of the directly designed optimized asymmetric HCR gears with the traditionally designed HCR gears that have the high (>2.0) transverse contact ratio and the full circular root fillet. Such gears are also used in aerospace gear transmissions [215].

10.2 Experimental Comparison

Experimental comparison of the traditionally designed symmetric gears and asymmetric directly designed gears was done by the Rotorcraft division of the Boeing Company [95]. Directly designed asymmetric gears were

TABLE 10.1

Comparison of Traditionally and Directly Designed Conventional Gears

Design Method	Traditional (Baseline)		Direct	
Tooth shape	Symmetric		Asymmetric	
Gear mesh	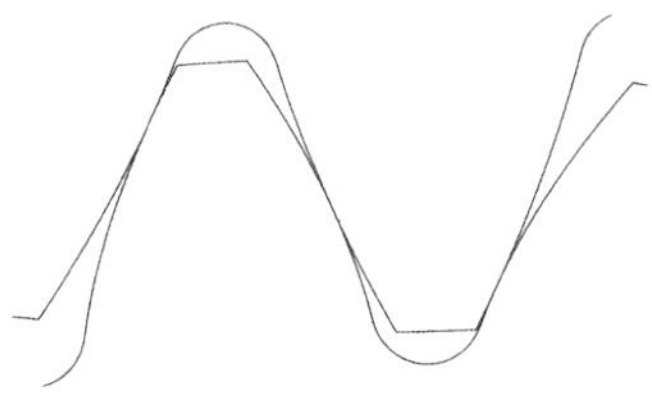		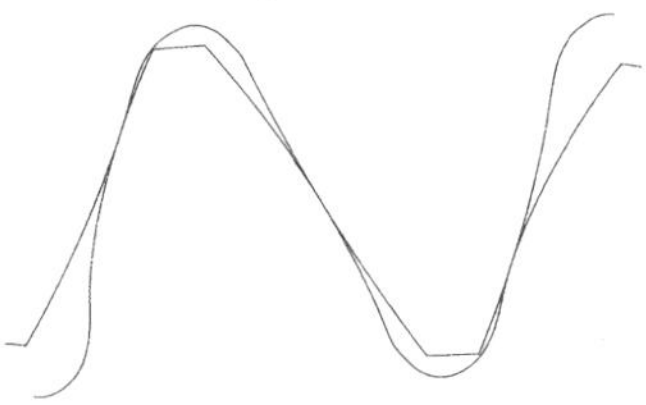	
Gears	Pinion	Gear	Pinion	Gear
Number of teeth	27	49	27	49
Module, mm	3.0		3.0	
X-shift	0.09	–0.09	N/A	N/A
Root fillet profile	Full Circle		Optimized	
Operating pressure angle	25°		32°/18°[a]	
Drive contact ratio	1.49		1.49/1.98[a]	
Pitch diameter (PD), mm	81.0	147.0	81.0	147.0
Outer diameter, mm	87.540	152.46	87.895	153.645
Root diameter, mm	74.285	138.962	73.685	139.488
Root clearance, mm	0.628	0.749	0.335	0.308
Tooth thickness at PD, mm	4.955	4.469	4.873	4.551
Tooth tip thickness, mm	1.543	1.797	1.13	1.13
Center distance, mm	114.0		114.0	
Face width, mm	30	30	30	30
Driving torque, Nm	300		300	
Bending stress, MPa	210	213	182 (–13%)	183 (–14%)
Contact stress, MPa	958		886 (–7.5%)	
Specific sliding velocity	0.258	0.230	0.228	0.228
Bearing load, N	8,172		8,734 (+7%)	

[a] Drive/coast flank.

analyzed to determine their bending and contact stresses relative to the symmetric involute gear tooth form, which is representative of helicopter main drive gears. Asymmetric and baseline symmetric toothed gear test specimens were designed, fabricated, and tested to experimentally determine their single tooth bending fatigue (STBF) strength and scuffing resistance. For fair comparison both gear sets had the circular (not optimized) root fillet. The gear test specimens are presented in Figure 10.1.

TABLE 10.2

Comparison of Traditionally and Directly Designed HCR Gears

Design Method	Traditional (Baseline)		Direct	
Tooth shape	Symmetric		Asymmetric	
Gear mesh	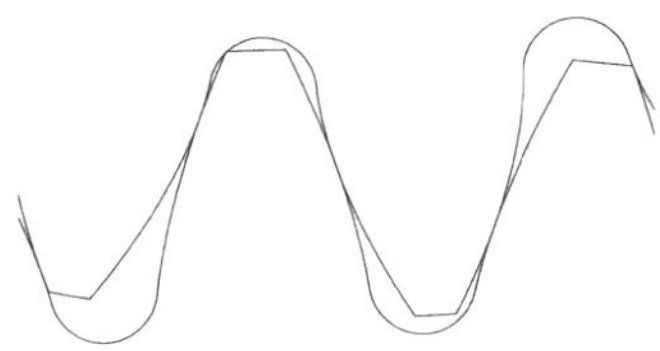		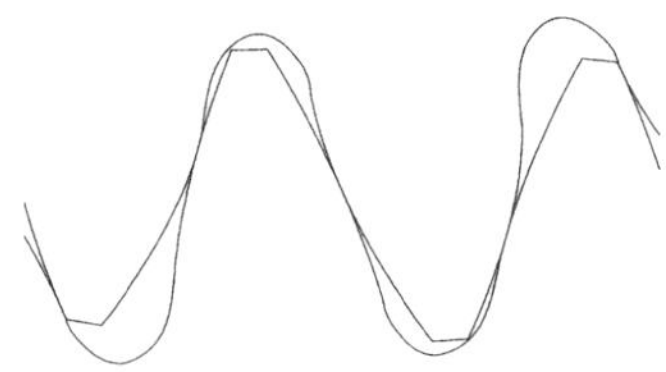	
Gear	Pinion	Gear	Pinion	Gear
Number of teeth	27	49	27	49
Module, mm	3.0		3.0	
X-shift	0.15	–0.15	N/A	N/A
Root fillet profile	Full circle		Optimized	
Operating pressure angle	20°		24°/16°[a]	
Drive contact ratio	2.04		2.04/2.48[a]	
Pitch diameter (PD), mm	81.0	147.0	81.0	147.0
Outer diameter, mm	89.40	153.60	89.576	154.963
Root diameter, mm	73.658	137.822	72.394	137.901
Root clearance, mm	0.371	0.389	0.322	0.262
Tooth thickness at PD, mm	5.040	4.384	5.020	4.404
Tooth tip thickness, mm	1.18	1.70	1.01	1.02
Center distance, mm	114.0		114.0	
Face width, mm	30	30	30	30
Driving torque, Nm	300		300	
Bending stress, MPa	147	150	126 (–14%)	126 (–16%)
Contact stress, MPa	824		774 (–6%)	
Specific sliding velocity	0.367	0.323	0.336	0.336
Bearing load, N	7,882		8,108 (+3%)	

[a] Drive/coast flank.

The objective of this work was to evaluate the potential benefits of asymmetric involute gear teeth for helicopter main transmission applications. This involved not only quantifying performance improvements achieved by these concepts, but evaluating the practicality of manufacturing gears with asymmetric teeth for aerospace applications.

Test specimen gears designed for this program were representative of helicopter main drive gears in diametral pitch, pressure angle, material, and processing. Standardized traditional toothed designs have been developed

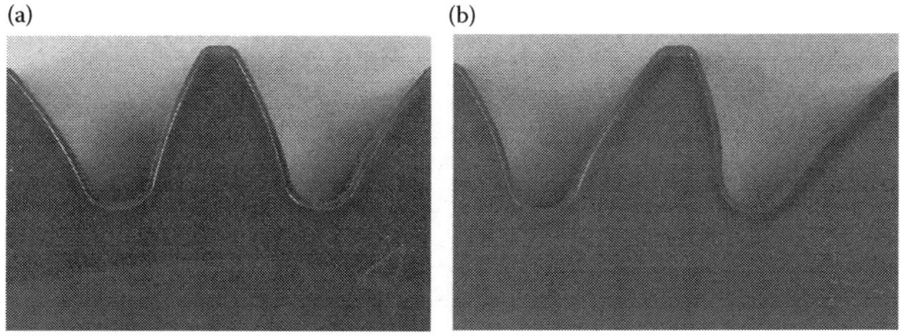

FIGURE 10.1
Test specimen gear tooth profiles: (a) baseline symmetric gear teeth, (b) asymmetric gear teeth. (Courtesy of Boeing Co., Philadelphia, Pennsylvania.)

for bending fatigue and scuffing test rigs that Boeing Rotorcraft uses for gear research. The standardized test specimen designs were modified to incorporate the asymmetric tooth configuration. Specimens of each type were manufactured using aerospace production techniques and requirements. A manufacturing approach was developed with a goal of reducing material and processing variability. The test specimen gear designs were analyzed to predict their bending and contact stresses, and compared to stresses predicted for the baseline test specimens.

The STBF test gears are 32-tooth gears with groups of 4 teeth removed per quadrant to allow for assembly into the STBF test fixture. For comparison, conventional symmetric involute gears and gears with asymmetric involute teeth were designed and tested. Both asymmetric toothed and conventional baseline specimens employ ground circular root fillets. The asymmetric gear tooth form for the STBF test specimens was nominally based on the standard STBF gear specimen. This enabled the asymmetric toothed specimen to fit the existing test fixture with only minor modifications for tooth load angle, and provided a direct comparison between asymmetric and conventional gears of the same diameter and face width. The gear parameters and finite element analysis (FEA) calculated bending stresses for the STBF test gears are presented in Table 10.3, and the scuffing test gears are within the design experience range of typical main transmission helicopter power gears. The gear parameters for the scuffing test gear specimens are presented in Table 10.4.

The asymmetric gear specimens and baseline symmetric circular fillet test gears were fabricated by Aero Gear (South Windsor, Connecticut). The specimens were fabricated from aerospace quality (AGMA Grade 3) 9310 steel with all pertinent records and certifications retained. All specimens were low pressure carburized and high pressure gas quenched. This heat treatment processes were performed at Solar Atmospheres (Souderton, Pennsylvania). The material for all specimens was from the same lot and the heat treat processes, grind stock removal, and shot peening processes for all

TABLE 10.3

STBF Test Gear Specimen Parameters

	Symmetric Gears (Baseline)	Asymmetric Gears
Number of teeth of both mating gears	32	32
Diametral pitch, 1/in	5.333	5.333
Pressure angle	25°	35°/15°[a]
Pitch diameter, in	6.000	6.000
Base diameter, in	5.4378	4.9149/5.7956[a]
Outside diameter, in	6.3975	6.3864
Root diameter, in	5.571	5.558
Form diameter, in	5.6939	5.6581/5.8110[a]
Circular tooth thickness, in	0.2895	0.2895
Face width, in	0.375	0.375
Torque, in-lb	5,000	5,000
Load application radius, in	3.06	3.06
Calculated maximum Bending stress, psi	57,887	54,703 (–5.5%)

[a] Drive/coast flank.

TABLE 10.4

Scuffing Test Gear Specimen Parameters

	Symmetric gears (Baseline)	Asymmetric gears
Number of teeth of both mating gears	30	30
Diametral pitch, 1/in	5.000	5.000
Pressure angle	25°	35°/18°[a]
Pitch diameter, in	6.000	6.000
Base diameter, in	5.4378	4.9149/5.7063[a]
Outside diameter, in	6.400 max	6.403 max
Root diameter, in	5.459 max	5.510
Form diameter, in	5.6864	5.6415/5.7607[a]
Circular tooth thickness, in	0.3096	0.3096
Face width, in	0.50	0.50
Drive contact ratio	1.417	1.25
Torque, in-lb	6,000	6,000
Calculated maximum Contact stress, psi	193,180	174,100 (–9.9%)

[a] Drive/coast flank.

specimens were identical. All gears were surface temper etch inspected and magnetic particle inspected after the completion of machining.

All specimens produced for this project were ground using conventional gear tooth form grinding equipment. The form grinding process is often used to grind symmetric gear teeth with circular fillets in helicopter main drives. The cubic boron nitride (CBN) form grinding wheels were produced from data shown on the engineering drawings for both the asymmetric gear teeth and baseline symmetric gear teeth. The CBN gear grinding setup is shown in Figure 9.2. Measurements of the gear teeth were carried out using conventional CMM gear checking equipment and software.

STBF tests were performed on nonrotating STBF test fixtures. These fixtures are loaded by Baldwin-Lima Hamilton IV-20 universal fatigue machines through a series of alignment fixtures and in-line load cells. These fatigue machines are capable of 18,000 lb (10,000 lb steady load and 8,000 lb alternating load).

For the STBF testing of the subject gears, pulsating fatigue load is applied to the tooth through the load link and fixture arrangement shown in Figure 10.2. The test gear teeth were cycled at approximately 1,200 cycles per minute. Prior to the start of testing, alignment of the fixture was verified with a strain-gaged baseline specimen. The specimen was instrumented with three strain gages across the face width and was used to align the fixture as well as correlate load applied to stress in the fillet of the tooth. For fatigue testing, each tested tooth is instrumented with a crack-wire, as seen in Figure 10.3. Upon failure of the crack-wire due to the presence of a fatigue crack, the test machine is triggered to shut down. The crack-wire is placed so that a crack length of 0.050 in. is detected. Magnetic particle inspection (MPI) is used to confirm the presence of a crack. Each tooth specimen was run continuously until failure or run-out. For this project, run-out was defined as 1.0×10^7 cycles.

Scuffing tests of asymmetric gear specimens and baseline specimens were conducted on a gear research test stand. The test stand is a split coupling torque design. The test gears are outboard of the main housing and can be

FIGURE 10.2
STBF test fixture with asymmetric gear installed. (Courtesy of Boeing Co., Philadelphia, Pennsylvania. From Brown, F.W., et al. Analysis and Testing of Gears with Asymmetric Involute Tooth Form and Optimized Fillet Form for Potential Application in Helicopter Main Drives, Gear Technology, June/July 2011, 46–55. With permission.)

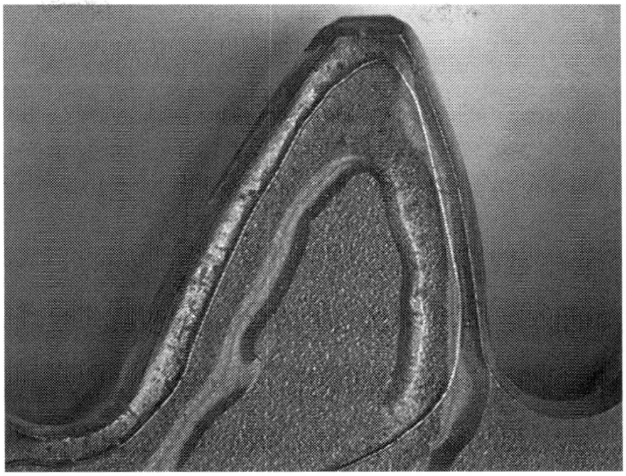

FIGURE 10.3
Asymmetric STBF test tooth with crack-wire installed. (Courtesy of Boeing Co., Philadelphia, Pennsylvania. From Brown, F.W., et al. Analysis and Testing of Gears with Asymmetric Involute Tooth Form and Optimized Fillet Form for Potential Application in Helicopter Main Drives, Gear Technology, June/July 2011, 46–55. With permission.)

FIGURE 10.4
Scuffing test rig with cover removed and test specimen gears installed. (Courtesy of Boeing Co., Philadelphia, Pennsylvania. From Brown, F.W., et al. Analysis and Testing of Gears with Asymmetric Involute Tooth Form and Optimized Fillet Form for Potential Application in Helicopter Main Drives, Gear Technology, June/July 2011, 46–55. With permission.)

quickly inspected or changed by removal of a simple cover (see Figure 10.4). A separate lubrication system serves the test specimen chamber, which was isolated from the test stand drive lubrication system. The lubricant supply to the test gears could be heated or cooled to supply lubricant at a constant temperature to the test gears. The test gears were subjected to a series of 15 min long, incrementally loaded runs. At the end of each 15-min run, a visual evaluation of the test gear teeth was conducted. If the condition of the gears did not meet the criteria for scuffing failure, the next higher incremental load was applied. This procedure was continued until a scuffing failure was observed. For purposes of this test program, a scuffing failure was declared when 25% of the available tooth contact surface exhibited visible evidence of radial scratch marks, characteristic of scuffing, on a minimum of ten teeth.

In the conclusion of the STBF tests, all crack locations were verified both visually and using MPI, as shown in Figure 10.5. Cracks were also opened to determine the origins and confirm the validity of the results. In Figure 10.6 the dark dashed line represents the extent of fatigue propagation, and the arrow indicates the fracture origin.

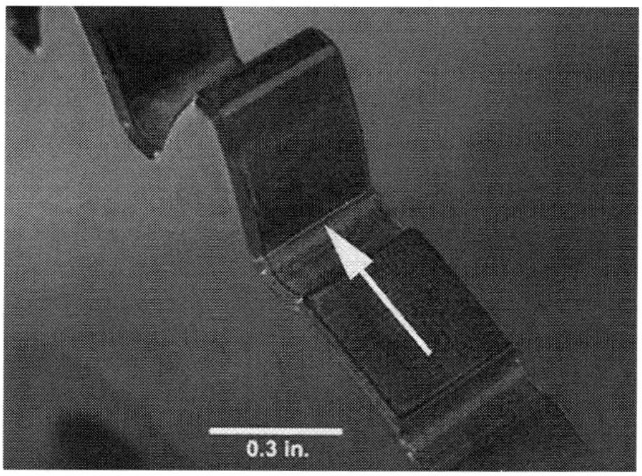

FIGURE 10.5
Cracked STBF test gear tooth showing MPI crack indication. (Courtesy of Boeing Co., Philadelphia, Pennsylvania. From Brown, F.W., et al. Analysis and Testing of Gears with Asymmetric Involute Tooth Form and Optimized Fillet Form for Potential Application in Helicopter Main Drives, Gear Technology, June/July 2011, 46–55. With permission.)

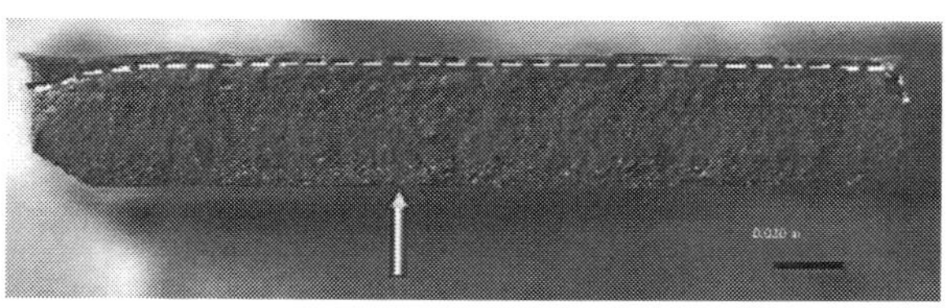

FIGURE 10.6
Fractograph of STBF test tooth. (Courtesy of Boeing Co., Philadelphia, Pennsylvania. From Brown, F.W., et al. Analysis and Testing of Gears with Asymmetric Involute Tooth Form and Optimized Fillet Form for Potential Application in Helicopter Main Drives, Gear Technology, June/July 2011, 46–55. With permission.)

The fatigue results of the STBF tests of the asymmetric tooth gears and the baseline specimens are presented in Figure 10.7. Curves for the asymmetric data were assumed to be parallel to the baseline curve.

Typical scuffing failures are shown in Figures 10.8 and 10.9. These figures show the vertical scratches indicative of a scuffing failure, associated with the breakdown of the separating lubricant film between the gears. Figure 10.9 shows the scuffing results for baseline and asymmetric gears. The 35° pressure angle asymmetric gears showed an improvement of approximately 25% in the mean scuffing load (torque) compared to the baseline symmetric tooth specimens. The mean 3-sigma levels are also shown, based on a population of eight baseline data points and six asymmetric data points.

The STBF test results shown in Figure 10.7 indicate that the asymmetric tooth gear design mean endurance limit was significantly higher, on the order of 16%, than the mean endurance limit of the baseline symmetric tooth design. It should be pointed out that there are relatively few data points, four failure points, and one run-out (included as a failure point in the data analysis), for the asymmetric tooth specimens. Nonetheless, the results of this testing indicate that asymmetric teeth offer an improvement in bending fatigue strength, although additional testing would serve to refine the magnitude of the improvement. It is interesting to note that the FE analysis of the asymmetric tooth STBF design predicted a 5.5% reduction in maximum bending stress compared to the baseline symmetric design.

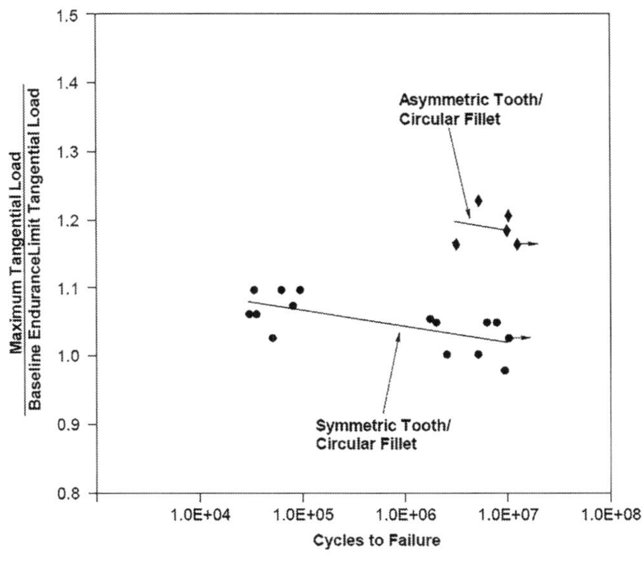

FIGURE 10.7
STBF test results for asymmetric gears with circular fillet along with baseline symmetric gears. (From Brown, F.W., et al. Analysis and Testing of Gears with Asymmetric Involute Tooth Form and Optimized Fillet Form for Potential Application in Helicopter Main Drives, Gear Technology, June/July 2011, 46–55. With permission.)

FIGURE 10.8
Scuffing failure of baseline test gear. (Courtesy of Boeing Co. Philadelphia, Pennsylvania. From Brown, F.W., et al. Analysis and Testing of Gears with Asymmetric Involute Tooth Form and Optimized Fillet Form for Potential Application in Helicopter Main Drives, Gear Technology, June/July 2011, 46–55. With permission.)

FIGURE 10.9
Close-up view of a representative scuffed tooth. (Courtesy of Boeing Co., Philadelphia, Pennsylvania. From Brown, F.W., et al. Analysis and Testing of Gears with Asymmetric Involute Tooth Form and Optimized Fillet Form for Potential Application in Helicopter Main Drives, Gear Technology, June/July 2011, 46–55. With permission.)

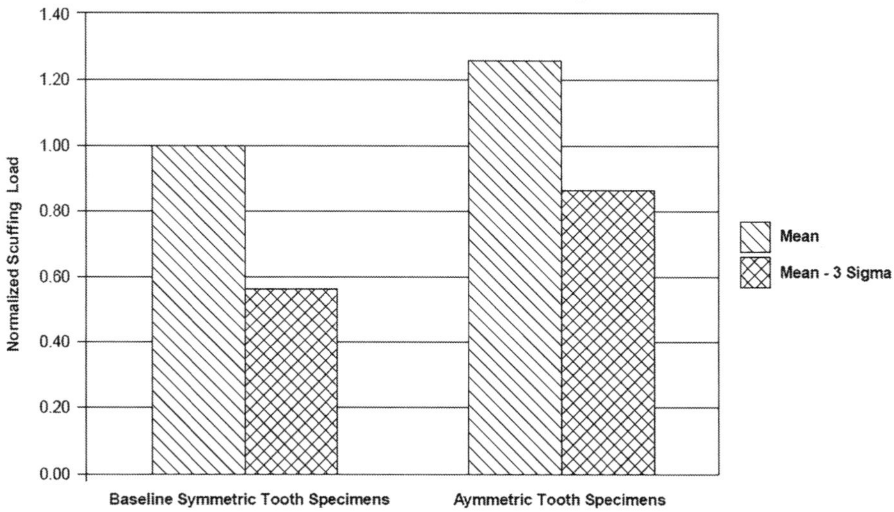

FIGURE 10.10
Results of baseline symmetric and asymmetric gear scuffing tests. (From Brown, F.W., et al. Analysis and Testing of Gears with Asymmetric Involute Tooth Form and Optimized Fillet Form for Potential Application in Helicopter Main Drives, Gear Technology, June/July 2011, 46–55. With permission.)

While not tested in this project, the combination of asymmetric teeth and optimized fillet geometry, in the same gear design, may offer improvements in tooth bending fatigue strength greater than either of the concepts taken individually. The decision was made early in this project to test each concept separately. The reasoning was that if one concept or the other proved to be impractical from a manufacturing standpoint, data of value would still be attained for the other concept. Since both concepts appear viable from a manufacturing standpoint, their combination in one gear design is worth further investigation.

The scuffing test results (Figure 10.10) indicated an improvement in mean scuffing load (torque) to failure of 25% for the asymmetric tooth gear specimens compared to the baseline symmetric tooth specimens. The improvement in calculated mean 3-sigma scuffing performance is even greater. Although based on limited testing, eight baseline points and six asymmetric tooth data points, this is a very significant improvement in scuffing resistance due to asymmetric gear tooth geometry. This improvement was in the primary drive direction of the asymmetric teeth. The opposite (coast) direction scuffing performance of the asymmetric teeth was not tested in this project. This improvement in scuffing resistance can be utilized to advantage in high-speed, scuffing-critical gear applications.

Test results demonstrated higher bending fatigue strength for the asymmetric tooth gears than for baseline designs. Scuffing resistance was significantly increased for the asymmetric tooth gears compared to a traditional symmetric involute tooth design.

11

Asymmetric Gear Applications

This chapter presents three examples of asymmetric gear application.

11.1 Gas Turbine Engine Gear Drive

Experimental application of spur asymmetric gears was done for a gear drive of the small size gas-turbine engine 9I56 made by The Kaluga Dvigatel plant (Kaluga, Russia) [216]. This engine is for the truck mounted electric generators. The main engine parameters are provided in Table 11.1.

The engine gearbox contains one helical gear stage. With a goal to explore the acoustic characteristics of the asymmetric gears, the serially made helical gears were used as a benchmark for comparison with the experimental spur asymmetric tooth gears (Figure 11.1).

The geometric data of the benchmark helical and asymmetric spur gears are shown in Table 11.2.

Two sets of asymmetric gears were made in order to test both the high (set #1) and low (set #2) pressure angle drive flanks for a comparison. The benchmark and experimental gears were made out of steel 12X2H4A, carburized, and case harden to achieve the tooth flank surface hardness Rc > 55.

In order to make the prototype gears quickly, their manufacturing process was altered compared to the mass-produced benchmark helical gears. Instead of using the protuberance hob for final machining of the tooth root and leaving a stock at the involute flanks for their subsequent grinding after heat treatment, the fly cutter was used for a final definition of the tooth root and roughing of the involute flanks. Then a preliminary tooth grinding was applied to create an event stock for a final grinding after heat treatment. Figure 11.2 shows the symmetric gear machining sequence and the fly cutter. The symmetric tooth fly cutter with the straight cutting edges was positioned with an offset S from a vertical axis for rough machining of the involute flanks of asymmetric gears.

Vibro-testing of the 5I56 engine gearbox with the benchmark helical and asymmetric spur gears was done to evaluate the acoustic characteristics of the asymmetric gears at the testing rig [14]. Three accelerometers were attached to the gearbox housing near the bearing locations, and an average vibration level for every gear set was registered at the maximum engine

TABLE 11.1

Main Parameters of 9I56 Gas-Turbine Engine

Power, hp	110
Turbine RPM	27,600
Output shaft RPM	6,000
Life, h	1200
Dimensions, mm	1034 × 577 × 660
Weight, kg	150

FIGURE 11.1
Experimental asymmetric tooth gear drive of the 9I56 gas-turbine engine; 1: driving pinion, 2: driven gear (From Kapelevich A.L. and R.E. Kleiss. Gear Technology, September/October 2002, 29–35. With permission.)

power and 27,600 RPM of the turbine shaft. A comparison criterion was a maximum vibration amplitude at the gear tooth mesh frequency.

Table 11.3 presents the results of the tolerance and stress analysis and acoustic testing.

The test results indicate that the asymmetric gear set #1 with the high 41° pressure angle drive flanks had a minimal vibration amplitude at tooth mesh frequency, lower than the benchmark helical gear set, and significantly lower than the asymmetric gear set #2 with the low 18° pressure angle drive flanks. This could be explained by a higher tooth tip deflection and the thicker elasto-hydrodynamic lubricant film thickness of the asymmetric gears with the high 41° pressure angle drive flanks that softens the tooth engagement impacts [14]. Besides, the gear stress analysis of the comparable gear sets has shown that the asymmetric gears with the high 41° pressure angle drive flanks has potentially greater load capacity.

TABLE 11.2

Geometric Data of Benchmark Helical and Experimental Asymmetric Spur Gears

Gear Pair	Benchmark Helical		Experimental Spur Asymmetric Sets #1 and #2	
Gear	pinion	gear	pinion	gear
Number of teeth	19	79	13	60
Normal module, mm	1.500		2.034	
Normal drive pressure angle	20°		41°(#1), 18°(#2)	
Normal coast pressure angle	20°		18°(#1), 41°(#2)	
Helix angle	14.5°		–	
X-shift coefficient	+0.300	−0.379	–	
Nominal pitch diameter (PD), mm	26.339	122.399	26.445	122.055
Drive base diameter, mm	24.654	114.570	19.958(#1), 25.151(#2)	92.116(#1), 116.081(#2)
Coast base diameter, mm	24.654	114.570	25.151(#1), 19.958(#2)	116.081(#1), 92.116(#2)
Tooth tip diameter, mm	30.128/30.228	123.937/124.137	30.790/30.990	125.270/125.470
Root diameter, mm	22.940/23.036	116.848/116.944	22.180/22.420	116.640/116.890
Tooth thickness at PD, mm	2.648/2.683	1.865/1.900	3.680/3.715	2.567/2.622
Face width, mm	26.90/27.00	21.90/22.00	26.90/27.00	21.90/22.00
Center distance, mm	74.250 ± 0.030		74.250 ± 0.030	
Accuracy per GOST 1643 [217][a]	6-5-5	6-5-5	6-5-5	6-5-5

[a] Close to the Q11-Q12 accuracy grade per the AGMA 2000-A88 standard.

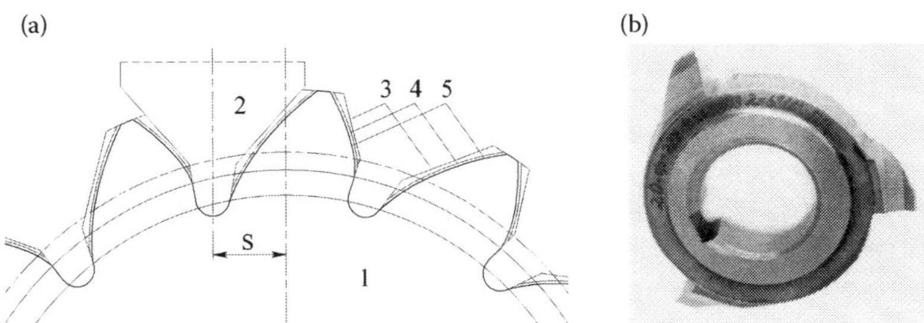

FIGURE 11.2
Asymmetric gear machining (a); 1: gear blank, 2: fly cutter (also the image b), 3: gear profile after roughing, 4: tooth flanks after preliminary grinding prior to heat treatment; 5: (thick lines) gear profile after heat treatment and final tooth flank grinding, 6: high pressure angle flank form diameter, 7: low pressure angle flank form diameter; *S*: fly cutter position offset.

TABLE 11.3

Tolerance and Stress Analysis and Acoustic Testing Results

Gear Pair	Benchmark Helical Set		Experimental Spur Asymmetric Sets #1 and #2	
Gear	pinion	gear	pinion	gear
Number of teeth	17	79	13	60
Normal module, mm	1.500		2.034	
Helix angle	14.5°		–	
Tolerance analysis results				
Operating transverse pressure angle	20.26/20.46°		40.97/41.05°(#1), 17.92/18.12°(#2)	
Operating transverse contact ratio	1.39/1.52		1.13/1.20(#1), 1.55/1.65(#2)	
Axial contact ratio	1.17		–	
Total contact ratio	2.56/2.69		1.13/1.20(#1), 1.55/1.65(#2)	
Normal backlash, mm	0.009/0.138		0.009/0.155(#1), 0.0110/0.196(#2)	
Radial clearance, mm	0.616/0.859	0.617/0.809	0.262/0.569	0.267/0.528
Stress analysis results				
Torque, Nm	38	175	38	175
RPM	27,600	6,000	27,600	6,000
Material	12X2H4A	12X2H4A	12X2H4A	12X2H4A
Max. Root bending stress, MPa	371	396	292(#1), 289(#2)	247(#1), 260(#2)
Max. Flank contact stress, MPa	1264		1123(#1), 1456(#2)	
Acoustic testing results				
Max. Vibration amplitude at tooth mesh frequency, g	7.7		2.7(#1) & 10.4(#2)	

11.2 Turboprop Engine Gearbox

Apparently the first application of asymmetric tooth gears in the aviation industry was for the TV7–117S turboprop engine gearbox [35,62,218]. The engine and gearbox were developed by Klimov Corporation (St. Petersburg, Russia) with the assistance of the Central Institute of Aviation Motors (CIAM, Moscow, Russia) for a commuter airplane Ilyushin Il-114 and produced by Chernyshev Enterprise (Moscow, Russia). The main characteristics of its gearbox are presented in Table 11.4.

The TV7–117S gearbox arrangement is shown in Figure 11.3. This arrangement was successfully used in an older generation of Russian turboprop engines AI-20 and AI-24, and it was able to provide high power transmission density for a required gear ratio. The first planetary differential stage has three planet gears. The second "star" type coaxial stage has five planet (idler)

TABLE 11.4

TV7–117S Turboprop Engine Data

Input turbine RPM	17,500
Output prop RPM	1,200
Total Gear Ratio	14.6:1
Overall Dimensions, mm:	
• Diameter	520
• Length	645
Gearbox weight, N	1,050
Maximum output power, hp	2,800

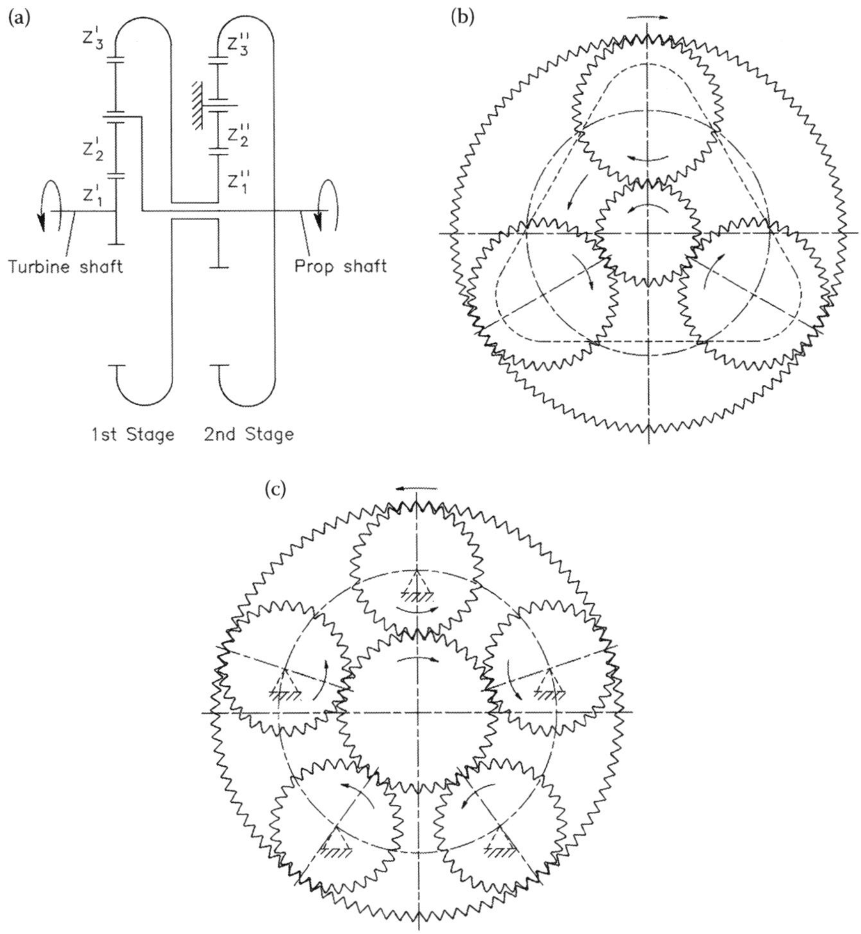

FIGURE 11.3
Gearbox arrangement (a), first (b) and second (c) stages with rotation directions (view from input shaft). ((a) from Novikov, A.S., et al., Gear Technology, January/February 2008, 60–65. With permission.)

gears and a stationary planet carrier. The first-stage sun gear is connected to the engine turbine shaft. Its ring gear is connected to the second-stage sun gear, and its planet carrier is connected to the second-stage ring gear and output propeller shaft. This arrangement makes it possible to transmit about 33% of the engine power through the first-stage carrier directly to the propeller shaft, bypassing the second stage. This allows a reduction in size and weight of the second stage that transmits only about 67% of the engine power from the first-stage ring gear to the second-stage sun gear, and then through the planet gears to the second-stage ring gear that is attached to the propeller shaft.

The gear geometry and accuracy parameters, and operating torques and stresses are presented in Table 11.5. All gears were made out of the forged blanks of the steel 20KH3MVF (EI-415). Its chemical composition includes the following: Fe, base material; C, 0.15–0.20%; S, <0.025%; P <0.030%; Si, 0.17%–0.37%; Mn, 0.25%–0.50%; Cr, 2.8%–3.3%; Mo, 0.35%–0.55%; W, 0.30%–0.50%; Co, 0.60%–0.85%; and Ni, <0.5%.

Protuberance hobs were applied for the sun and planet gear machining prior to carburizing and quenching. A custom protuberance hob (Figure 11.4) was used to provide a final cutting of the gear tooth root fillet, leaving the grinding stock only on the tooth involute flanks. This allowed avoiding possible grinding "hotspots" that could propagate initial cracks in the root area. Unground tooth root fillet remains residual compressive stress (about 300–600 MPa [219]) after heat treatment, which increases tooth bending strength. The protuberance gear hob profile parameters are shown in Table 11.6.

The first-stage sun gear tooth profile after the protuberance hobbing is shown in Figure 11.5.

The ring gear involute flanks were preliminarily machined with a special asymmetric tooth shaper cutter. Then the form disk mill cutter was used to cut root fillets. The first-stage ring gear tooth space profile after machining, and before carburizing and heat treatment, is shown in Figure 11.6. Its parameters are presented in Table 11.7.

Unlike that for the symmetric tooth gears, a proper positioning of the asymmetric tooth gear blanks relative to the cutting tool is critical. Otherwise, the drive flank of one gear can be positioned in contact with the coast profile of the mating gear, which will make a gear assembly impossible. After the tooth cutting, the gears are carburized and heat treated to achieve a tooth surface hardness of >59 HRC with the case depth of 0.6–1.0 mm. The tooth core hardness is 33–45 HRC. The MAAG HSS-30 and HSS-60 generating gear grinding machines were used for final gear tooth flank grinding of the sun and planet gears. Simultaneous processing of both flanks of asymmetric teeth required a special grinding machine setup. The tip/root relief profile modification was applied to the driving tooth flanks of the sun gears and both flanks of the planet gears (Figure 11.7). The flank modification parameters are presented in Table 11.8.

TABLE 11.5
Gear Data

Gear Geometry Data						
Stage	**First**			**Second**		
Gear	**Sun**	**Planet**	**Ring**	**Sun**	**Planet**	**Ring**
Number of gears	1	3	1	1	5	1
Nominal number of teeth	28	41	107	38	31	97
Module, mm	3.000	3.000	3.000	3.362	3.362	3.362
Nominal drive pressure angle,°	33	33 & 25	25	33	33&25	25
Nominal coast pressure angle,°	25	25 & 33	33	25	25&33	33
Nominal pitch diameter, mm	84.000	123.000	321.000	127.756	104.222	326.114
Drive base diameter, mm	70.448	103.156 & 111.476	290.925	107.145	87.408 & 94.457	295.560
Coast base diameter, mm	76.130	111.476 & 103.156	269.213	115.786	94.457 & 87.408	273.502
Tooth tip diameter, mm	90.02/90.16	128.44/128.60	323.88/324.11	134.07/134.23	110.93/111.07	329.67/329.90
Root diameter, mm	76.55/77.05	114.55/115.05	337.50/337.70	118.56/119.06	95.45/95.95	345.00/345.20
Tooth thickness at nominal pitch diameter, mm	4.773/4.814	4.325/4.365	0.667/0.621	4.972/5.018	5.253/5.299	1.104/1.059
Tooth tip radius, mm	0.20/0.40	0.20/0.40	0.30/0.50	0.20/0.40	0.20/0.40	0.30/0.50
Tooth tip land, mm	0.630/0.982	0.596/0.960	0.395/0.806	0.711/1.081	0.717/1.074	0.350/0.761
Face width, mm	34.75/35.00	31.75/32.00	25.48/26.00	37.75/38.00	34.75/35.00	27.48/28.00
Center distance, mm		103.50±0.01			116.00±0.01	
Accuracy and inspection parameters						
Accuracy grade per GOST 1643–81[217]	5-5-4B	5-5-4B	5-5-4B	5-5-4B	5-5-4B	5-5-4B

(Continued)

TABLE 11.5 (*Continued*)

Gear Data

Gear Geometry Data							
Stage		**First**			**Second**		
Gear		**Sun**	**Planet**	**Ring**	**Sun**	**Planet**	**Ring**
Runout tolerance, mm		0.016	0.016	0.022	0.022	0.016	0.022
Pitch variation, mm		±0.006	±0.006	±0.007	±0.007	±0.006	±0.007
Profile tolerance, mm		0.006	0.006	0.007	0.007	0.006	0.007
Lead tolerance, mm		0.0055	0.0055	0.0055	0.0055	0.0055	0.0055
Pin diameter, mm		6.000	6.000	6.000	7.000	7.000	7.000
Measurement over pins, mm		93.754/ 93.819	131.991/ 132.057	320.476/ 320.554	138.929/ 139.004	115.665/ 115.738	325.309/ 325.385
Tolerance Analysis Results							
Operating drive pressure angle,°		32.98/33.02	29.87/29.93		32.99/33.03	29.88/29.93	
Operating coast pressure angle,°		24.97/25.03	36.64/36.68		24.98/25.04	36.65/36.68	
Operating drive contact ratio		1.18/1.26	1.33/1.36		1.20/1.28	1.35/1.38	
Operating coast contact ratio		1.33/1.42	1.18/1.21		1.36/1.44	1.21/1.24	
Operating normal backlash, mm		0.196/0.322	0.197/.406		0.189/0.320	0.206/0.414	
Radial clearance, mm		0.651/ 1.033	0.936/ 1.308 0.953/ 1.376	0.921/ 1.159	0.978/ 1.356	0.948/ 1.336 0.898/ 1.321	0.936/ 1.164
Stress analysis results							
Maximum power, HP				2800			
RPM		17500	−11132	−3063	−3063	3755	1200
Torque per Mesh, Nm		374	548	1430	858	700	2191
Bending stress, MPa	Tension	240	257 280	298	289	318 345	371
	Compression	−384	−403 −306	−355	−470	−500 −385	−430
Contact stress, MPa		960	604		1043	809	

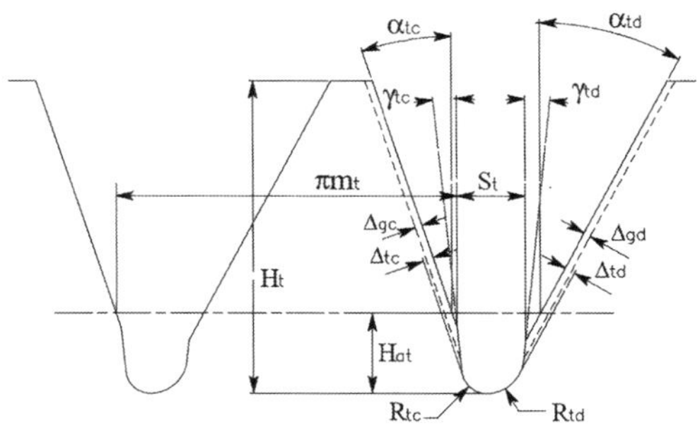

FIGURE 11.4
Protuberance gear hob profile.

TABLE 11.6
Protuberance Gear Hob Profile Data

Stage		**First**		**Second**	
Gear		**Sun**	**Planet**	**Sun**	**Planet**
Number of teeth		28	41	38	31
Hob module, mm	m_t	2.876	2.905	3.224	3.224
Hob tooth thickness at pitch line, mm	S_t	2.356	2.480	2.312	2.521
Hob tooth addendum, mm	H_{at}	2.124	2.157	1.858	2.124
Minimal whole depth, mm	H_t	7.80	7.80	8.20	8.20
Drive profile angle, °	α_{td}	29.0	30.0	29.0	29.0
Coast profile angle, °	α_{tc}	19.064	20.631	19.064	19.064
Drive protuberance angle, °	γ_{td}	6.0	6.0	6.0	6.0
Coast protuberance angle, °	γ_{tc}	6.0	6.0	6.0	6.0
Drive protuberance offset, mm	Δ_{td}	0.29	0.31	0.31	0.31
Coast protuberance offset, mm	Δ_{tc}	0.29	0.31	0.31	0.31
Drive grinding stock, mm	Δ_{gd}	0.17	0.20	0.20	0.20
Coast grinding stock, mm	Δ_{gc}	0.17	0.20	0.20	0.20
Drive side tip radius, mm	R_{td}	1.00	1.00	1.20	1.10
Coast side tip radius, mm	R_{tc}	0.65	0.70	0.80	0.80

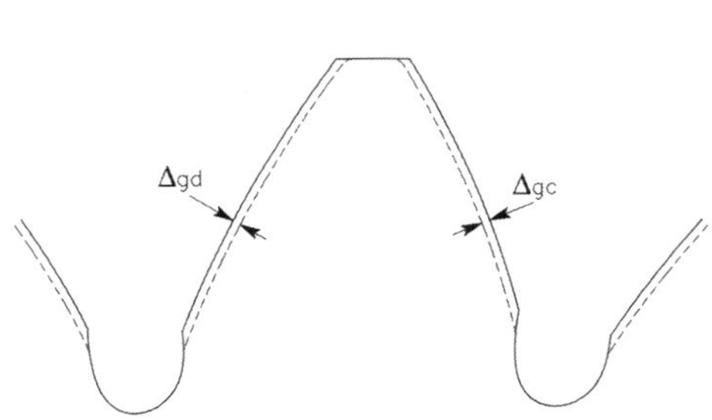

FIGURE 11.5
The first-stage sun gear tooth profile after the protuberance hobbing; Δgd and Δgc—drive and coast flank grinding stocks.

TABLE 11.7

Ring Gear Root Fillet Data

Stage		First	Second
Gear		**Ring**	**Ring**
Number of teeth		107	97
Fillet radius, mm	R	0.60/0.70	0.65/0.75
Fillet cut width, mm	B	1.20/1.40	1.30/1.50
Fillet angle, °	Φ	3.4	3.3
Drive flank fillet undercut, mm	Δ_d	0.06/0.08	0.06/0.08
Coast protuberance offset, mm	Δ_c	0.06/0.08	0.06/0.08
Drive grinding stock, mm	Δ_{gd}	0.19/0.20	0.19/0.20
Coast grinding stock, mm	Δ_{gc}	0.19/0.20	0.19/0.20

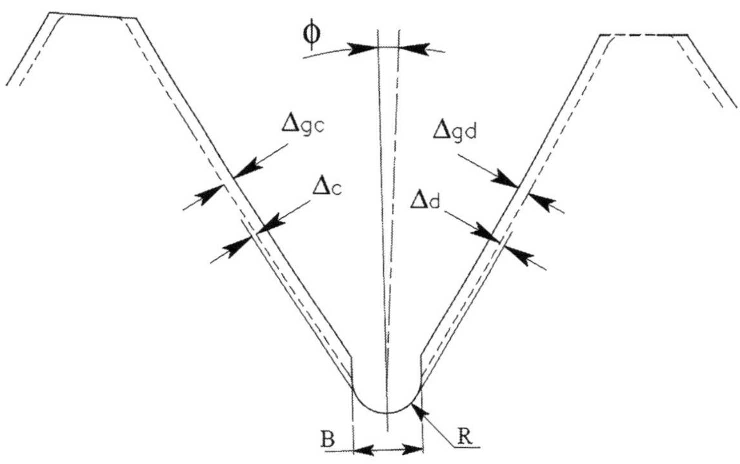

FIGURE 11.6
The first-stage ring gear tooth space profile after machining; Δgd and Δgc: drive and coast flank grinding stocks; Δd and Δc: drive and coast flank undercuts; R: root fillet radius; B: fillet cut width; φ: fillet angle.

The assembly of the gearbox includes the selection of planet gears and their initial angular orientation based on transmission error (TE) function of every gear. All planet gears were classified by a TE in several groups. Each group had planet gears with the similar TE. Then during assembly, position and orientation of each planet gear depended on its TE function profile, providing a better engagement of the driving flanks and load distribution between planet gears, and also reducing gear noise and vibration [13].

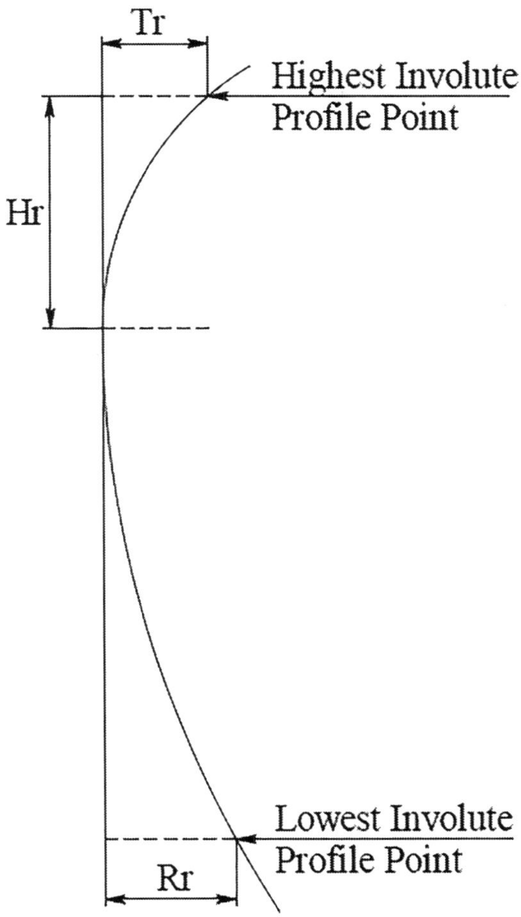

FIGURE 11.7
Tooth flack modification chart. *Hr*: tip relief height; *Tr*: tip relief depth; *Rr*: root relief depth.

TABLE 11.8
Tooth Flank Modification Parameters

Stage		**First**		**Second**	
Gear		**Sun**	**Planet**	**Sun**	**Planet**
Number of teeth		28	41	38	31
Tip relief height, mm	H_r	2.5/3.5	2.0/3.0	3.0/4.0	2.0/3.0
Tip relief depth, mm	T_r	0.006/0.010	0.002/0.006	0.008/0.012	0.003/0.008
Root relief depth, mm	R_r	0.008/0.014	0.002/0.006	0.010/0.017	0.002/0.008

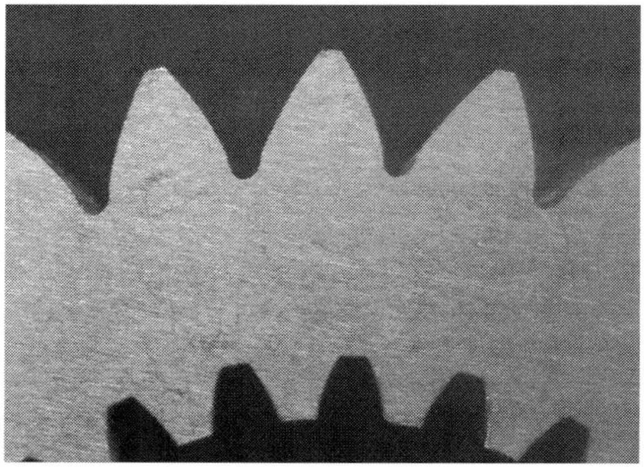

FIGURE 11.8
First-stage sun gear section. (Courtesy of Chernyshev Enterprise, Moscow, Russia.) (From Novikov, A.S., et al., Gear Technology, January/February 2008, 60–65. With permission.)

FIGURE 11.9
First-stage sun gear. (Courtesy of Chernyshev Enterprise, Moscow, Russia.) (From Novikov, A.S., et al., Gear Technology, January/February 2008, 60–65. With permission.)

Application of the asymmetric teeth provided a low weight to output torque ratio, a significantly reduced noise and vibration level, and cut-down duration and expense of operational development [218]. A section of the first-stage sun gear is shown in Figure 11.8. The photos of the gears and gear assemblies of the TV7–117S gearbox are shown in Figures 11.9–11.14. Figure 11.15 shows the Ilyushin Il-114 commuter airplane with the TV7–117S engines.

FIGURE 11.10
First-stage assembly. (Courtesy of Chernyshev Enterprise, Moscow, Russia.) (From Novikov, A.S., et al., Gear Technology, January/February 2008, 60–65. With permission.)

FIGURE 11.11
Second-stage sun gear. (Courtesy of Chernyshev Enterprise, Moscow, Russia.) (From Novikov, A.S., et al., Gear Technology, January/February 2008, 60–65. With permission.)

FIGURE 11.12
Second-stage planet gear carrier assembly. (Courtesy of Chernyshev Enterprise, Moscow, Russia.) (From Novikov, A.S., et al., Gear Technology, January/February 2008, 60–65. With permission.)

FIGURE 11.13
Second-stage ring gear. (Courtesy of Chernyshev Enterprise, Moscow, Russia.)

11.3 Modernization of Helicopter Gearbox

This section describes the modernization of the main gearbox for a *new Ukrainian helicopter* SMB-2 "Nadiya" (a deep modification of the Russian light multipurpose Mil-2 helicopter) with the asymmetric tooth gears [123]. This gearbox modernization was required because of replacement of the GTD-350

FIGURE 11.14
Assembled gearbox. (Courtesy of Chernyshev Enterprise, Moscow, Russia.) (From Novikov, A.S., et al., Gear Technology, January/February 2008, 60–65. With permission.)

FIGURE 11.15
Ilyushin Il-114 commuter airplane.

turboshaft engines with the new AI-450M engines, manufactured by the Motor Sich Company (Zaporozhye, Ukraine). In comparison to its predecessor, AI-450M has the 465 HP power output (vs. 400 HP of GTD-350), its fuel consumption is 27% lower and it is 25 kg lighter [220].

Instead of a complete redesign of the main gearbox, the task of transmitting of the increased engine power output was solved by implementation of asymmetric tooth gears in the most loaded second and third stages of the gearbox (Figure 11.16). Power of the two turboshaft engines is transmitted by their turbine shafts 1 through the bevel gears z_1, z_2 and then the spur gears z_3, z_4, and z_5 to the bull gear z_6 connected to the main propeller shaft 2. About 25% of one of the engine power is used to drive the tail propeller through the spur gear z_7 and the bevel gears z_8, z_9. The original bevel gear pairs had sufficient load capacity to withstand increased power of new engines and were not required any modernization. The second- and third-stage spur gears had to be redesigned to reduce contact stress level and to achieve sufficient tooth flank surface durability.

The original spur symmetric tooth gears were designed traditionally using preselected basic racks and their addendum modifications (X-shifts). The modernized spur gears with asymmetric teeth were constructed utilizing the Direct Gear Design method [107].

The main geometric parameters of the original and modernized spur gears are presented in Tables 11.9 and 11.10.

Parameters of asymmetric teeth were selected to satisfy the following requirements:

- Drive flank pressure angle should not exceed 32° to limit the maximum radial load and keep original bearings.
- Minimum drive flank contact ratio is 1.3.

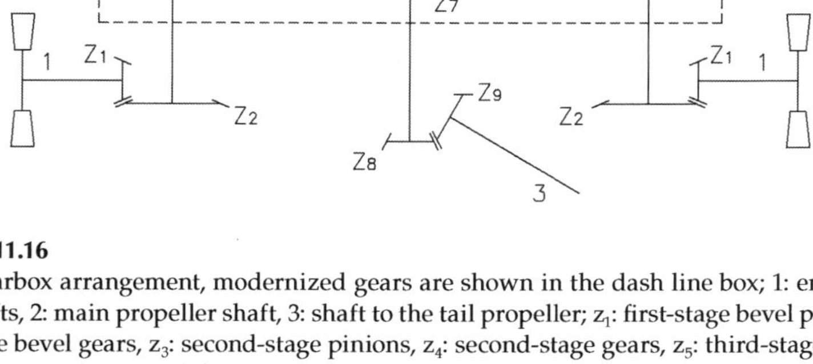

FIGURE 11.16
Main gearbox arrangement, modernized gears are shown in the dash line box; 1: engine turbine shafts, 2: main propeller shaft, 3: shaft to the tail propeller; z_1: first-stage bevel pinions, z_2: first-stage bevel gears, z_3: second-stage pinions, z_4: second-stage gears, z_5: third-stage pinions, z_6: third-stage gear, z_7: tail drive spur gear, z_8: tail drive bevel pinion, z_9: tail drive bevel gear. (Courtesy of Motor Sich JSC, Zaporozhye, Ukraine) (S.I. Shankin, A. A. Stupakov, I.L. Glikson, V.M. Ananiev, A.L. Kapelevich. Modernization of Main Helicopter Gearbox with Asymmetric Tooth Gears. Gear Solutions, March 2014, 46–49. With permission.)

TABLE 11.9

Geometric Parameters of Original and Modernized Gears of Second Stage

Gear Geometry	**Original**		**Modernized**	
Design Method	**Rack Generation**		**Direct Gear Design**	
Tooth Profile	**Symmetric**		**Asymmetric**	
Gear	pinion	gear	pinion	gear
Number of teeth	15	59	15	59
Center distance	206.504		206.504	
Basic rack module	5.500	5.500	N/A	N/A
Basic rack pressure angle	20°	20°	N/A	N/A
Addendum modification	0.5	0.0751	N/A	N/A
Operating module	5.581	5.581	5.581	5.581
Operating drive pressure angle	22.177°		32°	
Operating coast pressure angle	22.177°		20°	
Drive contact ratio	1.43		1.30	
Coast contact ratio	1.43		1.59	

TABLE 11.10

Geometric Parameters of Original and Modernized Gears of Third Stage

Gear Geometry	**Original**		**Modernized**	
Design Method	**Rack Generation**		**Direct Gear Design**	
Tooth Profile	**Symmetric**		**Asymmetric**	
Gear	pinion	gear	pinion	gear
Number of teeth	14	37	14	37
Center distance	206.504		206.504	
Basic rack module	8.000	8.000	N/A	N/A
Basic rack pressure angle	28°	28°	N/A	N/A
Addendum modification	0.5	0.0751	N/A	N/A
Operating module	8.098	8.098	8.098	8.098
Operating drive pressure angle	29.28°		32°	
Operating coast pressure angle	29.28°		20°	
Drive contact ratio	1.20		1.30	
Coast contact ratio	1.20		1.57	

- Minimum tooth tip thickness is 2.3 mm for the second-stage gears and 3.1 mm for the third-stage gears to avoid the harden through tooth tip.
- Minimum coast flank pressure angle in the second-stage is 20°, because the coast flanks of one of the gears are used to transmit torque to the tail propeller.
- The tooth root fillet is not ground after heat treatment. Its profile should not interfere with the grinding wheel trajectory.

Prior to heat treatment, the gear teeth were machined, leaving the grinding stock about 0.20mm on the tooth flanks and providing final root fillet profile. Unground tooth root fillet remains as residual compressive stress after heat treatment, which increases the tooth bending strength. This required the gear hobs with a protuberance. However, due to the low number of teeth of the second- and third-stage pinions, the protuberance hobs could not provide the required tooth root fillet surface finish. In order to solve this issue, the pinion flanks were machined with conventional hobs (without a protuberance) and then, before the gear heat treatment, the pinion root fillets were ground. The second-stage pinion tooth profiles before and after heat treatment are shown in Figure 11.17.

Prior to heat treatment, the driven gears of both the second and third stages were machined with the protuberance hobs. The third stage gear tooth profiles before and after heat treatment are shown in Figure 11.18.

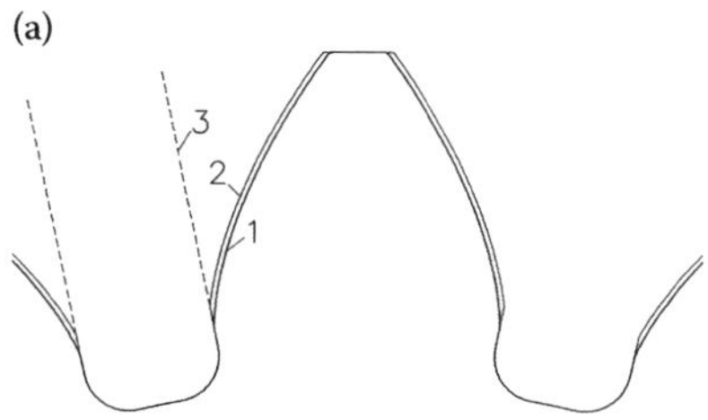

FIGURE 11.17
Second-stage pinion tooth profiles; (a) original symmetric, (b) asymmetric; 1: after heat treatment and flank grinding, 2: before treatment, 3: root fillet grinding wheel. (Courtesy of Motor Sich JSC, Zaporozhye, Ukraine) (S.I. Shankin, A. A. Stupakov, I.L. Glikson, V.M. Ananiev, A.L. Kapelevich Modernization of Main Helicopter Gearbox with Asymmetric Tooth Gears. Gear Solutions, March 2014, 46–49. With permission.)

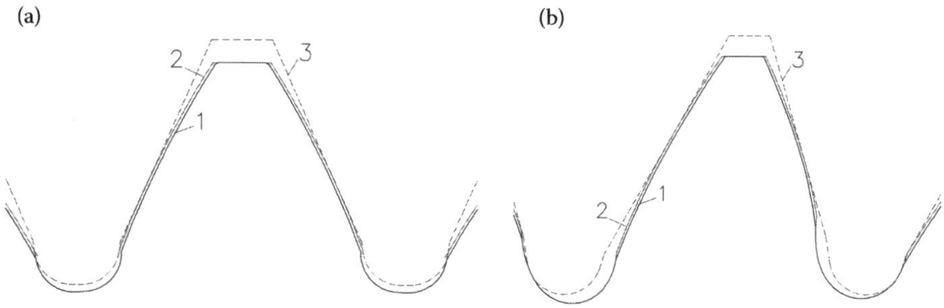

FIGURE 11.18
Third-stage gear tooth profiles; (a) original symmetric, (b) asymmetric; 1: after heat treatment and flank grinding, 2: before treatment, 3: profile of hob with protuberance. (Courtesy of Motor Sich JSC, Zaporozhye, Ukraine) (S.I. Shankin, A. A. Stupakov, I.L. Glikson, V.M. Ananiev, A.L. Kapelevich. Modernization of Main Helicopter Gearbox with Asymmetric Tooth Gears. Gear Solutions, March 2014, 46–49. With permission.)

TABLE 11.11
Stress Analysis Results of Original and Modernized Gears of Second Stage

Gear Geometry	Original			Modernized		
Design Method	**Rack Generation**			**Direct Gear Design**		
Tooth Profile	**Symmetric**			**Asymmetric**		
Gear	pinion		gear	pinion		gear
Number of teeth	15		59	15		59
Face width	58.0		55.0	58.0		55.0
Torque, Nm	1,255		4,936	1,255		4,936
Max. contact stress, MPa		1308			1186	
Contact stress reduction		–			9%	
Max. root bending stress, MPa	305		298	279		287
Root bending stress reduction	–		–	9%		4%

TABLE 11.12
Stress Analysis Results of Original and Modernized Gears of Third Stage

Gear Geometry	Original			Modernized		
Design Method	**Rack Generation**			**Direct Gear Design**		
Tooth Profile	**Symmetric**			**Asymmetric**		
Gear	pinion		gear	pinion		gear
Number of teeth	14		37	14		37
Face width	92.0		92.0	92.0		92.0
Torque, Nm	4,936		13,046	4,936		13,046
Max. contact stress, MPa		1513			1425	
Contact stress reduction		–			6%	
Max. root bending stress, MPa	367		371	352		364
Root bending stress reduction	–		–	4%		2%

The stress analysis results are shown in Tables 11.11 and 11.12.

All gears are made of the forged blanks of the steel 16H3NVFMB (VKS-5, DI-39). Its chemical composition includes the following: Fe, base material, C, 0.13%–0.19%, Si, 0.17%–0.37%, Mn, 0.50%–0.90%, Cr, 2.65%–3.25%, Ni, 0.4%–0.8%, S <0.025%, P <0.025%, Cu < 0.30%.

The gear blanks' position during machining must provide the asymmetric teeth pointed in certain direction (clockwise or counterclockwise). Otherwise, the drive flank of one gear will be positioned in contact with the coast profile of the mating gear, which would make gearbox assembly impossible.

After the tooth cutting, the gears are carburized and heat treated to achieve a tooth surface hardness of 59–60 HRC with a case depth of 1.2–1.4 mm. The core tooth hardness is 34–42 HRC.

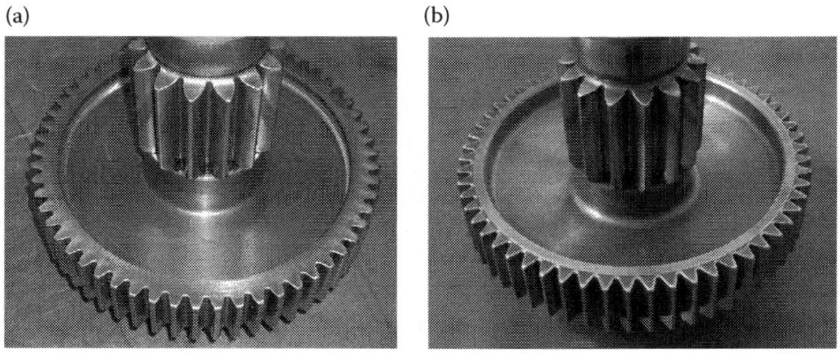

FIGURE 11.19
Compound gear; (a) original design with symmetric teeth, (b) modernized design with asymmetric teeth. (Courtesy of Motor Sich JSC, Zaporozhye, Ukraine) (Shankin, S.I., A. A. Stupakov, I.L. Glikson, V.M. Ananiev, A.L. Kapelevich. Modernization of Main Helicopter Gearbox with Asymmetric Tooth Gears. Gear Solutions, March 2014, 46–49. With permission.)

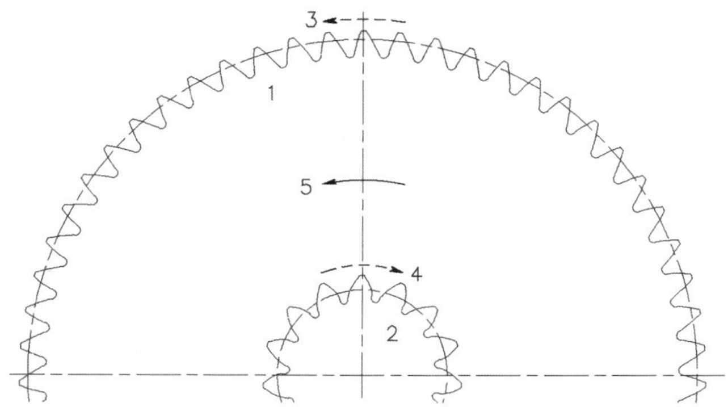

FIGURE 11.20
Asymmetric tooth slant directions of the compound gear; 1: driven second-stage gear, 2: driving third-stage pinion, 3: asymmetric tooth slant direction of the driven second-stage gear, 4: asymmetric tooth slant direction of the driving third-stage pinion, 5: compound gear rotation direction.

Final gear machining includes tooth grinding and honing. Asymmetric gears require a special setup for both these operations.

The compound gear that contains the second-stage gear and third-stage pinion is shown in Figure 11.19. In this compound gear, the second-stage gear is driven and the third-stage pinion is driving. For the asymmetric compound gear this means that asymmetric tooth slant directions of the gear and pinion should be opposite. Figure 11.20 illustrates the asymmetric tooth slant directions of the compound gear. The driven second-stage gear asymmetric tooth slant direction coincides with the compound gear rotation direction. The driving third-stage pinion asymmetric tooth slant direction is opposite to the compound gear rotation direction. This is a critical matter for proper gearbox assembly and operation.

Application of the asymmetric tooth gears in the most loaded second- and third spur gear stages of the SMB-2 "Nadiya" helicopter main gearbox resulted with contact stress reduction, achieving acceptable drive flank durability without increase of major gear dimensions, and utilizing more powerful and fuel-efficient AI-450M turboshaft engines without a complete redesigning of the helicopter gearbox.

Epilogue

The symmetry of the modern gear tooth shape is a status quo settled in the minds of engineers for centuries, like there is not any reason or possibility to change it. But there are reasons and possibilities to break this status quo. Many initially symmetric shapes of the essential mechanical components have evolved into asymmetric optimized shapes to achieve the best possible performance for many applications. There is no doubt that gears will follow suit for at least two reasons: a constant demand for gear drive performance improvement and a practical feasibility of modern gear fabrication technologies to cost-effectively produce virtually any required gear tooth shape.

This book described why asymmetric tooth gears should be considered for custom unidirectional gear drives and how to develop an asymmetric gearing that will deliver performance unattainable by the best symmetric tooth gears.

Asymmetric gear tooth proportions and designs are not covered by standards and are not limited by some tooth forming conditions, which are related to certain gear fabrication technologies, like a rack generation in the traditional gear design. This book suggests utilizing the alternative Direct Gear Design® method for asymmetric gear development that allows for deeper tooth geometry optimization to achieve maximum performance for a particular application. Besides the theoretical fundamentals, such as synthesis and analysis asymmetric gears, and areas of existence, the book demonstrated optimization approaches for each asymmetric tooth component including the drive and coast flank macro- and micro-geometry, and the root fillet to reduce contact and bending stresses, and a transmission error.

The book also described asymmetric gear fabrication methods and inspection parameter definition. It exhibited advantages of asymmetric tooth gears by analytical and experimental comparisons to the advanced existing symmetric tooth gears. It also contains examples of implementation of asymmetric tooth gears.

As an author, I would like to express my deep and sincere gratitude to readers of this book and encourage them to utilize its content for the practical application of asymmetric tooth gears.

Alexander L. Kapelevich

References

1. Burrows, M. and G. Sutton. 2013. Interacting gears synchronize propulsive leg movements in a jumping insect. *Science*, 341 (6151): 1254–1256.
2. Willis, R. 1841. *Principles of Mechanism*. London: John W. Parker, West Strand, Cambridge: J. & J.J. Deighton.
3. Willis, R. 1875. *Principles of Mechanism*, 2nd edition. New York: John Wiley & son.
4. Reuleaux, F. 1894. *The Constructor. A Hand-Book of Machine Design*. Philadelphia: H.H. Suplee.
5. Spooner, H.S. 1908. *Machine Drawing and Design for Beginners*. London: Longmans, Green.
6. Logue, C.H. 1910. *American Machinist Gear Book*. New York: McGraw-Hill Book Company.
7. Königl, K. 1912. Osterreichische patentschrift No. 62002.
8. Leutwiler, O.A., 1917. *Element of Machine Design*. London: McGraw-Hill Book Company. p. 134.
9. Niemann, G. and H. Glaubitz. 1950. Zahnfußfestigkeit geradverzahnter Stirnräder aus Stahl, *VDI-Zeitschrift, Band* 92: 923–932.
10. Vulgakov, E.B. 1974. *Gears with Improved Characteristics*. Moscow: Mashinostroenie.
11. Vulgakov, E.B. and G.V. Rivkin. 1976. Design of gears with asymmetric tooth profile. *Mashinovedenie*, 5: 35–39.
12. Vulgakov, E.B. and G.V. Rivkin. 1980. Generalized area of existence of involute gear mesh with asymmetric teeth. *Mashinovedenie*, 1: 55–59.
13. Vulgakov, E.B. (ed.) 1981. *Aviation Gearboxes. Handbook*. Moscow: Mashinostroenie.
14. Kapelevich, A.L. 1984. Research and development of geometry of modernized involute gears. PhD dissertation. Moscow State Technical University.
15. Rogachevsky, N.I. 1983. Measurement over pins (balls) of gears with asymmetric tooth profiles. *Mashinovedenie*, 5: 35–38.
16. Rinder, L. 1984. Schraubenverdichterläufer mit Evolventenflanken. *VDI Berichte*, 521: 181–201.
17. Bolotovsky, I.A., O.F. Vasil'eva and V.P. Kotelnikov. 1984. Involute gears with asymmetric teeth. *Vestnik Mashinostroeniya*, 4: 15–17.
18. Yoerkie, C.A. and A.G. Chory. 1984. Acoustic vibration characteristics of high contact ratio planetary gears. *The Journal of American Helicopter Society*, 40: 19–32.
19. Niemann, G. and H. Winter. 1985. *Maschinen-Elemente*, 2nd edition. Berlin: Springer-Verlag.
20. Kapelevich, A.L., A.I. Tolchenov and A.I. Eidinov. 1985. Application of the fly cutters in experimental production, *Aviatsionnaya Promyshlennost*, 3: 39–40.
21. Vulgakov, E.B. and A.L. Kapelevich. 1986. Asymmetric gear transmissions: Possible developments. *Vestnik Mashinostroeniya*, 4: 14–16.

22. Kapelevich, A.L. 1986. Measurement over pins of the gears with asymmetric teeth. *Mashinovedenie*, 6: 109–110.
23. Rinder, L. 1987. Sonderverzahnungen für Schraubenverdichterrotoren. *VDI Berichte*, 640: 137–150.
24. Kapelevich, A.L. 1987. Synthesis of asymmetric involute gearing gears. *Mashinovedenie*, 1: 62–67.
25. Andozhsky, V.D. and N.I. Rogachevsky. 1988. Involute gears with asymmetric tooth profile. *Vestnik Mashinostroeniya*, 10: 32–34.
26. Howe, D.C., C.V. Sundt and A.H. McKibbon. 1988. Advanced counter-rotating gearbox. NASA Contractor Report 180883.
27. Linke, H. and J. Borner. 1994. The influence of neighboring teeth on the tooth root capacity. *Proceedings of the International Gearing Conference*. Newcastle, UK, September 7–9: pp. 177–182.
28. Römhild, I. 1994. Evolventische und nichtevolventische Sondervenahnungen, Forschungsheft Forschungsvereinigung Antriebstechnik E.V., Forschungsvorhaben Nr. T 496.
29. Vulgakov, E.B. 1995. *Theory of Involute Gears*. Moscow: Mashinostroenie.
30. DiFrancesco, G. and S. Marini. 1997. Structural analysis of teeth with asymmetrical profiles. *Gear Technology*, 14 (4): 16–22.
31. DiFrancesco, G. and S. Marini. 1997. Structural analysis of asymmetrical teeth: reduction of size and weight. *Gear Technology*, 14(5): 47–51.
32. Erzhanov, N.M. 1998. Exploration of resistance to destruction of unsymmetrical teeth of cog wheels when bending. PhD dissertation. Taraz State University.
33. Venclík, J. 1998. Design of straight spur gears with asymmetric involute teeth with help of the method of generalized parameters. PhD dissertation of faculty of Mechanical Engineering Technical University in Brno.
34. Sarkisov, A.A. and Vulgakov E.B. 2000. The new gearbox of klimov corporation. *Aerospace Courier*, 2: 32–33.
35. Vulgakov, E.B. and A.L. Kapelevich. 2000. The gearbox for the TV7-117S turboprop engine. *Vestnik Mashinostroeniya*, 11: 13–17.
36. Litvin, F.L., Q. Lian and A.L. Kapelevich. 2000. Asymmetric modified gear drives: reduction of noise, localization of contact, simulation of meshing and stress analysis. *Computer Methods in Applied Mechanics and Engineering*, 188: 363–390.
37. Kapelevich, A.L. 2000. Geometry and design of involute spur gears with asymmetric teeth. *Mechanism and Machine Theory*, Elsevier. 35: 117–130.
38. Deng, G. and T. Nakanishi. 2001. Enhancement of bending load carrying capacity of gears using an asymmetric involute tooth. *Paper presented at the JSME International Conference on Motion and Transmissions (MPT2001-Fukuoka)*, Fukuoka, Japan.
39. Litvin, F.L., A. Fuentes, J.M. Hawkins and R.F. Handschuh. 2001. Design, generation and tooth contact analysis (tca) of asymmetric face gear drive with modified geometry. NASA/TM-2001-210614. ARL-TR-2373: 1–30.
40. Kleiss, R.E., A.L. Kapelevich and N.J. Kleiss Jr. 2001. New opportunities with molded gears. *Proceedings of the Annual Technical Meeting of the AGMA (01FTM9)*, Detroit.
41. Kapelevich, A.L. and R.E. Kleiss. 2002. Direct gear design for spur and helical gears. *Gear Technology*, 19(5): 29–35.

42. Sfantos, G.K., V. Spitas and T. Costopoulos. 2003. Optimized asymmetric gear tooth design. National Technical University of Athens (NTUA). Technical Report No. TR-SM-0307: 1–22.
43. Deng, G., T. Nakanishi and K. Inoue. 2003. Bending load capacity enhancement using an asymmetric tooth profile. *JSME International Journal*. Series C., 46 (3): 1171–1177.
44. Kapelevich, A.L. and Y.V. Shekhtman. 2003. Direct gear design: Bending stress minimization. *Gear Technology*, 20(5): 44–49.
45. Yang, S.C. 2005. Mathematical model of a helical gear with asymmetric involute teeth and its analysis. *International Journal of Advanced Manufacturing Technology*, 6(5): 448–456.
46. Brecher, C. and J. Schafer. 2005. Potentials of asymmetric tooth geometries for the optimization of involute cylindrical gears. *VDI Berichte*, 1904(I): 705–720.
47. Timofeyev. B.P. and D.A. Frolov. 2005. Calculation of geometric parameters of involute gears with asymmetric teeth. *Theory of Mechanisms and Machines*, 3(2): 15–21.
48. Karpat F. 2005. Analysis of involute spur gears with asymmetric teeth. PhD dissertation. Uludag University. Bursa-Turkey.
49. Herscovici, S. 2005. Megagears and unimegagears. *Gear Product News*, August: 22–26.
50. Kapelevich, A.L., Y.V. Shekhtman and T.M. McNamara. 2005. Turning an art into science. *Motion System Design*, August: 26–31.
51. Xiao, W.-Q., W. Li and M. Li. 2006. Finite element analysis of the tooth root bending stress of an unsymmetric gear with double pressure angles. *Journal of University of Science and Technology Beijing*, 28(6): 570–575.
52. Chira, F. 2006. On the elasticity of involute spur gears with asymmetric teeth. *Proceedings of the 15th International Conference on Manufacturing Systems—ICMaS*. Bucharest, Romania, October 26–27: 113–116.
53. Kargin, P.A. 2006. Synthesis of new self-locking gear drive arrangement. Izvestiya Vuzov. Severo-Kavkazskii Region. Russia. *Technical Sciences*, 1: 96–97.
54. Frolov, D.A. 2006. *Development of Geometry and Manufacturing Technology of Asymmetric Tooth Gears*. St. Petersburg, Russia: ITMO University.
55. Kapelevich A.L. and T.M. McNamara. 2006. Plastic gearing for small engine applications. *SAE International Conference*, San Antonio.
56. Yang, S.C. 2007. Study on an internal gear with asymmetric involute teeth. *Elsevier, Mechanism and Machine Theory*, 42: 977–994.
57. DiFrancesco, G. and S. Marini. 2007. Asymmetric teeth: Bending stress calculation. *Gear Technology*, 24 (2): 52–55.
58. Chira, F., V. Tisan and A. Dascalescu. 2007. Modelling of the asymmetric gears using applications in MATLAB and AutoLISP. Annals of the Oradea university. *Fascicle of Management and Technological Engineering*, VI(XVI): 1366–1371.
59. Shyue-Cheng Yang, S.C. 2007. Study on an internal gear with asymmetric involute teeth. *Mechanism and Machine Theory*, 42: 977–994.
60. Kumar, V.S., D.V. Muni and G. Muthuveerappan. 2008. Optimization of asymmetric spur gear drives to improve the bending load capacity. *Mechanism and Machine Theory*, 43: 829–858.
61. Kargin, P.A. 2008. Design and testing of self-locking gear. *Vestnik Mashinostroeniya*, 12: 3–6.

62. Novikov, A.S., A.G. Paikin, V.L. Dorofeyev, V.M. Ananiev and A.L. Kapelevich. 2008. Application of gears with asymmetric teeth in turboprop engine gearbox. *Gear Technology*, 25(1): 60–65.
63. Fetvaci, C. and E. Imrak. 2008. Mathematical model of a spur gear with asymmetric involute teeth and its cutting simulation. *An International Journal Based Design of Structures and Machines*. 36(1): 34–46.
64. Karpat, F. and S. Ekwaro-Osire. 2008. Dynamic analysis of high-contact-ratio spur gears with asymmetric teeth. *Proceedings of IMECE2008, 2008 ASME International Mechanical Engineering Congress and Exposition*. October 31-November 6, 2008. Boston, MA, USA: 1–7.
65. Karpat F., S. Ekwaro-Osire, K. Cavdar and F.C. Babalik. 2008. Dynamic analysis of involute spur gears with asymmetric teeth. *International Journal of Mechanical Sciences*, 50(12): 1598–1610.
66. Costopoulos, T. and V. Spitas, 2009. Reduction of gear fillet stresses by using one-sided involute asymmetric teeth. *Mechanism and Machine Theory*, 44: 1524–1534.
67. Kargin, P.A. 2009. Involute gears with asymmetric tooth profiles: Application, synthesis, fabrication. *Vestnik Mashinostroeniya*, 11: 3–6.
68. Kapelevich, A.L. and Y.V. Shekhtman. 2009. Tooth fillet profile optimization for gears with symmetric and asymmetric teeth. *Gear Technology*, 26(7): 73–79.
69. Kapelevich, A.L. and T.M. McNamara. 2009. A better way to design gears for medical devices. *Medical Design*, 9(2): 44–47.
70. Moya, J.L., A. Machado, A.M. Becerra Ferreiro, J.A. Velazquez and R. Goytisolo. 2009. Determining the Lewis factor for plastic spur gears with asymmetric teeth. *Proceedings of the ASME 2009 International Mechanical Engineering Congress & Exposition*. November 13–19, Lake Buena Vista, FL, USA: IMECE2009-10172.
71. Alipiev, O.L. and S.D. Antonov. 2010. Asymmetric involute –Lantern meshing formed by identical spur gears with a small number of teeth. *VDI—Berichte*, 2018: 925–939.
72. Moya, J.L., A.S. Machado, J.A. Velásquez, R. Goytisolo, A.E. Hernández, J.E. Fernández and J.M. Sierra. 2010. A study in asymmetric plastic spur gears. *Gear Solutions*, 8(4): 34–41.
73. Pedersen, N.L. 2010. Improving bending stress in spur gears using asymmetric gears and shape optimization. *Mechanism and Machine Theory*, 45: 1707–1720.
74. Sándor, R.N. and L. Mircea. 2010. Manufacture of gears with asymmetric teeth on CNC machine tools. Annals of the Oradea university. *Fascicle of Management and Technological Engineering*, IX(XIX): 3.212–3.218.
75. Fetvaci, C. 2010. Generation simulation of involute spur gears machined by pinion-type shaper cutters. *Strojniški vestnik—Journal of Mechanical Engineering*, 56(10): 644–652.
76. Fetvaci, C. 2010. Computer simulation of helical gears with asymmetric involute teeth. *Journal of the Faculty of Engineering and Architecture of Gazi University*, 25(3): 441–447.
77. Karpat, F. and S. Ekwaro-Osire. 2010. Efficiency of the high contact ratio involute spur gears with asymmetric teeth. *Proceedings of the ASME 2010 International Mechanical Engineering Congress & Exposition IMECE2010*. November 12–18, 2010, Vancouver, British Columbia, CA: 319–327.

78. Karpat, F. and S. Ekwaro-Osire. 2010. Influence of tip relief modification on the wear of spur gears with asymmetric teeth. *Tribology & Lubrication Technology*, 66(6): 50–60.
79. Kapelevich, A.L. and Y.V. Shekhtman. 2010. Area of existence of involute gears. *Gear Technology*, 27(1): 64–69.
80. Starzhinsky, V.E., N.N. Ishin and A.M. Goman. 2011. Perfection of quality parameters of plastic gears by using asymmetric tooth profile. *Machine Design*, 3(2): 109–114.
81. Alipiev, O.L. 2011. Geometric design of involute spur gear drives with symmetric and asymmetric teeth using the realized potential method. *Mechanism and Machine Theory*, 46: 10–32.
82. Simeonov, S. and G. Tsvetanov. 2011. Practical implementation and optimization of evolvent cylindrical toothed gears with asymmetric tooth profile. *International Virtual Journal MTM*, 6: 41–44.
83. Wang, S.C., G.R. Liu, G.Y. Zhang and L. Chen. 2011. Accurate bending strength analysis of the asymmetric gear using the novel ES-PIM with triangular mesh. *International Journal of Automotive and Mechanical Engineering*, 4: 373–397.
84. Wang, S., G.R. Liu, G.Y. Zhang, L. Chen. 2011. Design of asymmetric gear and accurate bending stress analysis using the ES-PIM with triangular mesh. *International Journal of Computational Methods*, 8(4): 759–772.
85. Li, N., W. Li, J.-Y. Han and N. Liu. 2011. Tooth profile design and meshing analysis on a helical gear with asymmetric involute teeth. *Journal of University of Science and Technology Beijing*, 33(7): 876–882.
86. Cárdenas Oliveros, J.A., D.R. Hernández Ochoa, G.A. Babilonia Sánchez and J.L. Moya Rodríguez. 2011. New method for geometrical design of spur gears with asymmetric involute-trochoid profile. *Revista Tecnica de la Facultad de Ingenieria Universidad del Zulia*, 34(2): 160–170.
87. Kumar, P., S. Saroj Mandal and S. Senthilvelan. 2011. Dedendum optimization of asymmetric spur gear. *15th National Conference on Machines and Mechanisms*. Chennai, Tamil Nadu, India. November 30–December 2. NaCoMM2011-156:1–8.
88. Sándor, R.N. and L. Mircea. 2011. Grinding gears with asymmetric teeth by using grinding wheel technology. *Academic Journal of Manufacturing Engineering*, 9(4): 94–99.
89. Sándor, R.N. and L. Mircea. 2011. Approach to the test of flexural strength of the teeth of gears with asymmetric teeth. annals of the Oradea university. *Fascicle of Management and Technological Engineering*, X(XX): 4.145–4.152.
90. Kargin, P.A. 2011. Transition curve of involute gear teeth with asymmetric profiles. *Russian Engineering Research*, 31(3): 200–202.
91. Dorofeev, V.L., K.B. Arnaudov and D.V. Dorofeev. 2011. Increase of strength of toothed gearings with help the gears with asymmetric teeth. *Eight International Congress "Machines, Technologies, Materials"*, 19–21 September: 83–87.
92. Wang, S., G. R. Liu, G. Y. Zhang and L. Chen. 2011. Accurate bending strength analysis of asymmetric gears using the novel ES-PIM with triangular mesh. *International Journal of Automotive and Mechanical Engineering (IJAME)*, 4: 373–396.
93. Sanders, A.A., D.R. Houser, A. Kahraman, J. Harianto, S. Shon. 2011. An Experimental investigation of the effect of tooth asymmetry and tooth root shape on root stresses and single tooth bending fatigue life of gear teeth. *Paper presented at the 11th International ASME Power Transmission and Gearing Conference*, Washington, DC.

94. Kapelevich, A.L. 2011. Measurement of directly designed gears with symmetric and asymmetric teeth. *Gear Technology,* 28(1): 60–65.
95. Brown, F.W., S.R. Davidson, D.B. Hanes, D.J. Weires and A.L. Kapelevich. 2011. Analysis and testing of gears with asymmetric involute tooth form and optimized fillet form for potential application in helicopter main drives. *Gear Technology,* 28(4): 46–55.
96. Stahl, K., B.R. Höhn, M. Otto and A. Froh. 2012. Normberechnung der geometrie und tragfahigkeit von asymmetrischen evolventenverzahnungen. *Konstruktion,* 10: 91.
97. Drágár, Z. and L. Kamondi. 2012. Deviation from standards in relation to cylindrical gears with external straight teeth. *Design of Machines and Structures,* 2(1): 25–34.
98. Dadhaniya, K.D., K.P. Hirpar and K.M. Vyas 2012. Effect of pressure angle on bending stress and deformation of asymmetric spur gear using FEA. *IOSR Journal of Engineering,* 2(4): 616–620.
99. Abdullah, M.Q. and M.J. Jweeg. 2012. Analytical solution of bending stress equation for symmetric and asymmetric involute gear teeth shapes with and without profile correction. *Innovative Systems Design and Engineering,* 3(6): 19–34.
100. Sándor, R.N. 2012. *Contributii la dimensionarea, testarea si executia rotilor dintate cu dinti asimetrici. teza de doctorat. din cluj-napoca centrul universitar nord din baia mare.* Facultatea De Inginerie. Romania.
101. Fuentes, A., I. Gonzalez-Perez, F.T. Sanchez-Marin and K. Hayasaka. 2012. On the behavior of asymmetric cylindrical gears in gear transmissions. *Proceedings of the FISITA 2012 World Automotive Congress, Lecture Notes in Electrical Engineering* 193: 143–150.
102. Wu, Y.R. and G.Y. You. 2012. Asymmetric gears design and performance analysis for external gear pumps. *International Journal of Computational Intelligence Research and Applications (IJCIRA),* 6(1): 1–8.
103. Karpat, F., S. Ekwaro-Osire and E. Karpat. 2012. A computer program for dynamic load simulation of spur gears with asymmetric and symmetric teeth. *World Journal of Mechanics,* 2(5): 239–245.
104. Kapelevich, A.L. 2012. Asymmetric gears: Parameter selection approach. *Gear Technology,* 29(4): 48–51.
105. Kapelevich A.L. and E. Taye. 2012. Self-locking gears: Design and potential applications. *Gear Solutions,* 10(5): 53–58.
106. Linke, H, J. Brechling, I. Römhild, R. Hess, D. Krüger, A. Klein-Hitpass and H. Blank. 2012. Tragfahigkeit von Zahnradern mit asymmetrischem Evolventenprofil. Part 1: 28–31 and Part 2: 34–37. *Antriebstechnik.* 11 and 12.
107. Kapelevich, A.L. 2013. *Direct Gear Design.* Boca Raton, FL: CRC Press.
108. Tian, X. and W. Li. 2013. Direct design and temperature field analysis for the asymmetric gear with a small number of teeth. *Journal of Harbin Engineering University,* 34(7): 907–912.
109. Li, N., W. Li and W.-Q. Xiao. 2013. Comparative analysis of root bending stress between spur gears with symmetric and asymmetric involutes. *Journal of University of Science and Technology Beijing,* 35(1): 120–125.
110. Seidler, R.T., S. Lieder and R. Ruhland. 2013. Asymmetrical bevel gear design. *Proceedings of the International Conference on Gears, Garching near Munich,* Germany, October 7–9, 2: 1415–1420.

111. Krüger, D., I. Romhild, H. Linke, J. Brechling and R. Hess. 2013. Increasing the load capacity of gear units by asymmetric gear tooth design. *Proceedings of the International Conference on Gears, Garching near Munich,* Germany. October 7–9. 2: 1327–1340.
112. Dorofeev, V.L., D.V. Dorofeev and K. Arnaudov. 2013. Selection of optimal geometry and stress calculation on wheel teeth with asymmetrical tooth profile. *Proceedings of the International Conference on Gears, Garching near Munich,* Germany. October 7–9. 2: 1341–1354.
113. Brecher, C., M. Brumm and J. Ingeli. 2013. Simulation and experimental analysis of the excitation behavior of asymmetric involute gears. *Proceedings of the International Conference on Gears, Garching near Munich,* Germany. October 7–9. 2: 1355–1366.
114. Masuyama, T., Y. Mimura and K. Inoue, 2013. Strength simulation of asymmetric involute tooth gears. *Proceedings of the International Conference on Gears, Garching near Munich,* Germany. October 7–9. 2: 1367–1378.
115. Sondur, V.B. and N.S. Dharashivkar. 2013. Theoretical and finite element analysis of load carrying capacity of asymmetric involute spur gears. *International Journal of Research in Aeronautical and Mechanical Engineering,* 1(3): 67–73.
116. Prajapati, J.M. and P.A. Vaghela. 2013. Objective function to optimize bending stress at critical section of asymmetric spur gear. *Journal of Mechanical and Civil Engineering (IOSR-JMCE),* 10(1): 58–65.
117. Flodin, A. and M. Andersson. 2013. Tooth root optimization of powder metal gears: reducing stress from bending and transient loads. *Gear Technology,* 30(4): 56–59.
118. Andersson. M. 2013. Optimizing the tooth root strength of sintered gears for a 6 speed manual automotive transmission. *Proceedings of EuroPM2013.* Gothenburg, Sweden. September 15–18, 2013. 2: 97–102.
119. Litzba, J. 2014. Holistic approach for development and optimization of helicopter gearboxes. *European Rotorcraft Forum 2014.* Southampton, UK, September 2–5.
120. Langheinrich, A. 2014. *Geometrie, beanspruchung und verformung asymmetrischer stirnradverzahnungen.* Doktor-Ingenieurs. Fakultät für Maschinenwesen der Technischen Universität München.
121. Mohan, N.A. and S. Senthilvelan. 2014. Preliminary bending fatigue performance evaluation of asymmetric composite gears. *Mechanism and Machine Theory,* 78: 92–104.
122. Thirumurugan, R. and C.C.C. Deepak. 2014. Design and analysis of asymmetrical spur gear drive for automobile gear box application. *Applied Mechanics and Materials,* 592–594: 2277–2281.
123. Shankin, S.I., A.A. Stupakov, I.L. Glikson, V.M. Ananiev and A.L. Kapelevich. 2014. Modernization of main helicopter gearbox with asymmetric tooth gears, *Gear Solutions,* 12(3): 46–49.
124. Vojtková, J. 2014. Effect of asymmetry on radii of curvature for spur gears with nonsymmetrical teeth. *Scientific Journal of Silesian University of Technology.* Series Transport 84: 47–51.
125. Olguner, S. and I.H. Filiz. 2014. A study on the design of asymmetric spur gears in gear pump applications. *Proceedings of International Gear Conference 2014.* Lyon. France. August 26–28. 1: 406–417.

126. Kapelevich, A.L. and Y.V. Shekhman. 2014. Fabrication of directly designed gears with symmetric and asymmetric teeth. *Gear Technology*, 31(8): 86–91.
127. Devendran, R.S., A. Vacca. 2014. Experimental characterization of external gear machines with asymmetric teeth profile. IFPE 2014, Int. Fluid Power Expo, March 4–8, Las Vegas, USA.
128. Deng, X., Hua, L. and X. Han. 2015. Research on the design and modification of asymmetric spur gear. Volume 2015. Article ID 897257: 1–13. http://dx.doi.org/10.1155/2015/897257.
129. Schumann, S. 2015. Möglichkeiten und grenzen asymmetrischer kegelradverzahnungen. Doktor-Ingenieurs. Fakultät Maschinenwesen der Technischen Universität Dresden.
130. Ewering, M., P. Michalke and M. Flinks. 2015. Design, test and certification from a practical point of view. *Proceedings of the International Conference on Gears*, Garching near Munich, Germany, October 5–7. 1: 261–271.
131. Langheinrich, A., M. Otto and K. Stahl. 2015. Geometry, strain and deformation of asymmetric spur gears. Proceedings of the International Conference on Gears, Garching near Munich, Germany, October 5–7. 1: 273–281.
132. Bhatia, S.S. and H.R. Chandrakar. 2015. Analysis of an asymmetric spur gear with an elliptical hole as stress relieving feature using ANSYS APDL as tool. *International Journal for Research in Applied Science & Engineering Technology (IJRASET)*, 3(VI): 963–972.
133. Banica, M. and F. Chira. 2011. Use genetic algorithms to design asymmetric involute gears. *Scientific Bulletin, Mechanics, Tribology, Machine Manufacturing Technology. Serie C*, XXV: 5–11.
134. Stadtfeld H.J. and J.K. Saewe. 2015. Non-involute gearing, function and manufacturing compared to established gear designs. *Gear Technology*, 32(1): 42–51.
135. Kapelevich, A.L. and T.M. McNamara. 2015. Advantages of direct gear design for automotive polymer gear drives. *Gear Solutions*, 13(12): 40–46.
136. Kapelevich, A.L. and S.D. Korosec. 2015. Involute spiral face coupling and gears: Design approach and manufacturing technique. *Gear Solutions*, 13(2): 33–37.
137. Mehr, A. 2015/2016. Benefits of asymmetry. *Liebherr Magazine 2015/2016—Gear Technology and Automation Systems*, 4–7.
138. Linke, H., J. Börner and R. Heß. 2016. *Cylindrical Gears: Calculation—Materials—Manufacturing*. Munich: Carl Hanser Verlag GmbH & Co. KG.
139. Weihmann, C. 2016. Oberflächenverdichten von gesinterten Zahnrädern mittels Profil-Querwalzen am Beispiel einer asymmetrischen Verzahnung für die Verwendung in den hochbelasteten Getriebestufen. Masterarbeit (Master's Thesis). Otto-Von-Guericke-Universität Magdeburg. Fakultät für Maschinenbau.
140. Mehr, A. 2016. The benefits of asymmetric gears. *Gear Solutions*, 14(5): 52–54.
141. Phalake, A.A., A.A. Utpat. 2016. Optimization of tooth fillet profile of spur gear with asymmetric tooth to analyze the bending stress concentration at root of tooth. *International Journal of Engineering Trends and Technology (IJETT)*, 36(3): 151–154.
142. Mertens, A.J., M. Kodeeswaran and S. Senthilvelan. 2016. Injection molded asymmetric spur gear tooth deflection: Numerical and experimental investigation. *Proceedings of the SPE ANTEC™ Plastics Technology Conference. Indianapolis*. May 23–25: 1670–1674.

143. Masuyama, T. and N. Miyazaki, 2016. Evaluation of load capacity of gears with an asymmetric tooth profile. *International Journal of Mechanical and Materials Engineering*, 11(11): 1–9.
144. Kapelevich, A.L. 2016. Analysis and optimization of asymmetric epicyclic gears. *Gear Solutions*, 14(8): 50–55.
145. Kapelevich, A.L. and Y.V. Shekhtman. 2016. Rating of asymmetric tooth gears. *Power Transmission Engineering*, 10(3): 40–45.
146. Goldfarb V.I. and N.A. Barmina (eds.). 2016. *Theory and Practice of Gearing and Transmissions. Direct Gear Design for Asymmetric Tooth Gears.* Mechanisms and Machine Science 34, New York: Springer International Publishing AG Switzerland. pp. 117–143.
147. Radzevich S. P. 2016. *Dudley's Handbook of Practical Gear Design and Manufacture*, 3rd Edition. Boca Raton, FL: CRC Press.
148. Kapelevich A.L. 2016. Pitch factor analysis for symmetric and asymmetric tooth gears. *Gear Solutions*, 14(12): 27–33.
149. Krüger. D. 2016. Auslegungsgrenzen, Grübchen- und Zahnfußtragfähigkeit asymmetrischer Stirnradverzahnungen. Doktor-Ingenieur. Von der Fakultät für Maschinenbau der Technischen Universität Chemnitz.
150. Flodin, A. 2016. Testing and validation of powder metal gears in a 6 speed manual transmission. Presented at WorldPM 2016 in Hamburg on October 12, 2016.
151. Flodin, A. 2016. Powder metal gear technology: A review of the state of the art. *Power Transmission Engineering*, 10(2): 38–43.
152. Winkel, O. and C. Fäustle, 2017. Sonderanwendungen beim verzahnungsfräsen. *GETPRO Kongress* 28–29. März 2017.
153. Mehr, A. 2017. Hard finishing of asymmetric tooth profiles—Solutions for series production. *Proceedings of MPT2017-Kyoto, the JSME International Conference on Motion and Power Transmissions.* February 28–March 3, Kyoto, Japan.
154. Masuyama, T., R. Kobayashi, N.Miyazaki, R. Hori and Y. Mimura. 2017. Fatigue strength simulation of gears with asymmetric tooth profile. *Proceedings of MPT2017-Kyoto, the JSME International Conference on Motion and Power Transmissions.* February 28–March 3. Kyoto. Japan. 1: 342–346.
155. Suciu, F., A. Dascalescu and M. Ungureanu. 2017. From design to manufacturing of asymmetric teeth gears using computer application. *IOP Conference Series: Materials Science and Engineering*, 200: 1–6.
156. Andersson, M. and Flodin, A. 2017. Redesigning and prototyping a six speed manual automotive transmission of powder metal gears. *Proceedings of MPT2017-Kyoto, the JSME International Conference on Motion and Power Transmissions.* February 28–March 3. Kyoto, Japan.
157. Kapelevich, A.L. and Y.V. Shekhtman. 2017. Analysis and optimization of contact ratio of asymmetric gear, *Gear Technology*, 34(2): 67–71.
158. Zhao, X. and A. Vacca. 2017. Formulation and optimization of involute spur gear in external gear pump. *Mechanism and Machine Theory*, 117: 114–132.
159. Kapelevich, A.L. and Y.V. Shekhman. 2017. Rating of helical asymmetric tooth gears. *Gear Technology*, 34(8): 78–81.
160. Kapelevich, A.L. and Y.V. Shekhman. 2018. Root fillet optimization of thin rim planet gears with asymmetric teeth. *Gear Solutions*, 16(3): 52–55.

161. Roth, Z. and J. Opferkuch. 2017. High load carrying tooth root fillet based on a Bézier curve. *Proceedings of the International Conference on Gears*, Garching near Munich, Germany, September 13–15.
162. Krueger D. 2017. Geometry, pitting and tooth root capacity of asymmetric gears. *Proceedings of the International Conference on Gears*, Garching near Munich, Germany, September 13–15.
163. Arnaudov K.B. and D.P. Karaivanov. 2017. Planetary gear trains. Bulgarian Academy of Sciences Publ. "Prof. Marin Drinov". Sofia. Bulgaria.
164. Kapelevich, A.L. and Y.V. Shekhman. 2018. Flank profile optimization for asymmetric tooth gears. *International Gear Conference 2018, Conference Proceedings*. August 27–29 2018, Lyon, France.
165. Townsend, D.P. 1967. *Dudley's Gear Handbook*, 2nd Edition. New York: McGraw Hill.
166. McVittie, D. 1993. The European rack shift coefficient "X" for Americans. *Gear Technology*. 10(4): 34–36.
167. Groman, M.B. 1962. The zones of involute mesh. *Vestnik Mashinostroeniya*, 12: 12–17.
168. Goldfarb, V.I. and A.A. Tkachev. 2005. New approach to computerized design of spur and helical gears. *Gear Technology*, 22(1): 26–32.
169. Vulgakov, E.B. and A.L. Kapelevich. 1980. Area of existence of helical gearing. *Vestnik Mashinostroeniya*, 7: 9–11.
170. Vulgakov, E.B. and L.M. Vasina. 1978. *Involute Gears in Generalized Parameters*. Moscow: Mashinostroenie.
171. Vulgakov, E.B. and A.L. Kapelevich. 1982. Helical gear mesh possibilities. *Vestnik Mashinostroeniya*, 3: 12–14.
172. NASA/SP-8100. 1974. *Liquid Rocket Engine Turbopump Gears*. NASA.
173. Henry Z.S. 1995. Bell Helicopter Advanced Rotorcraft Transmission (ART) Program, Contract Report ARL-CR-238, NASA.
174. Anderson N.E. and S.H. Loewenthal, 1984. Efficiency of nonstandard and high contact ratio involute spur gears. AVSCOM Technical Report 84-C-9, NASA.
175. Standard ISO 6336. 2006. Calculation of load capacity of spur and helical gears. ISO copyright office. Case postale 56, CH-1211 Geneva 20, Switzerland.
176. Inoue, K. and T. Masuyama. 2006. Possibilities of fatigue strength simulation in reliability design of carburized gears. *Paper presented at the International Conference "Power Transmissions' 06"*. Novi Sad, Serbia.
177. Pedersen, N.L. 2010. Improving bending stress in spur gears using asymmetric gears and shape optimization. *Mechanism and Machine Theory*, 45: 1707–1720.
178. Der Hovanesian J. et al. 1989. Gear root stress optimization using photoelastic optimization techniques. *SAE transactions*, 97(4): 748–755.
179. Pulley F.T. et al. 2000. Method for producing and controlling a fillet on a gear. US Patent No. 6164880.
180. Spitas, V., T. Costopoulos and C. Spitas. 2005. Increasing the strength of standard involute gear teeth with novel circular root fillet design. *American Journal of Applied Sciences*, 2(6): 1058–1064.
181. Roth, M.A., Z. Roth and K. Paetzold. 2013. Developing a bionic gear root fillet contour. *Proceedings of the International Conference on Gears*, Garching near Munich, Germany. October 7–9. 2: 1315–1326.

182. Hansong, X., J.W. Zu and W. Zaton. 2005. Fillet shape optimization for gear teeth. *ASME Power Transmission and Gearing Conference*, September 24–28. Long Beach, CA.
183. Rastrigin, L.A. (ed.). 1969. *Random Search Theory and Application*. Riga: Zinatne.
184. Joachim, F.J., J. Börner and N. Kurz. 2012. How to minimize power losses in transmissions, axles and steering systems. *Gear Technology*, 29(6): 58–66.
185. Standard ANSI/AGMA 2001-D04. 2004. *Fundamental Rating Factors and Calculation Methods for Involute Spur and Helical Gear Teeth*. ISBN: 1-55589-839-4 American Gear Manufacturers Association, 500 Montgomery Street, Suite 350, Alexandria, Virginia 22314.
186. Popper, J.B. 1961. Cooperating Wedges including mating worms, US Patent No. 2973660.
187. Munster N.S. and G.V. Tzarev. 1968. Self-locking cylindrical gears, theory mechanisms and machines. *Publications of Tashkent Polytechnic Institute*, 30A: 3–6.
188. Iskhakov T.G. 1969. Self-locking in gear mechanisms. *Publications of the Kazan Aviation Institute*, 105(105): 3–15.
189. Timofeev G.A. and V.V. Panukhin. 2003. Self-locking criteria analysis, *Vestnik Mashinostroenia*, 9: 3–8.
190. Taye, E. 2009. Actuator with self-locking helical gears for a continuously variable valve lift system, US Patent No. US2009/0283062 A1.
191. Adams, C.E. 1986. *Plastic Gearing*. New York: Marcel Dekker, Inc.
192. Starzhinsky, V.E., E.V. Shalobaev et al. 1998. *Plastic Gears in Instrument Mechanisms*. Saint-Petersburg—Gomel: MPRI NASB.
193. Standard ANSI/AGMA 1006-A97. 1997. *Tooth Proportions for Plastic Gears*. AGMA. American Gear Manufacturers Association 1500 King Street, Suite 201, Alexandria, Virginia 22314.
194. Standard ANSI/AGMA 1106-A97. 1997. *Tooth Proportions for Plastic Gears (Metric Addition)*. AGMA. American Gear Manufacturers Association 1500 King Street, Suite 201, Alexandria, Virginia 22314.
195. Standard ANSI/AGMA 920-A01. 2001. *Materials for Plastic Gears*. AGMA. American Gear Manufacturers Association 1500 King Street, Suite 201, Alexandria, Virginia 22314.
196. VDI-Standard: VDI 2736. Parts 1–4. VDI-Gesellschaft Produkt- und Prozessgestaltung (GPP) Fachbereich Getriebe und Maschinenelemente.
197. Mao, K., W. Liebig, C.J. Hooke and D. Walton. 2010. Polymer gear surface thermal wear and its performance prediction. *Tribology International*, 43(1–2): 433–439.
198. ANSI/ASME Y14.7. 1971. *Gear Drawing Standard—Part 1: For Spur, Spur, Helical, Double Helical and Rack*. New York: ASME International.
199. ANSI/ASME Y14.7. 1971. *Gear Drawing Standard—Part 2: For Bevel and Hypoid Gears*. New York: ASME International.
200. American Standard ANSI/AGMA 2015-1-A01. 2002. *Accuracy Classification System - Tangential Measurements for Cylindrical Gears*. Alexandria, VA: AGMA.
201. American Standard ANSI/AGMA 2015-2-A06. 2006. *Accuracy Classification System—Radial Measurements for Cylindrical Gears*. Alexandria, VA: AGMA.
202. International Standard ISO 1328-1. 1995. Cylindrical gears—ISO system of accuracy—Part 1: Definitions and allowable values of deviations relevant to corresponding flanks of gear teeth. ISO.

203. International Standard ISO 1328-2. 1997. Cylindrical gears—ISO system of accuracy—Part 2: Definitions and allowable values of deviations relevant to radial composite deviations and runout information. ISO.
204. Zou, Z. and, E. Morse. 2001. Statistical tolerance analysis using gapspace. *This paper was presented at the 7th CIRP International Seminar on Computer Aided Tolerancing*. ENS de Cachan. France.
205. Creveling, C.M. 1996. *Tolerance Design: A Handbook for Developing Optimal Specifications*. Upper Saddle River, NJ: Prentice Hall, p. 422.
206. Cox, N.D. 1986. *How to Perform Statistical Tolerance Analysis*. Milwaukee: American Society for Quality Control.
207. Saari, O.E. 1954. How to calculate exact wheel profiles for form grinding helical gear teeth. *American Machinist*, 13: 172–175.
208. Kapelevich, A.L., A.I. Tolchenov and A.I. Eidinov. 1985. Application of the fly cutters in experimental production, *Aviatsionnaya Promyshlennost*, 3: 39–40.
209. Flodin, A. and M. Andersson. 2012. Tooth root optimization of powder metal gears—Reducing stress from bending and transient loads. *Paper presented at the Powder Metal World Congress and Exhibition (PM2012)*. Yokohama. Japan.
210. Walsh, S.F. 1993. Shrinkage and warpage prediction for injection molded components. *Journal of Reinforced Plastics and Composites*, 12(7): 769–777.
211. Autodesk® Moldflow® plastic injection molding simulation software. http://usa.autodesk.com/moldflow/.
212. Gear Design Solutions. 2015. Dontyne systems brochure. *Release*, 5(2): 1–12.
213. Nezhurin, I.P. 1961. Calculation of the measurement over pins of the helical gears with odd number of teeth, *Vestnik Mashinostroeniya*, 2: 14–17.
214. Standard ANSI/AGMA 2000-A88. 1988. *Gear Classification and Inspection Handbook*. Alexandria, VA: AGMA.
215. Coy, J.J., D.P. Townsend and E.V. Zaretsky. 1985. *Gearing*. NASA Reference Publication 1152. AVSCOM Technical Report 84-C-15. NASA.
216. Small-sized gas-turbine engines of series 9I56. http://kadvi.esy.es/product/malorazmernie-gdt/.
217. GOST 1643-81. 1981. *Tolerances for Cylindrical Gears*. Moscow: Interstate Standard. IPK.
218. Sarkisov, A.A. and Vulgakov E.B. 2000. The new gearbox of Klimov Corporation. *Aerospace Courier*, 2: 32–33.
219. Yeliseev, Y.S., V.V. Krymov, I.P. Nezhurin, V.S. Novikov and N.M. Ryzhov. 2001. *Fabrication of Gas Turbine Engine Gears*. Moscow: Vysshaya Shkola.
220. Mi-2MSB - Ukrainian Combat Variant of the Mi-2 Helicopter. http://www.defence24.com/265849,mi-2msb-ukrainian-combat-variant-of-the-mi-2-helicopter.

Index

Page numbers followed by *f* indicate figures; those followed by *t* indicate tables.

S

T